330.1
BACO 1

A First Course in
Econometric Theory

SOMERVILLE COLLEGE
LIBRARY

A First Course in Econometric Theory

ROBERT BACON

OXFORD UNIVERSITY PRESS
1988

Oxford University Press, Walton Street, Oxford OX2 6DP
Oxford New York Toronto
Delhi Bombay Calcutta Madras Karachi
Petaling Jaya Singapore Hong Kong Tokyo
Nairobi Dar es Salaam Cape Town
Melbourne Auckland
and associated companies in
Beirut Berlin Ibadan Nicosia

Oxford is a trade mark of Oxford University Press

Published in the United States
by Oxford University Press, New York

© Robert Bacon 1988

All rights reserved. No part of this publication may be reproduced, stored in a retrieval system, or transmitted, in any form or by any means, electronic, mechanical, photocopying, recording, or otherwise, without the prior permission of Oxford University Press

This book is sold subject to the condition that it shall not, by way of trade or otherwise, be lent, re-sold, hired out or otherwise circulated without the publisher's prior consent in any form of binding or cover other than that in which it is published and without a similar condition including this condition being imposed on the subsequent purchaser

British Library Cataloguing in Publication Data
Bacon, Robert—1942—
A first course in econometric theory.
1. Econometrics.
I. Title.
330'.028
ISBN 0-19-877278-5
ISBN 0-19-877277-7 (pbk)

Library of Congress Cataloging in Publication Data
Bacon, Robert William.
A first course in econometric theory.
Includes index.
1. Econometrics. I. Title.
HB139.B33 1988 330'.028 87-35035
ISBN 0-19-877278-5
ISBN 0-19-877277-7 (pbk.)

Phototypeset by Macmillan India Ltd., Bangalore 25.

Printed and bound in
Great Britain by Biddles Ltd,
Guildford and Kings Lynn

Preface

I make no apology for presenting yet another textbook on econometric theory. There are many ways of teaching the subject and my belief is that different approaches suit different types of student. My approach has been to stress the understanding of the technical details of the subject at the expense of generality, in the belief that if we really understand the simplest cases then we can hope to progress to general cases without loss of comprehension.

This book is the reflection of several years' teaching econometric theory to undergraduates. Often the detailed questioning of my inadequate explanations of topics has resulted in my understanding the subject better, and I believe that this has improved the text. Were some of my students now to read this book they would see a reflection of these valuable interchanges. In a genuine sense such students have made it possible for me to wish to and to be able to produce this book. It is dedicated to them.

R. B.

Lincoln College,
Oxford,
January 1987

Contents

List of Figures   x
List of Tables   xi
Abbreviations   xii

1. **Introduction**   1
   1.1 The meaning and uses of econometrics   1
   1.2 About the book   3
   1.3 Some useful text books   4

2. **The Basic Model and Least Squares Estimation**   6
   2.1 Introduction   6
   2.2 The method of ordinary least squares   10
       Example 2.1. A least squares estimation   15
   2.3 Least squares and the scale of variables   17
   2.4 Least squares and transformations of the equation   18
   2.5 Residuals and goodness of fit   20
       Example 2.2. Goodness of fit of least squares   23
   2.6 Diagrammatic representation of least squares   23
   2.7 OLS and the model without an intercept   26
   2.8 OLS in a model with no slope coefficient   28
   Problems 2   29
   Answers 2   31

3. **The Properties of Least Squares Estimation**   39
   3.1 The Gauss–Markov theorem   40
   3.2 Estimation of the variances of OLS estimators   50
       Example 3.1. Variances and covariances of least squares   53
   3.3 Least squares and forecasting   53
       Example 3.2. Forecasting from least squares   57
   Problems 3   58
   Answers 3   62

4. **Multiple Regression**   76
   4.1 The model and assumptions   76
   4.2 The Gauss–Markov theorem   77
   4.3 Multiple regression and goodness of fit   81
       Example 4.1. A multiple regression   85
   4.4 Multiple regression and the problem of multicollinearity   87
   4.5 Estimation with parameter restrictions   89
       Example 4.2. Multiple regression subject to a linear restriction   96

	4.6	Forecasting with multiple regressions	97
		Example 4.3. Forecasting from a multiple regression	98
	4.7	The many-variable generalization	98
	Problems 4		99
	Answers 4		102
5.	Hypothesis Testing		115
	5.1	Introduction	115
	5.2	Hypothesis testing	118
	5.3	Test statistics	122
		Example 5.1 Hypothesis tests on a multiple regression	126
	5.4	Relations between test statistics	130
		Example 5.2. 't' scores and partial correlations	135
	5.5	Confidence intervals	135
		Example 5.3. Confidence interval for a regression coefficient	136
	5.6	Tests for structural change	138
	5.7	Tests for predicted values	141
	Problems 5		144
	Answers 5		146
6.	Maximum Likelihood Estimation and Inference		154
	6.1	Introduction	154
	6.2	Estimation and inference with independent stochastic explanatory variables	155
	6.3	Asymptotic properties of estimators	157
	6.4	Maximum likelihood estimation	159
	6.5	Large sample test statistics	162
		Example 6.1. ML estimation and tests	167
	Problems 6		169
	Answers 6		170
7.	Generalized Least Squares: Serial Correlation and Heteroskedasticity		176
	7.1	Heteroskedasticity	178
		Example 7.1. Tests for heteroskedasticity	190
	7.2	Serial correlation	191
		Example 7.2. Tests for serial correlation	197
		Example 7.3. Forecasts with serially correlated errors	199
	7.3	Seemingly unrelated regressions	200
	Problems 7		204
	Answers 7		207
8.	Specification and Measurement Errors		219
	8.1	Omitted variables	219
	8.2	Wrongly included variables	221
	8.3	Measurement error	223

	8.4	Random measurement errors and probability limits	230
	8.5	Testing for specification and measurement errors	232
		Example 8.1. A test for measurement error	236
	Problems 8		237
	Answers 8		240
9.	Simultaneous Equations		247
	9.1	Identification and prior information	252
	9.2	The estimation of simultaneous equations	260
	9.3	Simultaneous equations and forecasting	267
		Example 9.1. The estimation of a system of equations	269
	Problems 9		271
	Answers 9		273
10.	Dynamic Models		278
	10.1	Finite lag models	278
	10.2	Infinite lags	285
	10.3	Estimation with lagged dependent variables	289
		Example 10.1. Estimation of lag structures	294
	Problems 10		297
	Answers 10		299
11.	Applying Econometric Theory		303
	11.1	Introduction	303
	11.2	Measures of econometric performance	304
	11.3	Multicollinearity: lack of data variability	308
	11.4	Modelling strategy	313
	11.5	Final thoughts	315
Index			317

List of Figures

2.1	The Deterministic Relation between Consumption and Income	8
2.2	Scattergram of Consumption against Income	9
2.3	Time Series Plot of Consumption and Income	9
2.4	Scattergram of Y against X	11
2.5	Horizontal and Vertical Measures of Closeness to a Line	12
2.6	Data Points and the True Line	24
2.7	Scattergram of Actual Versus Fitted Values	25
2.8	Time Series Plot of Actual and Fitted Values	26
2.9	Time Series Plot of Residuals	27
3.1	Schematic Representation of a Repeated Samples Experiment	40
3.2	The Influence of the Spread of X on the Variance of the Estimator	52
3.3	The Effect of the Size of the Independent Variable on the Variance of the Predictor	55
5.1	The Normal Distribution	117
5.2	5% Critical Regions on an $N(0, 1)$	119
5.3	Hypothetical and Actual Distributions for an Estimated Parameter	120
5.4	A One-tailed Test for a Two-sided Hypothesis	121
5.5	A Two-tailed Test for a Two-sided Hypothesis	122
5.6	Confidence Intervals for a Single Parameter	137
7.1	Heteroskedastic Scatter of Points	179
7.2	Plot of Estimated SEE Versus $\tilde{y}$	185
7.3	Bounds for a Durbin–Watson Test	196
10.1	A Polynomial Lag Pattern	279
11.1	Critical Values of t for Different d.f.	309

List of Tables

2.1	UK Consumption and Income 1961–1982	7
2.2	Residuals, Actual and Fitted Values	24
3.1	Forecasts from a Least Squares Estimation	58
3.2	The Probabilities of Values of a Bivariate Distribution	63
4.1	UK Unemployment 1961–1982	85
4.2	Matrix of Pairwise Correlations	86
6.1	Restricted and Unrestricted ML Estimates	174
7.1	Results for Two-sample Test for Variance Equality	190
10.1	Coefficients from Unrestricted OLS	294
10.2	Coefficients from a Quadratic Restriction	295
10.3	Coefficients from a Rational Lag Restriction	296

Abbreviations

AH	alternative hypothesis
a.p.c	average propensity to consume
AR	autoregressive process
BLUE	best linear unbiased estimator
d.f.	degrees of freedom
DWS	Durbin Watson Statistic
Eff	efficiency
ESSQ	explained sum of squares
FIML	full information maximum likelihood
GLS	generalized least squares
i.i.d.	independently identically distributed
ILS	indirect least squares
IV	instrumental variable
LDV	lagged dependent variables
LM	Lagrange multiplier
LR	likelihood ratio
MA	moving average
ML	maximum likelihood
m.p.c.	marginal propensity to consume
MSE	mean square error
NH	null hypothesis
OLS	ordinary least squares
PDL	polynomial distributed lag
plim	probability limit
RLS	restricted least squares
RSSQ	residual sum of squares
SEE	standard error of estimate
SURE	seemingly unrelated regressions estimation
TSLS	two-stage least squares
TSSQ	total sum of squares
UMPU	uniformly most powerful unbiased

1
Introduction

1.1 The meaning and uses of econometrics

The study of econometric theory has become over the last decades an essential part of every economist's training, and it has been required to be introduced into that training at progressively earlier stages. This is because a large part of 'applied economics'—the study of actual data and facts—is no longer merely a cataloguing of instances but has as a crucial aspect the evaluation and quantification of economic hypotheses. We need only to think of disagreements over whether the money supply affects inflation, or of attempts to quantify the effect of inflation of the savings ratio, to see that economists are interested in the actual magnitudes of economic relationships. Theoretical economics may postulate that there is a relationship between two variables, e.g. that income affects consumption, but applied economics demands both evidence that such a relationship exists in the real world and, if so, quantification, for example of the marginal and average propensities to consume. The study of the methodology of how to quantify such relationships from economic data is known as econometrics. Of course econometrics is just a sub-branch of general statistical theory but because the problems typically faced by the economist require only a small range of all current statistical techniques (utilizing very special variations of these techniques) it is convenient to isolate the methods used by economists under the heading of econometrics. In fact, as economists have examined more varied economic models their statistical requirements have increased and it is no longer easy to characterize econometrics as the use of one or two statistical tools in the study of economics. Pragmatically, econometrics is defined as those techniques of statistical analysis which economists use. However, in this book, which is designed merely as a first course in the study of such techniques, only a few techniques are investigated. The use of statistical (i.e. data-related) methods of analysis can have one or more of several possible goals. There are five main uses to which a study of the methods of analysing data can be put.

    1. *Estimation.* The most common demand made upon economic data is that we should be able to derive from it values of the constants (parameters) of the economic model which we believe has generated that data. In the case of the consumption function, where our data would be the consumption and income levels for an economy over a number of years, we would wish to obtain a value (estimate) for the marginal propensity to consume (m.p.c.). Estimation is concerned to produce merely the value rather than what we do with the value.

2. *Inference.* In practical terms we are rarely content to accept an estimated value of a parameter at face value. We recognize that the data may be untypical or rather scanty, and accordingly we are prepared to take the estimate as an approximation to the true value of the unknown parameter. We believe that income affects consumption, but when our statistical technique reveals that based on the last ten quarters data the estimate of the m.p.c. is 0.89 (say), we do not believe that this number is the true value generating the levels of consumption. Rather we hope that it will be a good approximation to the truth. In the spirit of such an understanding (which we formalize in some detail in later chapters) we are naturally led to ask questions about our estimates. For example, is the estimate (given that it is an approximation) compatible with some hypothetical value of the true parameter? A very simple hypothesis might be that m.p.c. is zero, i.e. that consumption is unaffected by income. In order to support the Keynesian approach we would wish to know whether our estimated value was compatible with a true value of zero (hoping and expecting to find that it was not). Clearly if the estimated value were incompatible with the true value's being zero then we should be led to argue that the data supported the alternative hypothesis that income did have an effect upon consumption. This technique of trying to infer something about the value of the true relationship from the data lies at the heart of all discussions of whether a given theory is supported by the data or not.

3. *Forecasting.* Once we are convinced that a given theory is supported by the (known) data over a certain historical period then it is natural to wish to use our knowledge of the model gained from estimating it to forecast future behaviour. Clearly, much economic policy is based in making forecasts about economic variables for some period about which we are at present ignorant. The estimated model is used to bring our knowledge to bear in a systematic way. For example, companies constantly need to forecast the sales of their products or the prices of their inputs over the next few years or governments need to forecast the growth of output and the rate of inflation before they can decide how to design their policies. These examples serve to illustrate how important the exercise of forecasting is today and therefore how necessary it is to know how best to use knowledge about the economy gathered from the past to produce a forecast (estimate) of the future.

4. *Regulation and control.* Our discussion of the use of models for forecasting highlighted the fact that once an estimate based on the past could be used to construct a forecast of the future, we could see that action based on such a forecast might follow. The study of how to relate the decisions to be made to the estimated model through which we see the decisions working is one aspect of control theory which poses some particular problems for the econometrician. Although this area is potentially very important we will in fact pay it virtually no attention since it is not at present studied or used in a way that is amenable to the elementary methods presented in this book.

5. *Experimental design.* For certain types of economic model we may be able to collect the data ourselves from sample surveys (e.g. of household expenditures on various goods). Since such surveys are very expensive to mount it is essential that they give the most useful estimates possible at lowest cost. To achieve this our knowledge of how data will be utilized to produce estimates can be harnessed to show us how best to design the survey,—which income ranges of households to sample, and so on. Again this is a subject of great practical importance, but since it has received little special attention from economists we shall not specifically deal with it.

These various goals are clearly interlinked, but for teaching purposes it is very convenient to separate them. Although the third and fourth goals are practically of the most importance and the second goal is more important than the first, the amount of space devoted to each topic is inversely proportional to its importance. This is not as odd as it might at first appear because, as we have indicated, each branch of the subject forms the basis for the next. We cannot make inferences or forecasts without first knowing how to estimate the equation. Indeed once the correct estimation technique is established for any situation then the extensions to inference and to forecasting are in all cases fairly similar, so that the greater part of our study is concerned with how to vary the method of estimation in various situations that can arise.

1.2 About the book

The purpose of this textbook is to give the reader a thorough grounding in the central ideas and techniques of econometric theory while using a low level of mathematics. There are a large number of proofs to follow and theoretical problems to solve, but a uniform standard of mathematics has been assumed throughout the book. The avoidance of matrix algebra is designed to make the text accessible to those who have not taken courses in matrix algebra or who find its interpretation difficult. A price has had to be paid for this—no result generalizable to any number of variables has been proved—and instead the reader is given a full account of simple cases and the generalizations have been indicated. Anybody wishing to master econometric theory will certainly need to progress to another text after this but it may be that the insights gained from detailed analysis of simple cases will be helpful in more complex cases.

The book has also taken only the 'big' themes of econometrics for protracted treatment—this allows us to see the links between parts of the subject and to have an overview without being overwhelmed by detail, but again a price is paid: many specialized and useful techniques and problems are not covered. Again the reader will need to move on to other texts for such insights. At the end of this chapter brief notes on some currently useful texts are given.

The main ideas of this book can easily be covered in a single lecture course but an important aspect of the book is contained in the problems (which are supplied with detailed answers). Many of these problems illustrate important extensions of arguments of the text or important technical developments. It is strongly recommended that readers make serious attempts to solve problems before turning to the answers for help—only by doing this will the important aspects of the problem become fully apparent.

I have deliberately not used references to primary material, even though many theorems and problems can be traced to such sources. Students in a first course rarely wish to use primary sources; good secondary sources, such as those recommended at the end of this chapter, are much more accessible and also contain references to primary sources.

1.3 Some useful textbooks

There are enormous ranges of textbooks available, from the very simple to the extremely rigorous and mathematically demanding. The ones recommended here are those which I envisage being useful to a student who is committed to wading through this book. Omission from this list does not imply that I view the book as 'less good' than others, merely that its complementarity with this book is less marked.

Peter Kennedy, *A Guide to Econometrics*, 2nd edition (Blackwell, 1985). This is a unique book in that it presents no proof and relatively few formulae and yet describes and explains a very wide range of econometric problems and techniques. The ease of reading allows the user to gain a quick overview on the topics it addresses. It also contains a large bibliography.

G. S. Maddala, *Econometrics* (McGraw Hill, 1977). This book also presents virtually no proofs in the main text and uses scalar algebra and not matrix algebra to give important results. It covers a very wide body of material and is worth consulting on any problem the reader may have come across elsewhere. It is a reference book using a simple mathematical level.

A. S. Goldberger, *Econometric Theory* (Wiley, 1964). This is one of the classic texts of econometrics. It uses matrix algebra very extensively in its systematic treatment of the proofs of the central results in econometrics. Its scope in terms of topics covered is perhaps less than some other books but its uniform style is very valuable.

G. G. Judge, William E. Griffiths, R. Carter Hill, and Tsoung-Chao Lee, *The Theory and Practice of Econometrics* (Wiley, 1980). This is a compendium of econometric results and proofs which covers an enormous range of material, and which has an equally extensive bibliography. The level of mathematics is based on matrix algebra so that it is considerably more difficult to use than Maddala.

D. B. Owen, *Handbook of Statistical Tables* (Addison Wesley, 1962). Most econometric textbooks include standard statistical tables and I have not thought it necessary to include them here. However, if a wider range of tables is needed then this reference book by Owen contains most tables that will be required.

2
The Basic Model and Least Squares Estimation

2.1 Introduction

We discussed in Chapter 1 the need to estimate from a set of data the parameters of the economic model that generated the data. To make this idea more concrete let us begin with an actual set of data on macro-economic consumption and disposable income for the UK published in the 1984 edition of *Economic Trends*.[1] Table 2.1 gives figures at constant 1980 market prices from 1961 to 1982. We can assume for the moment that our theoretical model has been decided upon before seeing the data and that the particular measures of the concepts used (i.e. GNP at constant prices as a measure of income) are those we would have ideally chosen (i.e. are those generated by the theory itself).

Let us further assume that the data is generated by a simple linear relationship between income and consumption in the same year. If we were still working with purely theoretical models we would write this perhaps as:

$$C = a + bY, \qquad (2.1)$$

where a and b are constants (parameters) and C stands for the consumption at a point in time and Y for income at the same point in time. Graphically we would represent (2.1) as in figure 2.1. There are several features of this model that are noteworthy. The interpretation of the parameters is straightforward —if income were zero the level of consumption would be a, and for every unit that income increases consumption increases by b units. This implies that the average propensity to consume is less than the marginal propensity if a is greater than zero. The equation is additive in the parameters, that is a and b enter additively rather than (say) multiplicatively as in (2.2):

$$C = aY^b. \qquad (2.2)$$

This property is dictated of course by economic theory but it does have important implications for estimation as we shall see later. The variables themselves are in the levels and are not in logs or rates of change or any other transformation. Although such transformations do not in themselves present

---

[1] It is essential to give the source of data precisely since even official government figures are revised, often for several years after their first publication.

TABLE 2.1. UK Consumption and Income 1961–1982

Year	Consumption[a]	Disposable income[b]
1961	89,628	98,153
1962	91,653	99,087
1963	95,894	103,764
1964	98,806	107,617
1965	100,389	110,359
1966	102,212	112,846
1967	104,739	114,443
1968	107,804	116,468
1969	108,401	117,553
1970	111,168	122,123
1971	114,719	123,687
1972	121,519	134,511
1973	127,734	143,870
1974	125,552	142,713
1975	124,824	142,629
1976	125,097	141,639
1977	124,646	139,318
1978	131,485	149,602
1979	137,863	158,295
1980	136,890	160,620
1981	137,063	156,630
1982	138,865	155,627

Source: Economic Trends, 1984.

[a] Consumption is consumer's expenditure at 1980 market prices in £ million.
[b] Income is Real Personal Disposable Income at 1980 market prices in £ million.

difficulties for estimation, it is important to start with that transformation which corresponds to our theory.

Now that we have a clear view of the process which generates the data we can turn to the central question of how to estimate the unknown parameters a and b from the known data. In practice it would be sensible next to plot the data on a graph in order to see whether our assumption about the true model is roughly supported by the evidence. This elementary step can often suggest all sorts of subsidiary hypotheses and can reveal gross inadequacy of the model very quickly and should never be omitted. There are two common

8 Basic Model and Least Squares Estimation

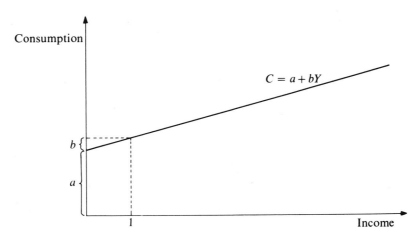

FIG. 2.1 The Deterministic Relation between Consumption and Income

presentations of the data used by econometricians for data of this type. The first form is the 'scattergram', in which each variable is assigned to its own axis and the pairs of values for different years are plotted on the graph. This is illustrated by figure 2.2. The second presentation, which is commonly used when the observations can be arranged in some natural sequence (here the index being the year), is to plot against the index on the same graph. This type of time series plot, shown in figure 2.3, conveys different information from the scattergram and is used as a complementary presentation rather than as an alternative.

The data as plotted in figure 2.2 do appear to follow a relation of the type we postulated. Consumption grows with income, the slope of the function appears to be constant, and there appears to be a positive intercept. All of this is revealed at a glance but one other feature of the data is also self-evident. The data do not obey a relation of the type (2.1) exactly. No straight line could be drawn which passes through all twenty-two points. Clearly our model is inadequate to describe the data exactly, even though it is evident that it is a good approximation. This phenomenon, which would have occurred whatever the actual data chosen, forces us to make a crucial adjustment to the simple model we started with. We take the basic model and add to it an extra term U which in effect incorporates all that part of consumption which is not 'explained' by our model.

$$C = a + bY + U. \qquad (2.3)$$

The interpretation of this new term is crucial to the study of econometrics.

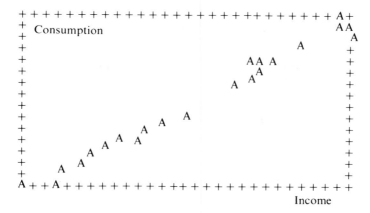

FIG. 2.2 Scattergram of Consumption against income

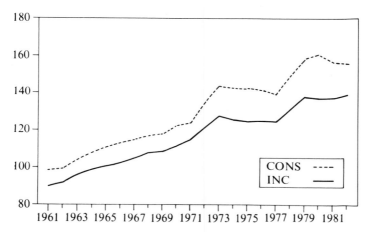

FIG. 2.3 Time Series Plot of Consumption and Income (£000 million)

The justification for its existence is that, as well as the dominant income factor, many small factors will affect consumption at any point in time. Those small influences, such as the state of expectations or the degree of uncertainty about the future, unforeseen events during the period, and so on, are such that we cannot perhaps measure them or indeed even know what they are. Nevertheless there can be many such factors all affecting consumption. The totality of this effect we call the *disturbance* term because it disturbs our central relationship from being exactly true at any time. An alternative name for U is the *equation error* or *error term* because its existence denotes an error in our specification of the precise determinants of consumption. A third name for U is the *random* or *stochastic* component of the equation. To the econometrician concerned with a particular theory it appears that the extra effects on consumption are random (i.e. unpredictable). Of course, were the econometrician to have a complete understanding of the model, this interpretation could not hold.

The essential reformation of the model of the determination of consumption into a deterministic component (that caused by the level of income) and a stochastic component (those factors unknown to the econometrician) allows us to link actual data to theoretical economic models. The next step in the process is to find a method of deriving values for the parameters of the theoretical model from the data when the data do not conform exactly to our model. The most important technique for solving this problem is commonly called the *least squares* method and we now present this for a general relationship in section 2.2.

2.2 The method of ordinary least squares

We now need to develop a model which is valid whatever data is being considered and for this we need some standardized notation.

Let Y_t be the tth observation on the dependent (determined) variable and let X_t be the tth observation on the independent (explanatory or determining) variable. Then the general single (explanatory) variable linear model can be written:

$$Y_t = \alpha + \beta X_t + U_t. \tag{2.4}$$

The device of subscripting the equation by the index t which runs from 1 to T (the total number of observations) allows us to capture in one expression the relationship for all the different data points.

The values α and β are the parameters of the theoretical model which have to be estimated from the data and U_t is the error term for observation t. This model includes an intercept as well as a slope term and therefore does not assume strict proportionality between the variables X and Y. We shall see that if we wish to insist on strict proportionality (e.g. that the a.p.c. = m.p.c. for a consumption function) the equation (2.4) must be modified.

It is necessary to say a little more about terminology. The way of writing the equation indicates that 'causality' flows from X to Y and not vice versa. Accordingly we can say that Y is an *endogenous* variable while X is *exogenous* as far as the equation is concerned. We shall meet situations later where the determining variable is itself endogenous to a larger (multi-equation) model.

The problem of estimation is that, given the T data points—remembering that a point Y_t, X_t is a single piece of information and that we do not observe U_t—we need to obtain estimates of the parameters α and β. A natural method which occurs to us on looking at data, such as that given in figure 2.2, is to fit by eye a line which is as close as possible to the data. This 'curve-fitting' approach raises two distinct difficulties.

1. The data is not always so helpful as in figure 2.2 and hence the choice of 'best' line is open to some ambiguity if a formal criterion of closeness is not used. For example, with a scattergram such as figure 2.4 it would be easy for different econometricians to choose different lines to fit the data if purely visual means were used.

2. More fundamentally the fact that a line is chosen to fit closely to a set of points does not automatically tell us that the estimated parameters describing that line are the 'best' estimates of the unknown parameters. We have not yet shown that the two approaches are equivalent.

We begin with the establishment of a formal criterion of estimation and then examine the properties of the resulting estimators. If we wish the estimated line to be as close as possible to the data it is necessary to give a definition of closeness. It might seem obvious that we should take the distance of any point from the line as its 'closeness'. However even here there are two difficulties. First, distance can be measured vertically, or horizontally, or at any angle (including 90°) to the line. Visual fits do not always distinguish sharply enough between these concepts. For example, in figure 2.5 we show the scatter fitted with a line where, first, vertical and, second, horizontal distances have been indicated. The points are clearly much closer when distances are measured horizontally than when they are measured vertically (and they would be closest when measured orthogonally, i.e. at 90° to the line). In fact, we choose to measure distances vertically, that is in the direction of the variable whose behaviour we are trying to explain. There is a good reason for this as we shall see later. The second difficulty is to distinguish

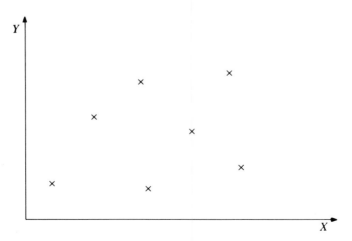

FIG. 2.4   Scattergram of Y against X

12 Basic Model and Least Squares Estimation

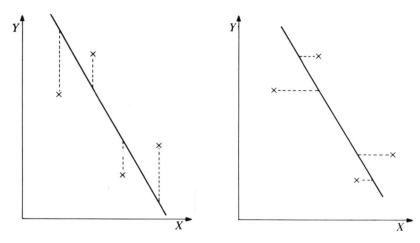

FIG. 2.5 Horizontal and Vertical Measures of Closeness to a Line

points above and below the line. We would obviously wish to argue that a point one unit above the line (in the Y direction) is equidistant from it as a point one unit below the line. However any algebraic formulation would distinguish these as positive and negative values. In a model where we are trying to fit the line as closely as possible to the points, large negative deviations would make for a small measure of closeness. This is unacceptable and in order to avoid such a difficulty we need to make deviations in either direction equally bad. This could be achieved by taking just the absolute value of the deviations or the square of the deviation or the absolute value of the cubed deviation, and so on. There are an infinite number of possible measure. For manipulative purposes absolute values are very cumbersome.[1] It was a natural choice from early on in the study of statistics to use the *squared* distance from the line as the measure of how close it was to a point. Hence we use as a principle of estimation the choice of *that line which minimizes the sum of the squared deviations between it and the data (measured in the Y direction)*. We can now give a formal expression of this principle. The data is generated by a process described by (2.4):

$$Y_t = \alpha + \beta X_t + U_t$$

$$t = 1 \ldots T.$$

Accordingly we would wish to consider equations of the form:

$$\alpha^+ + \beta^+ X_t \qquad (2.5)$$

---

[1] They rule out the use of calculus techniques for optimization and require programming techniques instead.

where α^+ and β^+ can be any values we wish. The set of all possible α^+ and β^+ generates all possible lines relating X_t to the value of (2.5), which we can call the fitted value Y_t^+ (Y_t^+ being the value of the line at the point X_t for a given α^+ and β^+). Our estimation principle is then to choose those values for α^+ and β^+ which minimize the sum of the squared distance between the line Y_t^+ and the actual data Y_t at that same value of X_t. The problem is: choose α^+, β^+ to minimize

$$(Y_1 - Y_1^+)^2 + (Y_2 - Y_2^+)^2 + \ldots + (Y_T - Y_T^+)^2 \tag{2.6}$$

where each Y_t^+ is a function of α^+ and β^+. The notation above, even using dots to indicate that all values of the index t are to be included, is somewhat cumbersome and there is a more compact form using the summation operator Σ. The equation can be written:

$$\sum_{t=1}^{t=T} (Y_t - Y_t^+)^2 \tag{2.7}$$

(the subscripts and superscripts indicate the range of values over which the sum is to be taken, the index being understood just to take integer values). Substituting (2.5) into (2.7) the problem is to minimize:

$$S(\alpha^+, \beta^+) \triangleq \sum_{t=1}^{t=T} (Y_t - \alpha^+ - \beta^+ X_t)^2 \tag{2.8}$$

(where the *symbol* $\triangleq$ stands for 'is defined equal to'). Here we have a standard minimization problem in two variables. The necessary conditions to obtain a minimum are obtained by partially differentiating a function and setting the derivatives equal to zero. Here we need to solve:

$$\frac{\partial S(\alpha^+, \beta^+)}{\partial \alpha^+} = 0$$

$$\frac{\partial S(\alpha^+, \beta^+)}{\partial \beta^+} = 0. \tag{2.9}$$

Differentiating (2.8) requires us to differentiate each term of the sum and then add up the individual components. We obtain the so-called 'normal equations':

$$\frac{\partial S}{\partial \alpha^+} = -2\Sigma(Y_t - \alpha^+ - \beta^+ X_t) = 0 \tag{2.10}$$

$$\frac{\partial S}{\partial \beta^+} = -2\Sigma X_t(Y_t - \alpha^+ - \beta^+ X_t) = 0. \tag{2.11}$$

From (2.10) we obtain

$$T\alpha^+ = \Sigma Y_t - \beta^+ \Sigma X_t \tag{2.12}$$

(remembering that a constant summed over all t is T times the constant). This equation is usually written

$$\alpha^+ = \bar{Y} - \beta^+ \bar{X} \tag{2.13}$$

where $\bar{Y} = \Sigma Y_t / T$ (the mean of the Y_t) etc. We can substitute (2.13), which holds at the minimum, into (2.11), to obtain:

$$\Sigma X_t (Y_t - \bar{Y} + \beta^+ \bar{X} - \beta^+ X_t) = 0. \tag{2.14}$$

Using the notation below to express the variables in *deviations* from their means—

$$y_t = Y_t - \bar{Y}$$
$$x_t = X_t - \bar{X}$$

—we can write (2.14)

$$\Sigma(x_t + \bar{X}) y_t - \beta^+ x_t) = 0. \tag{2.15}$$

However the properties of summations are useful to reduce this expression further. We do not prove (but rather leave it to the reader as a simple exercise) but merely state that for any constant K and variable W_t

(i) $$\sum_{1}^{T} K W_t = K \Sigma W_t \tag{2.16}$$

(ii) $$\Sigma w_t = 0$$

where $$w_t = W_t - \bar{W}. \tag{2.17}$$

Using these results (2.15) becomes

$$\Sigma x_t (y_t - \beta^+ x_t) = 0 \tag{2.18}$$

∴ $$\beta^+ = \Sigma x_t y_t / \Sigma x_t^2 \tag{2.19}$$

and from (2.13):

$$\alpha^+ = \bar{Y} - \bar{X} \frac{(\Sigma x_t y_t)}{(\Sigma x_t^2)}. \tag{2.20}$$

The equations (2.19 and 2.20) yield the 'optimal' estimators for the parameters of a line which passes as close as possible (in a squared deviation sense) to the set of data points (Y_t, X_t). To denote that these are optimal values we write them with a 'hat', $\hat{\alpha}, \hat{\beta}$. Strictly speaking, as we noted above, only the necessary conditions have been derived and we should check that the second-order conditions for a minimum are satisfied at the values $\hat{\alpha}, \hat{\beta}$ (see problem 2.2). The functions (2.19) and (2.20) are known as the (ordinary) least squares estimators[1] and the procedure is sometimes called the *regression* of Y on X.

---

[1] We distinguish an *estimator*, the formulae used in fitting data, from an *estimate*, which is a particular outcome or value of an estimator.

Result 2.1.

For the model:
$$Y_t = \alpha + \beta X_t + U_t$$
the least squares estimators of the parameters are given by:
$$\hat{\beta} = \Sigma x_t y_t / \Sigma x_t^2$$
$$\hat{\alpha} = \bar{Y} - \bar{X} \cdot (\Sigma x_t y_t / \Sigma x_t^2)$$

Example 2.1. A least squares estimation

We can apply these results to the data given in table 2.1. Using the numbers in *raw form* (e.g. consumption in 1961 is $89,628,000,000) we can obtain the following values for the least squares estimates:

$$\hat{\alpha} = 15{,}750{,}146{,}000$$
$$\hat{\beta} = 0.775173.$$

Thus the line:
$$15.75 \times 10^9 + 0.775 \text{ Income}$$
is the 'best fit' to the twenty-two points, where we take squared deviations between consumption and the line as a measure of 'closeness'.

There are several important features of these estimators.

1. The values of the estimators can be calculated and do not depend in any way on any unobservable variable, nor do they require any assumption to be made. They result from a purely mechanistic 'best' fitting procedure.

2. The estimates have economic dimensions as we shall show below in detail. $\hat{\alpha}$ is measured in the same units as Y while $\hat{\beta}$ is in Y units per unit of X.

3. The estimated line passed through the point of means $\bar{Y}$, $\bar{X}$ as (2.13) reveals. This is a property of the particular model we are studying rather than a property of any least squares type estimator.

4. There is a relationship between the estimated parameter values and the true values.
Since:
$$Y_t = \alpha + \beta X_t + U_t$$
$$\therefore \quad \bar{Y} = \alpha + \beta \bar{X} + \bar{U} \qquad (2.21)$$

16 Basic Model and Least Squares Estimation

where
$$\bar{U} = \Sigma U_t / T$$

$$\therefore \quad y_t = \beta x_t + u_t \quad (2.22)$$

where
$$u_t = U_t - \bar{U}.$$

Substituting (2.22) into (2.19) we have

$$\hat{\beta} = \frac{\Sigma(\beta x_t^2 + x_t u_t)}{\Sigma x_t^2} \quad (2.23)$$

$$\therefore \quad \hat{\beta} = \beta + \frac{\Sigma x_t u_t}{\Sigma x_t^2}. \quad (2.24)$$

Similarly:

$$\hat{\alpha} = \alpha - \frac{(\Sigma x_t u_t)}{(\Sigma x_t^2)} \cdot \bar{X} + \bar{U} \quad (2.25)$$

These equations show that the estimated values are related to the true values but that *in general the two are not equal*. Since Σx_t^2 is necessarily positive (except if all the X_t are equal[1]) then only if $\Sigma x_t u_t$ were equal to zero would $\hat{\beta}$ equal β, and for $\hat{\alpha}$ to equal α we would also require $\bar{U}$ to equal zero. Since the U_t are by construction unknown we would have no way of telling whether these conditions held or not for any particular set of data. Thus we have the very important result that for least squares estimation (and for all other methods of estimation) we cannot guarantee that the values of the estimators are equal to the true parameter values for a given set of data.

5. Since the estimates minimize the total of the squared deviations they also, for the given value of T, minimize the average squared deviation.

Having given the general formulae for deriving estimators for a line and using them to construct an estimate of 'least squares' fit for a particular set of data it is helpful to review what has been achieved. So far we have shown how to find general expressions for the estimators of a line such that the average (squared) distance between that line and a set of data points is minimized. At this stage, and without making some assumptions about the nature of the error term, no more can be said about the line. In particular we cannot use the property of the least squares fit by itself to tell us anything about the 'true' relation between Y and X. The relationship between the 'true' line and the estimated line is dealt with in Chapter 3.

From a practical point of view there are a number of issues that arise even if we use the least squares estimation technique purely in this 'curve-fitting' fashion, that is to find the relationship which is as close as possible to a given set of data. First, since economic data can often be presented in different units

---

[1] In this case the OLS estimator does not exist i.e. there is no unique solution to the minimization problem (see problem 2.2).

at different times (e.g. pounds weight or kilograms, dollars or millions of dollars, etc.), it is essential to understand how the properties of least squares fitting are affected by the *scaling* of the data. Second, the same underlying relationship can be presented in different transformations and this also requires analysis of the properties of the estimates applied to transformations of the same behavioural relation. Third, as has been revealed, it is often desired to omit the intercept term from a relationship so this case must be treated.

These three basic variations are dealt with in this chapter while the important generalization to models with more than one explanatory variable is left until Chapter 4.

A final issue of practical importance is to define a measure of 'goodness of fit', so that one equation can be compared to other studies in order to see which has the closer relationship to the data. This theme is taken up in section 2.5.

2.3 Least squares and the scale of variables

So far we have discussed our model, paying no attention to the dimensions of the variables, but this ignores the fact that in reality the same data can often be presented in various units. Further, as we saw in example 2.1, the handling of data holding many figures (most of which were zeros) is very cumbersome. There are two issues involved in changing the scale of data. First if we had made an *exact* scale change, measuring in millions so that the 1961 consumption figure was 89,628, how would the value of the least squares estimates be affected? The second problem is, supposing that for simplicity we not only rescale the series but we round the data, so that the number of significant figures decreases, what effect will this have on our estimates (e.g. if we rounded our consumption figure in 1961 to 89, with the units now being thousands of millions of pounds)? The former problem we can now give a straightforward answer to, but we postpone a treatment of the second until we deal systematically with all measurement error problems in Chapter 8.

We take a simple economic example where the units of the variables are not the same (unlike the consumption function). Consider the demand for meat per week for a household t where the quantity demanded Y_t, measured in kilograms per week, is to be related to the price per kilogram X_t, measured in pence, by the linear equation

$$Y_t = \alpha + \beta X_t + U_t. \tag{2.26}$$

The dimensions of both sides of the equation must be the same so that α, βX_t, and U_t are all in kilograms per week. Since X_t is in pence per kilogram, units of β are in squared kilograms per penny per week.

18 Basic Model and Least Squares Estimation

Let us now measure prices not in pence per kilo but in pounds per kilo and we shall denote the same data (on X_t) measured in £ as X_t^*

$$\therefore \qquad X_t^* = \lambda X_t$$

where $\lambda = 0.01$. Let us denote the new regression by

$$Y_t = \alpha + \beta^* X_t^* + U_t. \qquad (2.27)$$

We know that if X^* is one-hundredth part of X then, to keep the equation the same, the scale of β^* should be one hundred times β:

$$\beta^* = \beta/\lambda. \qquad (2.28)$$

We can check whether the least squares estimators also share this scaling property. Carrying out least squares estimation of α and β^* we have, using our standard formulae on the transformed variables (Y, X^*),

$$\hat{\hat{\beta}}^* = \Sigma x_t^* y_t / \Sigma x_t^{*2} \qquad (2.29)$$

$$\hat{\hat{\alpha}} = \bar{Y} - \bar{X}^* \cdot \hat{\hat{\beta}}^*. \qquad (2.30)$$

Since $x_t^* = \lambda x_t$ and $\bar{X}^* = \lambda \bar{X}$ it is easy to show that:

$$\hat{\hat{\beta}}^* = \hat{\beta}/\lambda \qquad (2.31)$$

$$\hat{\hat{\alpha}} = \hat{\alpha}. \qquad (2.32)$$

These results are very important. First, they show that least squares has the essential property of changing the estimated parameters by exactly the same percentage as the theoretical parameters would change if the units in which the independent variable was measured were changed. Second, it shows that if we knew the estimated value of the regression parameter where the data was measured in pounds (say) we could immediately calculate the estimated value of the parameter for the same data rescaled in any units by just making the inverse scaling to that of the data.

Similar results hold for the rescaling of the dependent variable and for simultaneous rescaling of both variables (see problem 2.4).

2.4 Least squares and transformations of the equation

Let us suppose that our basic model is expressed as usual in the form (2.4):

$$Y_t = \alpha + \beta X_t + U_t$$

i.e. that the level of Y depends on the level of X. However, a different formulation of the same model is that the difference between Y and X depends on the level of X; i.e.

$$Y_t - X_t = A + BX_t + U_t. \qquad (2.33)$$

Now it is evident that $A = \alpha$, $B = \beta - 1$, but the important empirical issue is how the two sets of ordinary least squares estimators are related. Applying our standard formulae to (2.33) and denoting

$$Y_t - X_t \text{ by } Y_t^*$$

$$\therefore \quad \hat{B} = \Sigma y_t^* x_t / \Sigma x_t^2. \tag{2.34}$$

However it is easy to show that

$$y_t^* = y_t - x_t \tag{2.35}$$

and hence that

$$\hat{B} = \hat{\beta} - 1. \tag{2.36}$$

Similarly:

$$\hat{A} = \hat{\alpha}. \tag{2.37}$$

Hence the simple transformation of the equation by an equivalent addition to both sides does not change the estimation of the *basic* (structural) parameters.

Further linear transformations are possible, e.g. we could subtract a constant K from both sides

$$(Y_t - K) = (\alpha - K) + \beta X_t + U_t. \tag{2.38}$$

If we define a new dependent variable Y_t^* where

$$Y_t^* = Y_t - K \tag{2.39}$$

then OLS applied to the transformed equation yields

$$\hat{\hat{\beta}} = \Sigma y_t \, x_t / \Sigma x_t^2 \tag{2.40}$$

$$\hat{A} = \bar{Y}^* - \hat{\hat{\beta}} \bar{X} \tag{2.41}$$

Simple manipulation produces the result that

$$\hat{\hat{\beta}} = \hat{\beta}$$

and

$$\hat{A} = \hat{\alpha} - K.$$

The 'structural' parameters estimated in a roundabout way, by applying OLS to the transformed equation and then 'detransforming' the answers, are identical to estimates obtained by direct application of OLS. This property however does not apply when the transformation is not 'additive'.

It is important to see that our basic model is 'additive' in the *parameters*— the equation (2.4) is a linear or additive function of α and β. This does not imply that it must be linear in the variables. For example, the economic model might relate the rate of change of prices to the level of unemployment —here the dependent variable is the rate of change of prices. Any other non-linear transformations of the variables can be accommodated within this framework provided the model is linear in the parameters. Some models are of course naturally non-linear in parameters, e.g. a constant elasticity Engel curve where Q is quantity of a good and Y is income: $Q = AY^\alpha$. Sometimes it is possible to make a transformation to the equation which renders it linear in parameters—in this case the log transform $\log Q = \log A + \alpha \log Y$. This has become linear in α but the intercept $\log A$ is not a linear transform of A; however given an estimate of $\log A$ we can clearly recover an estimate of A.

Finally there are some equations which cannot be made linear in this way, e.g.

$$Y_t = \alpha + AX_t^\beta.$$

This cannot be estimated using the standard OLS approach but advanced techniques are available to handle so called 'non-linear least squares' problems.

2.5 Residuals and goodness of fit

Once we have estimated the parameters of our model it will obviously be useful to have a measure of how close our estimated line is to the data. The natural criterion is the value of the sum of squared deviations that we minimized. That is, we can use $S(\hat{\alpha}, \hat{\beta})$ to describe the goodness of fit. We denote the deviations between the data (in the Y direction) and the estimated line by $\hat{U}_t$, which we call the *residual*:

then
$$\hat{U}_t = Y_t - \hat{Y}_t = Y_t - \hat{\alpha} - \hat{\beta} X_t \tag{2.42}$$

$$S(\hat{\alpha}, \hat{\beta}) = \sum_{}^{T} \hat{U}_t^2. \tag{2.43}$$

This important value is known as the *residual sum of squares* (RSSQ). It is purely a function of the data (being expressible solely in terms of Y_t and X_t). It can be seen that the measure has the dimension of the square of units of Y and also that it will increase without limit as the number of observations in the data set increases. In order to give a natural dimension to the measure of goodness of fit it is often preferred to divide the RSSQ by the number of observations in which it is based (T) and then to take the square root to reduce it to a measure with the same dimension as that of Y. There is one further refinement that is more usually incorporated and that is to divide the RSSQ not by the number of observations but rather by the number of observations less the number of parameters we have estimated from the data in order to construct the residuals. This number is called the *degrees of freedom* (d.f.) and leads to a measure generally known as the standard error of estimate (SEE):

$$\text{SEE} = \sqrt{\frac{1}{(T-K)} \cdot \sum \hat{U}_t^2} \tag{2.44}$$

where K is the number of estimated parameters (here 2). There are some useful alternative expressions for the SEE which do not require us to directly calculate the residuals but which use the basic data (and the estimated parameter values). Substituting in from (2.42), (2.13), and (2.19) we have:

$$\text{SEE} = \sqrt{\Sigma(y_t - \hat{\beta} x_t)^2 / (T-2)}$$

$$= \sqrt{\left\{ \Sigma y_t^2 - \frac{(\Sigma x_t y_t)^2}{\Sigma x_t^2} \right\} / (T-2)} \qquad (2.45)$$

This measure of goodness of fit is certainly very useful not only in assessing the adequacy of our fitted equation—it could be compared with the mean value of Y or the range of variation of Y in order to show exactly how important or unimportant the residual term is on average—but it can also be used to compare the adequacy of one estimated set of data with another, e.g. consumption functions, over different time periods. However, if we wished to compare equations which were for different dependent variables, e.g. a consumption function and a Phillips curve, the residuals would be measured in different units and so would not be directly comparable. Here it would be useful to have a dimensionless measure of goodness of fit. Such a measure is provided by the squared correlation coefficient, which can be derived in a number of ways. In the theory of statistics the association between two variables Y and X is often measured by the statistic

$$r^2 = \frac{(\Sigma x_t y_t)^2}{(\Sigma x_t^2)(\Sigma y_t^2)}. \qquad (2.46)$$

It is obvious that this statistic is dimensionless—if we were to change the units of measurement of Y or X, the value would remain the same. The value of the squared correlation coefficient can be shown to lie between zero and unity and so any particular value can be compared to the possible extremes (see problem 2.1).

Hence we can measure the 'goodness of fit' by a statistic lying within a known range and which is dimensionless. For some purposes the fact that r^2 is always positive conceals an important aspect of the relationship between Y and X. This can be a positive or a negative association and r^2 does not distinguish between them—hence the (unsquared) correlation coefficient r may be used to summarize the goodness of fit. It is defined by:

$$r = \Sigma xy / \{(x_t^2)^{\frac{1}{2}} (\Sigma y_t^2)^{\frac{1}{2}}\} \qquad (2.47)$$

The correlation coefficient has the same sign as the numerator and hence can be seen to vary between -1 and $+1$.

Describing the model by the correlation between independent and dependent variables is straightforward when there is only one independent variable but for the many-variable case some generalization incorporating all the variables is needed. A different definition of goodness of fit, which allows such

22 Basic Model and Least Squares Estimation

a generalization, is the use of the correlation between the actual dependent variable (Y) and the fitted value ($\hat{Y}$).

$$r^2 = \frac{(\Sigma y\hat{y})^2}{\Sigma \hat{y}^2 \Sigma y^2} \qquad (2.48)$$

It can be seen that in the single explanatory variable case the OLS fitted values $\hat{Y}$ and the independent variable X each have the same correlation with the dependent variable.

A third way to approach the goodness of fit is to focus on the dependent variable Y. This takes different values in the sample data and a completely successful regression would be able to reproduce these with the fitted value $\hat{Y}$. If we take as a measure of the actual variation of the data the total sum of squares (TSSQ) defined as

$$\text{TSSQ} = \Sigma(Y_t - \bar{Y})^2 \qquad (2.49)$$

and define an analogous concept for the fitted values (around their means) which we call the 'explained sum of squares' (ESSQ):

$$\text{ESSQ} = (\hat{Y}_t - \bar{\hat{Y}})^2 \qquad (2.50)$$

then it can be shown that these two concepts are linked by the equation:

$$\text{TSSQ} = \text{ESSQ} + \text{RSSQ} \qquad (2.51)$$

where RSSQ is as before the sum of squared deviations between actual and fitted values.

Since $\Sigma(Y_t - \bar{Y})\}^2 = \Sigma\{(Y_t - \hat{Y}_t) + (\hat{Y}_t - \bar{Y})\}^2$ and using the fact that by OLS $\bar{Y} = \bar{\hat{Y}}$ (by 2.13), therefore $\text{TSSQ} = \text{RSSQ} + \text{ESSQ} + 2\Sigma(Y_t - \hat{Y}_t)(\hat{Y}_t - \bar{Y})$. To obtain our result we need to prove that the last term is identically zero. Replacing $\hat{Y}_t$ by $(\hat{\alpha} + \hat{\beta} X_t)$ we have

$$\Sigma(Y_t - \hat{Y}_t)(\hat{Y}_t - \bar{Y}) = \Sigma(Y_t - \hat{\alpha} - \hat{\beta} X_t)(\hat{\alpha} + \hat{\beta} X_t - \bar{Y}).$$

Using (2.13) to replace $\hat{\alpha}$ by $\bar{Y} - \hat{\beta}\bar{X}$ gives this as

$$= \Sigma(y_t - \hat{\beta} x_t)(\hat{\beta} x_t).$$

When $\hat{\beta}$ is replaced by its OLS value (2.19) this expression does go to zero. This is an important by-product in that it is equivalent to showing that for OLS:

$$\Sigma x_t \hat{U}_t = \Sigma \hat{Y}_t \hat{U}_t = 0 \qquad (2.52)$$

—the *estimated* residual has zero correlation with the independent variable and with the fitted value of the dependent variable. Given that we have established that (2.51) is a property of any OLS estimation (whatever the true error terms) we can then derive a measure of goodness of fit from it. Since we are attempting to make ESSQ as near to TSSQ as possible (to find a line to fit

Basic Model and Least Squares Estimation 23

the data as closely as possible) the ratio of the two is used as a measure:

$$R^2 = \text{ESSQ/TSSQ}. \qquad (2.53)$$

It can again easily be shown that in the single-variable case this measure of goodness of fit (the explained variation in Y) is equal to the squared correlations (2.46) and (2.48) (see problem 2.12). The decomposition (2.51) allows us to write

$$R^2 = 1 - (\text{RSSQ/TSSQ}) \qquad (2.54)$$

Hence, for a given set of data (since the TSSQ is invariant with respect to the actual line chosen) we can see that by minimizing RSSQ (the least squares principle) we have maximized the goodness of fit as defined by (2.54). A final word of interpretation on this last measure may be useful—we see that the ratio of explained sum of squares to total sum of squares (both in deviations from means) lies between zero and unity. It is clear that if $\hat{Y}_t = Y_t$ for all t there is indeed a perfect fit and the measure would attain its upper bound. The lower bound is attained when $\hat{Y}_t = \bar{\hat{Y}} = \bar{Y}$ for all t; that is, when $\hat{\beta} = 0$. If the data produces a zero-estimated slope then the fitted line is purely the intercept (which is then the mean of Y_t). However this criterion of goodness of fit only counts deviations from the mean of Y as being 'relevant'—in effect it is measuring the degree to which this model improves on a model which related Y solely to its mean value—and hence the ESSQ and R^2 would be zero. (See below for a discussion of models without slopes.)

Example 2.2. Goodness of fit of least squares

These measures of goodness of fit can be applied to the data of table 2.1. (measured in £ million).

(2.43):   RSSQ $= 4.77 \times 10^7$

(2.44):   SEE $= 1544$ (d.f. $= 20$) (£ million)

(2.46) or (2.48) or (2.54): $r^2 = 0.991$

                           $\bar{Y} = £116{,}225$ million

The actual values of the dependent variable (Y), the fitted values ($\hat{Y}$), and the residuals ($\hat{U}$) are given in table 2.2.

2.6 Diagrammatic representation of least squares

The basic model (2.4) is considered to be generated by a systematic component ($\alpha + \beta X_t$) and a stochastic component U_t. The data points (Y_t, X_t) can be plotted together with the 'true' line as in figure 2.6. The vertical distance

24 Basic Model and Least Squares Estimation

TABLE 2.2. Residuals, Actual, and Fitted values (in £ million)

Year	Residual ($\hat{U}$)	Actual (Y)	Fitted ($\hat{Y}$)
1961	−2,207	89,628	91,835
1962	−906	91,653	92,559
1963	−291	95,894	96,185
1964	−365	98,806	99,171
1965	−908	100,389	101,297
1966	−1,013	102,212	103,225
1967	276	104,739	104,463
1968	1771	107,804	106,033
1969	152	108,401	106,874
1970	751	111,168	110,417
1971	3,090	114,719	111,629
1972	1,500	121,519	120,019
1973	460	127,734	127,274
1974	−825	125,552	126,377
1975	−1,488	124,824	126,312
1976	−448	125,097	125,545
1977	900	124,646	123,746
1978	−233	131,485	131,718
1979	−593	137,863	138,456
1980	−3,368	136,890	140,258
1981	−103	137,063	137,166
1982	2,477	138,865	136,388

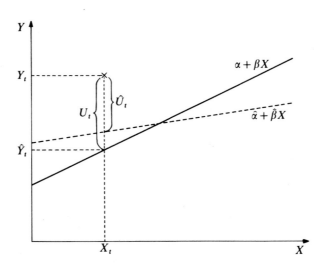

FIG. 2.6 Data Points and the True Line

between the line and the value of Y_t for the associated value of X_t is the random term U_t. Both the true line and random term are unknowable. However, OLS produces an estimate of the line ($\hat{Y}_t = \hat{\alpha} + \hat{\beta}_t$) which may or may not coincide with the true line. The vertical distance between the value of Y and the estimated line at a given value of $X(\hat{Y})$ is the residual or estimated error term ($\hat{U}$).

Least squares, by construction, minimizes the average squared residual (since it minimizes their sum) and so it is clear that if unusually large U_t occur in our data sample OLS may pull the estimated line away from the true line in order to minimize the squared residuals. The total effect then clearly depends on the pattern of the actual U_t.

As well as giving the estimated line (i.e. the value of $\hat{\alpha} + \hat{\beta} X$ for *all* values of X, whether or not these occurred in the sample) it is useful to show the relation between the actual data points and the estimated values. This can be done in several ways, each of which may give an insight as to the adequacy of the model. The first method is to plot the bivariate scattergram between the actual Y_t and the fitted values $\hat{Y}_t(\hat{\alpha} + \hat{\beta} X_t)$. This is shown in figure 2.7 for our estimated consumption function.

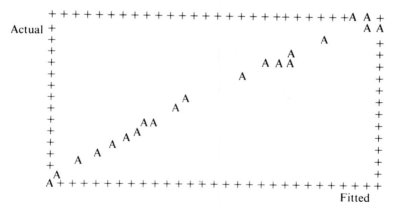

FIG. 2.7 Scattergram of Actual Versus Fitted Values

In this presentation if there were a perfect fit all the points would lie on the 45° line from the origin. Now the actual relation between Y_t and $\hat{Y}_t$ obeys the equation:

$$Y_t = \hat{Y}_t + \hat{U}_t \qquad (2.55)$$

and a regression of Y_t on $\hat{Y}_t$ does indeed have a zero intercept and unit slope (see problem 2.13). Hence we see that the points will be scattered around the perfect fit line with the sum of errors zero (by the properties of the least square fit). However *patterns* of errors may still show up, for example, larger variation at high values of Y_t, which might cast doubt on the adequacy of the model to incorporate all systematic factors.

The second way of plotting the estimated data is to plot both the actual values and the fitted values on the vertical axis with the index (usually time) on the horizontal axis. The vertical gap between the two series is then the residual. This approach is shown in figure 2.8. It concentrates on showing how much of the actual Y level (at a particular t) can be attributed to the constant ($\hat{\alpha}$), how much to the independent variable ($\hat{\beta}X_t$) and how much to the residual ($\hat{U}_t$). It is a particularly good way of seeing whether the model can 'explain' sudden jumps in the level of Y_t or, for example, only responds to them after a delay.

The third way to present the results is to concentrate on the residuals and to plot them against the time index. This information is of course available in figure 2.8 but since the level of the fitted line is moving up and down it can be hard to judge whether the gap between it and the actual values has any pattern. Figure 2.9 plots the residuals from our example. This graph will have the property that the average residual ($\hat{U}_t$) is zero from the least squares fit, but it can reveal patterns in the residuals which may suggest that there are similar patterns in the true errors. This in turn might suggest that some systematic factor could be identified which would improve the model.

2.7 OLS and the model without an intercept

So far we have considered a model in which Y is related to X in such a way that, as X tends to zero, Y tends to a non-zero value (α). Often in economic

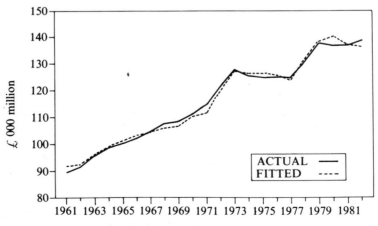

FIG. 2.8 Time Series Plot of Actual and Fitted Values

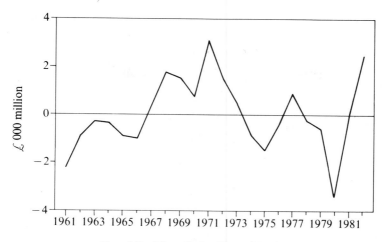

FIG. 2.9 Time Series Plot of Residuals

theory relationships between variables are known to be strictly proportional so that we need to write:

$$Y_t = \beta X_t + U_t. \tag{2.56}$$

If we apply the least squares fitting criterion we wish to minimize,

$$S(\beta^+) = \Sigma(Y_t - \beta^+ X_t)^2 \tag{2.57}$$

and the optimal value is given (see problem 2.14) by:

$$\hat{\beta} = \Sigma Y_t X_t / \Sigma X_t^2. \tag{2.58}$$

It is useful to notice that the difference in estimators (2.19) and (2.58)—which correspond to the same model with or without an intercept—is not in structure but solely depends on whether the data is in deviation form (2.19) or in 'raw' form (2.58). The residuals and residual sum of squares and standard error of estimate follow as before:

$$\hat{Y}_t = \hat{\beta} X_t \tag{2.59}$$

$$\hat{U}_t = Y_t - \hat{Y}_t$$

$$\text{RSSQ} = \Sigma \hat{U}_t^2$$

$$\text{SEE} = \frac{1}{T-1} \Sigma \hat{U}_t^2$$

On the measurement of goodness of fit there is a need to redefine our measures. The measures (2.46), (2.48), and (2.53) are all based on variables measured in deviations from the mean. If (2.46) and (2.48) are used then, by the Cauchy–Schwarz inequality (problem 2.1) they will still lie between zero and one; however the relationship (2.51) does not hold if TSSQ and ESSQ are still in deviations from the mean. This implies that if we attempted to obtain a value of R^2 from (2.54) it would not necessarily lie between zero and one (see problem 2.7). At the same time the interpretation of using a goodness of fit statistic to describe how well the model has described variations in Y around its mean is less acceptable if the intercept is known to be zero. It is possible however to replace all these measures by their analogues in raw data form:

$$\tilde{r}^2 = \frac{(\Sigma X_t Y_t)^2}{\Sigma X_t^2 \Sigma Y_t^2} \qquad (2.60)$$

or

$$\tilde{r}^2 = \frac{(\Sigma Y_t \hat{Y}_t)^2}{\Sigma Y_t^2 \Sigma \hat{Y}_t^2}. \qquad (2.61)$$

These clearly lie between zero and one by the Cauchy–Schwarz inequality (problem 2.1).

Define:
$$\widetilde{\text{TSSQ}} = \Sigma Y_t^2 \qquad (2.62)$$

and
$$\widetilde{\text{ESSQ}} = \Sigma \hat{Y}_t^2 \qquad (2.63)$$

∴
$$\widetilde{\text{TSSQ}} = \widetilde{\text{ESSQ}} + \text{RSSQ} \qquad (2.64)$$

$$\tilde{R}^2 = \widetilde{\text{ESSQ}}/\widetilde{\text{TSSQ}} \qquad (2.65)$$

∴
$$\tilde{R}^2 = 1 - \text{RSSQ}/\widetilde{\text{TSSQ}} \qquad (2.66)$$

It can be shown that (2.60), (2.61), and (2.66) all yield the same measure of goodness of fit (problem 2.15).

2.8 OLS in a model with no slope coefficient

A degenerate but not uninteresting case of our models is where no measured X variable is thought to affect Y. Hence Y fluctuates around an intercept term with variations in the stochastic term. The model can be written:

$$Y_t = \alpha + U_t. \qquad (2.67)$$

The OLS estimator of $\hat{\alpha}$ turns out to be $\bar{Y}$. Hence we can see that measuring the performance of a general model relative to the mean of Y_t is equivalent to comparing the general model to one in which the only variable to 'explain' Y_t is the intercept. Hence the usual R^2 is a measure of the superiority of a model including X (and an intercept) to a model excluding X but including the intercept.

Another useful way of looking at this model is to rewrite it in the form:

$$Y_t = \beta X_t^o + U_t \qquad (2.68)$$

where X_t^o is the 'unit' variable (i.e. $X_t^o = 1$ for all t). Using OLS on the model without an intercept we have the OLS estimator:

$$\hat{\beta} = \frac{\Sigma Y_t X_t^o}{\Sigma X_t^{o2}} \qquad (2.69)$$

which on substitution yields $\hat{\beta} = \bar{Y}$ (which was the value of $\hat{\alpha}$ in (2.67). Hence the intercept can be treated as if it were a variable with a particular set of values. In this case an estimator does exist even though all the X_t have the same value. This is in contrast to the model (2.4) where if X_t was constant then OLS did not work. We can see that this is not surprising—a model with an intercept (a unit variable) *and* a variable which is constant has in fact related Y to a given variable twice. There is obviously no way to find a unique combination of $\hat{\alpha}$ and $\hat{\beta}$ that fits best—infinitely many combinations would have the same residual sum of squares. This is an example of the problem of 'perfect' multicollinearity, which is treated in detail in Chapter 4.

Problems 2

2.1 Prove the Cauchy–Schwarz inequality:

$$(\Sigma a_t^2 \Sigma b_t^2) \geq (\Sigma a_t b_t)^2.$$

(Try for two observations first before going to the general case.)

2.2 Show that the OLS formulae (2.13) and (2.19) generally satisfy the necessary (second-order) conditions for the sum of squares to be minimized, and discuss conditions under which the condition would not be satisfied.

2.3 For the model (2.4) show that the residuals from OLS obey the condition: $\Sigma \hat{U}_t = 0$. Does this imply that $\Sigma U_t = 0$?

2.4 In the equation (2.4) we change the scale of measurement of Y_t replacing it by Y_t^* where $Y_t^* = \mu Y_t$. What relation is there between the estimated coefficients of the regression of Y_t^* on X_t and that of Y_t on X_t? How is the goodness of fit affected by the rescaling? Generalize your results to the case where both dependent and independent variables are rescaled.

2.5 For the model $Y_t = \beta X_t + U_t$ derive the least squares estimator when deviations are measured in the X direction. Under what circumstances does the value of this estimator equal that of the conventional estimator $\hat{\beta}$? Which estimator gives a better fit?

2.6 For the basic model (2.4) the data is transformed into deviations from arbitrary (known) values Y^+, X^+. Show how the values of the estimators from a regression between untransformed data and from a regression between transformed data are related.

2.7 Show that the formula (2.54), when applied to a model without an intercept, can lead to a measure of goodness of fit that is negative.

2.8 Can the following equations be converted into forms that are linear in their parameters?

(i) $\qquad Y_t = A \cdot \exp(b\, X_t) \cdot U_t$

(ii) $\qquad Y_t = \exp(-b\, X_t) + U_t$

(iii) $\qquad Y_t = \exp(a + b\, X_t + U_t)$

(iv) $\qquad Y_t = a/(b - X_t) + U_t.$

2.9 The model $Y_t = \alpha + \beta X_t + U_t$ is estimated by OLS. The data on the dependent variable is also available by its two sub-components Y_{1t} and Y_{2t} (where $Y_t = Y_{1t} + Y_{2t}$). Show that if the two components are separately each regressed on X_t, the sums of the estimated parameters are equal to those that would be obtained from a regression using the aggregate data.

2.10 The model $Y_t = \alpha + \beta X_t + U_t$ is rewritten in the form

$$Y_t - X_t = \alpha + (\beta - 1)X_t + U_t.$$

Show that the values of the OLS estimators of α and β are identical whichever equation is used. Are the two measures of goodness of fit (SEE and R^2) also equal for the two regressions?

2.11 A variable X_t is divided into components X_{it} such that $\Sigma X_{it} = X_t$ (for all t). If each component X_{it} is separately regressed on the total $X_t (t = 1 \ldots T)$ to give parameter estimates $\hat{\alpha}_i$, $\hat{\beta}_i$, show that $\Sigma \hat{\alpha}_i = 0$, $\Sigma \hat{\beta}_i = 1$.

2.12 The following model is estimated by OLS: $Y_t = \alpha + \beta X_t + U_t$. Show that these measures of goodness of fit are equal:

(i) $\qquad (\Sigma x_t\, y_t)^2 / (\Sigma x_t^2 \cdot \Sigma y_t^2),$

(ii) $\qquad \hat{\beta}(\Sigma x_t y_t)/(\Sigma y_t^2),$

(iii) $\qquad (\Sigma y_t\, \hat{y}_t)^2 / \{(\Sigma y_t^2) \cdot (\Sigma \hat{y}_t^2)\},$

(iv) $\qquad 1 - (\Sigma \hat{U}_t^2 / \Sigma y_t^2).$

2.13 Show that the regression of Y_t on the fitted value $\hat{Y}_t$ has a zero intercept and a unit slope.

2.14 For the model $Y_t = \beta X_t + U_t$, derive the ordinary least squares estimator for β. What are the necessary conditions on the data for the estimator to exist?

2.15 In the model $Y_t = \beta X_t + U_t$, which is estimated by OLS, does the estimated line pass through the point of means $(\bar{Y}, \bar{X})$?

2.16 For the model $Y_t = \beta X_t + U_t$, show that the following measures of goodness of fit are equal:

(i) $\qquad (\Sigma X_t\, Y_t)^2 / (\Sigma X_t^2) \cdot (\Sigma Y_t^2)$

(ii) $\qquad (\Sigma Y_t\, \hat{Y}_t)^2 / (\Sigma Y_t^2) \cdot (\Sigma \hat{Y}_t^2),$ (where $\hat{Y}_t$ is the fitted value)

(iii) $$1 - RSSQ/TSSQ^*$$

(where TSSQ* is the total sum of squares about the origin).

2.17 There is a true model $Y_t = \beta X_t + U_t$. This is estimated from the augmented model $Y_t = \alpha + \beta X_t + U_t^*$. Under what circumstances will the estimator of β from the augmented model coincide with that which would have been obtained from the true model? Show that, if the augmented model is estimated subject to the restriction that the value of the estimator of the intercept is zero, the value of the estimator for β will always be equal to that obtained from the true model.

2.18 In the model $Y_t = \beta X_t + U_t$ show, without using calculus, that the estimator $\hat{\beta} = (\Sigma X_t Y_t)/(\Sigma X_t^2)$ minimizes the sum of the squared residuals. (Hint: use the technique of 'completing the square', i.e. construct an expression for the sum of squares which can be expressed as a perfect square of a function of the estimator.)

Answers 2

2.1 From the inequality cancel the term $a_t^2 b_t^2$ from both sides. What is left can be written:

$$\sum_{v,t}^{v \neq t} (a_t b_v - a_v b_t)^2.$$

This expression is strictly positive unless $a_t/b_t = a_v/b_v$ for all v and t. Hence, unless one series is an exact linear multiple of the other, the inequality holds.

2.2 The conditions for a minimum of a function are given by the values of the second derivatives (see standard texts on calculus). In the case of the sum of squares function $S(\hat{\alpha}, \hat{\beta})$ we require:

(i) $\partial^2 S/\partial \hat{\alpha}^2 > 0$

(ii) $\partial^2 S/\partial \hat{\beta}^2 > 0$

(iii) $(\partial^2 S/\partial \hat{\alpha}^2) \cdot (\partial^2 S/\partial \hat{\beta}^2) > (\partial^2 S/\partial \hat{\alpha} \partial \hat{\beta})^2$.

From the normal equations (2.10) and (2.11) we obtain

$$\partial^2 S/\partial \hat{\alpha}^2 = 2T$$
$$\partial^2 S/\partial \hat{\beta}^2 = 2\Sigma X_t^2$$
$$\partial^2 S/\partial \hat{\alpha} \partial \hat{\beta} = 2\Sigma X_t.$$

Hence the first two expressions are positive at the optimum whatever the values of the data. The third condition requires $T \Sigma X_t^2 > (\Sigma X_t)^2$. This is strictly positive, by the Cauchy–Schwarz inequality (see problem 2.1), unless X_t is constant for all t. In such a case the condition holds with equality and there is no turning point in the sum of squares function. Expressing the model

32   Basic Model and Least Squares Estimation

as a two-variable case where one of the variables is the 'unit' variable (see (2.67)), the condition is equivalent to imposing the restriction that the variables are not perfectly 'multicollinear'.

2.3 From the definitions and the values of the OLS estimators we have $\hat{U}_t = Y_t - \hat{\alpha} - \hat{\beta}X_t$. Hence

$$\Sigma \hat{U}_t = \Sigma Y_t - T(\bar{Y} - \hat{\beta}\bar{X}) - \hat{\beta}(\Sigma X_t)$$
$$= \Sigma y_t - \hat{\beta}\Sigma x_t = 0$$

from the properties of sums of deviations around means.

There is of course no reason why the sum of the true errors should be zero in any particular sample; indeed if this were generally to hold (for any size sample) the only way it could do so would be for every error to be zero.

2.4   The original model

$$Y_t = \alpha + \beta X_t + U_t$$

becomes

$$Y_t^* = \mu\alpha + \mu\beta X_t + \mu U_t$$
$$= A + BX_t + E_t.$$

By OLS on the transformed data we have:

$$\hat{B} = \Sigma x_t \cdot y_t^* / \Sigma x_t^2 = \mu\hat{\beta}$$
$$\hat{A} = \bar{Y}^* - \hat{B}\bar{X} = \mu\hat{\alpha}$$

where $\hat{\alpha}$ and $\hat{\beta}$ are given by the OLS estimators (2.19) and (2.13). The goodness of fit for the transformed model is:

$$R^{*2} = (\Sigma x_t y_t^*)^2 / (\Sigma x_t^2) \cdot (\Sigma y_t^{*2})$$
$$= R^2;$$
$$SEE^* = (\Sigma \hat{U}_t^2 / T - 2)^{\frac{1}{2}} = \mu \, SEE.$$

Hence R^2 is unaffected by change of scale but the SEE is affected by the same scale as the dependent variable. Generalizing to the case where both Y_t and X_t are rescaled we have $Y_t^* = \mu Y_t$, $X_t^* = \gamma X_t$ and therefore the transformed model is $Y_t^* = \mu\alpha + (\mu/\gamma)\beta X_t + \mu U_t$ (where we relabel the coefficients A and B). By OLS we obtain:

$$\hat{B} = (\mu/\gamma)\hat{\beta}, \qquad A = \mu\hat{\alpha}$$
$$R^{*2} = R^2, \qquad SEE^* = \mu \, SEE.$$

Scaling affects the estimated coefficients in exactly the same way as it affects the relation of the true coefficients, while the R^2 goodness of fit is unaffected by changes in the scale of either variable.

2.5 Rewriting the equation we have:

$$X_t = \frac{Y_t}{\beta} - \frac{U_t}{\beta}$$

and hence

$$S(\beta^*) = \Sigma(X_t - Y_t/\beta^*)^2.$$

Using

$$\frac{\partial S(\beta^*)}{\partial \beta^*} = 0$$

We obtain

$$\beta^* = \Sigma Y_t^2 / \Sigma X_t Y_t.$$

We can see that on any of our definitions of the correlation coefficient type goodness of fit (R^{*2}) e.g.

$$R^{*2} = (\Sigma X_t \hat{X}_t)^2 / [(\Sigma X_t^2)(\Sigma \hat{X}_t^2)]$$

where

$$\hat{X}_t = Y_t/\beta^*$$

∴

$$R^{*2} = R^2.$$

It can be shown that the standard error of estimate is

$$SEE^* = \left\{ \frac{(\Sigma X_t^2)(\Sigma Y_t^2) - (\Sigma X_t Y_t)^2}{(\Sigma Y_t^2) \cdot (T-1)} \right\}.$$

The SEE for the original model has the same form except that the term (ΣY_t^2) in the denominator is replaced by (X_t^2) so that if the squared variation in Y is greater than that in X, the minimization of the sum of squares in the X direction gives the closer fit.

We could alternatively carry out a direct regression of X on Y to give an estimator for $(1/\beta)$. It can be seen that $(1/\beta) = (\Sigma X_t Y_t)/(\Sigma Y_t^2)$ which, on inverting, gives the same estimator for β as the above procedure.

2.6 Write the model

$$Y_t^\circ = A + BX_t^\circ + U_t$$

where

$$Y_t^\circ = Y_t - Y^+ \quad \text{etc.}$$

We see that

$$A = (\alpha + \beta X^+ - Y^+) \quad \text{and} \quad B = \beta.$$

Now

$$\hat{B} = (\Sigma y_t^\circ x_t^\circ)/(\Sigma x_t^{\circ 2})$$

$$\hat{A} = \bar{Y}^\circ - \hat{B} \bar{X}^\circ$$

From definitions we have:

$$y_t^\circ = y_t \quad \text{etc.}$$

∴

$$\hat{B} = \hat{\beta}$$

$$\hat{A} = \bar{Y} - Y^+ - \hat{\beta}(\bar{X} - X^+) = \hat{\alpha} - Y^+ + \beta X^+.$$

Hence the estimators based on the transformed data give the same values for

34 Basic Model and Least Squares Estimation

the structural parameters (α, β) as would a regression using the basic data (once the transformation has been corrected for).

2.7 From the formula:

$$R^2 = 1 - \text{RSSQ}/\text{TSSQ}$$

where
$$\text{RSSQ} = \Sigma(Y_t - \hat{\beta} X_t)^2$$
$$= \Sigma(Y_t^2) - (\Sigma X_t Y_t)^2/(\Sigma X_t^2)$$

and
$$\text{TSSQ} = \Sigma(Y_t - \bar{Y})^2.$$

Hence
$$R^2 = 1 - (1 - r^{*2})g$$

where
$$g = (\Sigma Y_t^2)/\Sigma(Y_t - \bar{Y})^2$$
$$= \Sigma(Y_t/\bar{Y})^2/\{\Sigma(Y_t/\bar{Y})^2 - T\}$$

and
$$r^{*2} = \Sigma(X_t Y_t)^2/\{(\Sigma Y_t^2)(\Sigma X_t^2)\}.$$

The value of g has an upper limit at plus infinity (when all the Y_t are equal) and a lower limit at unity when the variation of the standardized variables is very large. Since r^{*2} varies between zero and unity (Cauchy–Schwarz inequality) it follows that a low value of r^{*2} and a high value of g (which are independent variables) can produce a negative value of the standard measure of goodness of fit.

2.8

(i) Taking logs:
$$\log Y_t = \text{Log } A + b X_t + \exp U_t$$
so this is linear in b and $\log A$ (but not A).

(ii) This cannot be made linear because the error term is already additive—any transformation would upset the additivity.

(iii) Take logs:
$$\log Y_t = a + B X_t + U_t.$$

(iv) This also cannot be made linear because of the additivity of the error term.

2.9 From the regression using the aggregate data we have:

$$\hat{\beta} = \Sigma(x_t y_t)/(\Sigma x_t^2), \quad \hat{\alpha} = \bar{Y} - \hat{\beta} \bar{X}.$$

From the two disaggregated regressions we have:

$$\hat{\beta}_i = (\Sigma x_t \cdot y_{it})/\Sigma(x_t^2)$$
$$\hat{\alpha}_i = \bar{Y}_i - \hat{\beta}_i \bar{X}.$$

Hence
$$(\Sigma \hat{\beta}_i) = \Sigma(y_{1t} + y_{2t}) x_t/\Sigma(x_t^2) = \hat{\beta}$$

and
$$(\Sigma \hat{\alpha}_i) = \bar{Y}_1 + \bar{Y}_2 - (\Sigma \hat{\beta}_i \bar{X}) = \hat{\alpha}.$$

A useful implication of this result is that, given the aggregate parameters and $(I-1)$ of the disaggregated parameter sets, we can infer the values of the final set. It can also be seen that this result does not generalize to the case where Y is regressed separately on disaggregated components of X.

2.10 Writing the new equation:

$$Y_t^* = A + BX_t + U_t$$
$$B = (\Sigma y_t^* \cdot x_t)/(\Sigma x_t^2) = \hat{\beta} - 1$$
$$\hat{A} = \bar{Y}^* - B\bar{X} = \bar{Y} - \bar{X} - (\hat{\beta} - 1)\bar{X} = \hat{\alpha}$$

where $\hat{\alpha}, \hat{\beta}$ are the OLS estimators for the untransformed model.

$$\text{SEE}^* = \Sigma y_t^{*2} - (\Sigma y_t^* \cdot x_t)^2/\Sigma x_t^2$$
$$= \Sigma y_t^2 - (\Sigma y_t \cdot x_t)^2/\Sigma x_t^2$$
$$= \text{SEE}$$

(using $y_t^* = y_t - x_t$).

$$R^{*2} = (\Sigma y_t^* \cdot x_t)^2/(\Sigma y_t^{*2})(\Sigma x_t^2)$$
$$= \frac{(\Sigma y_t \cdot x_t)^2 + M}{(\Sigma y_t^2)(\Sigma x_t^2) + M}$$

where
$$M = (\Sigma x_t)^2 - 2(\Sigma y_t x_t)(\Sigma x_t).$$

Hence
$$R^{*2} > R^2 \quad \text{if} \quad M > 0$$
$$R^{*2} < R^2 \quad \text{if} \quad M < 0.$$

(From the Cauchy–Schwarz inequality we see that M cannot take a value to make R^{*2} negative.) But the condition that $M > 0$ is equivalent to having for the untransformed model: $\hat{\beta} < 1/2$. Hence if $\hat{\beta}$ is less than one-half, the transformation will increase the value of the squared correlation coefficient. It can easily be seen, using an extension of this argument, that whatever the value of R^2, we can find a value of R^{*2} as close as we like to unity by choosing K sufficiently large or small and using the transformation $Y_t - KX_t = \alpha + (\beta - K)X_t + U_t$. The estimates of the structural parameters are identical once the transformation has been taken into account.

2.11 From the standard formulae for OLS estimators:

$$\hat{\beta}_i = (\Sigma x_{it} \cdot x_t)/(\Sigma x_t^2)$$
$$\hat{\alpha}_i = \bar{X}_i - \hat{\beta}_i \bar{X}.$$

Summing over components we see

$$\Sigma \hat{\beta}_i = 1 \qquad \left(\text{since } \sum_i x_{it} \cdot x_t = x_t^2\right)$$

and $\quad \Sigma \hat{u}_i = 0 \quad \left(\text{since } \sum_i X_i = \bar{X}\right).$

2.12 Comparing (i) and (ii): substituting the formula for the estimator $\hat{\beta}$ into (ii) (i.e. $\hat{\beta}=(\Sigma x_t y_t)/(\Sigma x_t^2)$) we immediately obtain (i). Comparing (iii) and (i): replace $\hat{y}_t$ by $(\hat{\beta} x_t)$ in (iii) and then cancel $\hat{\beta}^2$ in numerator and denominator to obtain (i). Comparing (iv) and (i): first replace $\hat{U}_t$ by $(y_t - \hat{\beta} x_t)$ and then replace $\hat{\beta}$ by the formula for the OLS estimator. This then simplifies to (i).

2.13 Writing the relation between actual and fitted values:

$$Y_t = A + B \hat{Y}_t + E_t$$

we have
$$\hat{B} = (\Sigma y_t \cdot \hat{y}_t)/(\Sigma \hat{y}_t^2)$$
$$= \hat{\beta}(\Sigma y_t \cdot x_t)/\hat{\beta}^2(\Sigma x_t^2)$$

(using) $\quad \hat{y}_t = \hat{\beta} x_t.$

Hence, by substituting in the formula for $\hat{\beta}$:

$$\hat{B} = 1.$$

Similarly $\quad \hat{A} = \bar{Y} - \hat{B} \bar{\hat{Y}}$

but $\quad \bar{\hat{Y}} = \Sigma(\hat{\alpha} + \hat{\beta} X_t)/T = \bar{Y} - \hat{\beta}\bar{X} + \hat{\beta}\bar{X}_t$
$\quad = \bar{Y}.$

Hence $\quad \hat{A} = 0.$

2.14 Define the sum of squared deviations function

$$S(\beta^+) = \Sigma(Y_t - \beta^+ X_t)^2$$

and minimize with reference to β^+

$$\therefore \quad \frac{\partial S}{\partial \beta^+} = -2\Sigma(Y_t - \beta^+ X_t)X_t = 0.$$

Hence at the minimum

$$\beta^+ = \Sigma Y_t X_t / \Sigma X_t^2.$$

The value of the second derivative where the first derivative is zero must be positive for it to be a minimum. But:

$$\frac{\partial^2 S}{\partial \beta^{+2}} = \Sigma X_t^2$$

which is strictly positive unless all the X values are zero. Hence, providing at least one of the values of the independent variable is non-zero, the OLS estimator exists (i.e. there is a minimum to the sum of squares function).

2.15 Since from the formula for OLS in a model without an intercept:
$$\hat{\beta}=(\Sigma X_t Y_t)/(\Sigma X_t^2)$$
for the regression line to pass through the point of means for the fitted equation it must satisfy the condition:
$$\bar{Y}=\bar{X}(\Sigma X_t Y_t)/(\Sigma X_t^2).$$
It is clear that in general this condition is not satisfied and hence the property that OLS passes through the point of means is true only when the equation includes an intercept.

2.16 Comparing (i) and (ii), Replace $\hat{Y}_t$ by $\hat{\beta} X_t$ in (ii) and cancel $\hat{\beta}^2$ in numerator and denominator to yield (i). Comparing (i) and (iii), since
$$1-\text{RSSQ}/\text{TSSQ}^* = 1 - \Sigma(Y_t-\hat{\beta} X_t)^2/\Sigma Y_t^2$$
$$= 1 - \frac{\{\Sigma Y_t^2 - (\Sigma X_t Y_t)^2/(\Sigma X_t^2)\}}{\Sigma Y_t^2}$$
$$= (\Sigma X_t Y_t)^2/\{(\Sigma Y_t^2)(\Sigma X_t^2)\}$$
which is measure (i). These measures of goodness of fit take as their point of comparison the model $Y_t = U_t$ (i.e. zero intercept as well as zero slope).

2.17 We must minimize the sum of squares subject to the restriction. This can be done either by substituting the restriction into the equation and then minimizing (in which case we return to the basic model without an intercept), or else by using a Lagrangian.

Define: $S(\alpha^*, \beta^*, \lambda^*) = \Sigma(Y_t - \alpha^* - \beta^* X_t)^2 + \lambda^* \alpha^*.$

The first-order conditions for a minimum are:
$$\frac{\partial S}{\partial \alpha^*} = -2\Sigma(Y_t - \alpha^* - \beta^* X_t) + \lambda^* = 0$$
$$\frac{\partial S}{\partial \beta^*} = -2\Sigma(Y_t - \alpha^* - \beta^* X_t) X_t = 0$$
$$\frac{\partial S}{\partial \lambda^*} = \alpha^* = 0$$

Substituting the third condition into the second we obtain the standard normal equation for a model without an intercept and hence:
$$\beta^* = \Sigma(Y_t \cdot X_t)/(X_t^2)$$
$$\alpha^* = 0.$$

We also have from the first equation:
$$\lambda^* = 2\Sigma(Y_t - \beta^* X_t).$$

38 Basic Model and Least Squares Estimation

If λ^* is zero then we know that the constraint would not have been binding, i.e. the results would have been the same in the absence of a constraint. This occurs if $\bar{Y} = \bar{X}(\Sigma Y_t \cdot X_t)/(\Sigma X_t^2)$ (*). By substitution we can show that if this condition holds $\Sigma(x_t \cdot y_t)/\Sigma(x_t^2) = \Sigma(Y_t \cdot X_t)/\Sigma(X_t^2)$. This result is implicit in problem 2.15 where it was shown that the model without an intercept passes through the point of means only if the condition (*) holds. The value of λ^* indicates the extent to which the constraint is binding, since we can see that $\lambda^* = 2T\alpha^*$, i.e. it is proportional to the value that would have occurred in the model with an intercept, had the intercept not been constrained to equal zero.

2.18 The sum of squares for the model is:

$$S(\beta^*) = (Y_t - \beta^* X_t)^2$$
$$= \alpha_0 \beta^{*2} + \alpha_1 \beta^* + \alpha_2 \quad \text{(i)}$$

where
$$\alpha_0 = \Sigma X_t^2, \quad \alpha_1 = -2\Sigma X_t Y_t, \quad \alpha_2 = \Sigma Y_t^2.$$

The function (i) can be written as:

$$S = \alpha_0 \left(\beta^{*2} + \frac{\alpha_1 \beta^*}{\alpha_0} + \frac{\alpha_1^2}{4\alpha_0^2} \right) - \left(\frac{\alpha_1^2}{4\alpha_0} - \alpha_2 \right)$$

$$= \alpha_0 \left[\left(\beta^* + \frac{\alpha_1}{2\alpha_0} \right)^2 - \left(\frac{\alpha_1^2}{4\alpha_0} - \alpha_2 \right) \right].$$

Since only the first term is a function of β^*, S will be minimized when the first term is zero (being always positive or zero). Hence the minimum is achieved at

$$\beta^* = -\alpha_1/2\alpha_0$$
$$\therefore \quad \beta^* = \Sigma X_t Y_t / \Sigma X_t^2.$$

3
The Properties of Least Squares Estimation

In Chapter 2 we have shown how to derive an estimator for the parameters of a one-variable model. These ordinary least squares estimators by construction have the property that they minimize the sum of the squared residuals between the data and the estimated regression line. The line is, in a particular sense, the 'closest' possible to the data. However this property of the method of OLS, as we have already mentioned, does not guarantee that the estimated line (parameters) is near to the true line (parameters) and we are certainly interested in the ability of OLS to tell us something about the true line or economic process which has generated the data. It is evident that no method of estimation could be devised which would inevitably give very accurate values—the nature of the error term is such that there could occur a set of U_t in the given observation period that are untypical (e.g. several very large positive values). OLS will then pull the estimated line towards these observed points and away from the systematic true line. Recognizing this limitation, we require a method that taken in all circumstances would 'on average' produce the most accurate estimates of the parameters. In order to develop this line of analysis it is necessary to define terms. The key here is the mention of 'all circumstances'. This is most easily understood by reference to a controlled experiment of the type used in sciences rather than in social sciences. Imagine that the investigator wants to know the exact values of the parameters in a linear equation linking the input of fertilizer to the yield of wheat obtained from plots of land treated with different levels of the fertilizer. Symbolically we have the relation:

$$Y_t = \alpha + \beta X_t + U_t \tag{3.1}$$

where Y_t is yield on plot t, X_t is application of fertilizer to plot t, and there are T plots with different levels of input. The error term U_t expresses all the other unmeasured factors on plot t which affect the yield on that plot (e.g. soil variation, micro-climatic variations, variation in the uniformity of sowing, and so on). Such an experiment could be carried out on a strip of land divided into T plots, the data recorded, and OLS estimates of α and β obtained. However we can imagine that next to the first strip of land a second strip could be treated with exactly the same set of input levels—see figure 3.1. However, even with the same X level the error would be different and so the resulting yield would be different.

We can represent this by a generalized expression

$$_\tau Y_t = \alpha + \beta X_t + {}_\tau U_t \tag{3.2}$$

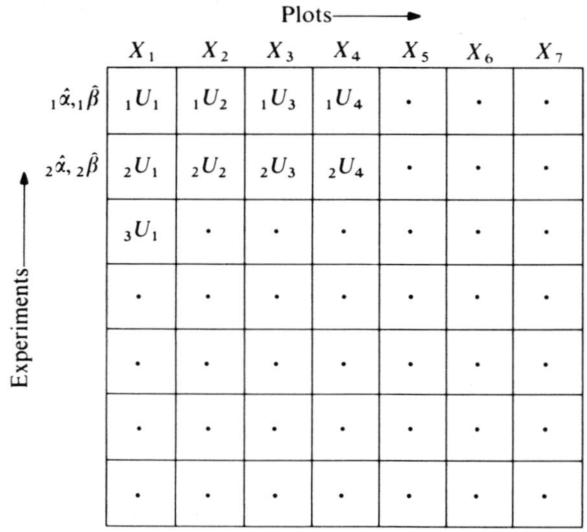

40 Properties of Least Squares Estimation

	X_1	X_2	X_3	X_4	X_5	X_6	X_7
$_1\hat{\alpha}, _1\hat{\beta}$	$_1U_1$	$_1U_2$	$_1U_3$	$_1U_4$	.	.	.
$_2\hat{\alpha}, _2\hat{\beta}$	$_2U_1$	$_2U_2$	$_2U_3$	$_2U_4$	.	.	.
	$_3U_1$	.	.	.	.	.	.
	.	.	.	.	.	.	.
	.	.	.	.	.	.	.
	.	.	.	.	.	.	.
	.	.	.	.	.	.	.

(Plots →, Experiments ↑)

FIG. 3.1 Schematic Representation of a Repeated Samples Experiment

where τ is the experiment or sample number. The parameters α and β are universal for this model and therefore are identical for all samples. The set of X_t (the values from X_1 to X_T) are assumed to be the same in every sample ('X is fixed in repeated sampling') so that for a given level of X the variation in Y is determined by the variation in the error for that level of X. Conceptually we could carry out several experiments (all with the same set of X values) and for each one obtain an OLS estimate of the parameters α and β. These would themselves be distinguished by the experiment number—$_\tau\hat{\alpha}$ and $_\tau\hat{\beta}$. Having defined this framework we can then perform the purely conceptual experiment of considering the set of samples which contain all possible values of the error terms. This is purely a theoretical construct but it does allow us to analyse the 'typical' or 'average' outcome.

3.1 The Gauss–Markov theorem

In order to derive any interesting properties for the method of OLS in the 'typical' situation it is necessary to make some very particular and important assumptions about the error term and the model itself. For the model we are considering there are five assumptions made:

(i) $E_\tau(_\tau U_t) = 0$ for all t \hfill (3.3)

(ii) $E_\tau(_\tau U_t^2) = 0$ for all t \hfill (3.4)

(iii) $E_\tau({}_\tau U_t \, {}_\tau U_s) = 0$ for all $s, t, s \neq t$ (3.5)

(iv) X_t is fixed (with respect to τ) in repeated samples (3.6)

(v) $\dfrac{1}{T} \Sigma x_t^2 > 0$ (3.7)

The symbol E, known as the expectation operator, is the average of a random variable over all possible outcomes of the variable where each outcome is weighted by its frequency (probability of occurring). For a random variable r which can take any of a (discrete) set of values $r_i (i = 1, 2, \ldots)$ and for which the probability of a value r_i is given by $f(r_i)$ its expected value of the variable is

$$E(r_i) = \sum_{i=1}^{\infty} f(r_i) r_i \qquad (3.8)$$

This is clearly just the population mean of the variable. The general properties of expectations are dealt with in problem 3.1, but it is important to know the following results. If a and b are any two random variables then

$$E(a+b) = E(a) + E(b) \qquad (3.9)$$

('the average of a sum is the sum of the averages'). Further if k is a constant then

$$E(ka) = kE(a) \qquad (3.10)$$

However it is *not generally true* that the analogous property to (3.9) holds for the product of two random variables, i.e.

$$E(ab) \neq E(a) E(b) \qquad (3.11)$$

When the expected value of the product is equal to the product of the expected values

$$E(ab) = E(a) E(b) \qquad (3.12)$$

then the variables are said to be *independent*. We can now turn to the interpretation of the assumptions:

Assumption (i) means that the typical or repeated sample average value of the error term is zero for every value X, that is the stochastic term is not affected by factors that would be persistently positive (or negative), for a given value of X.

Assumption (ii) means that the average squared distance of the stochastic terms is the same at every level of X. There is no tendency for certain values of X to be associated with stochastic terms that have a particularly large (or small) range or variation between the repeated samples. The errors are said to be *homoskedastic*.

Assumption (iii) is to be interpreted as saying that for any pairs of values of the X variable there is no tendency for the associated stochastic terms to be related, i.e. no tendency for a high value of U_t to be always associated with a higher than average (or lower than average) value of U_s. The error terms are pairwise independent.

Assumption (iv) we have already met in our discussion of the idea of the 'repeated trial'.

Finally assumption (v), as we saw in Chapter 2, is required in order to allow a unique solution for an ordinary least squares estimation to be derived.

It is extremely important to understand the role of these assumptions in econometrics. They are in no sense meant to be a description of reality or even necessarily a reasonable approximation in certain situations. They are a set of conditions which, *if true*, allow us to claim certain properties for OLS estimators. As we shall see, a large part of econometrics is addressed to the question of how we should proceed if one or more of these assumptions does not hold.

If the assumptions are correct then the *Gauss–Markov* theorem follows. This says that the *ordinary least squares estimator* (OLS) is the *best linear* (*in Y*) *unbiased estimator*. We shall first define these terms, next prove the theorem, and then discuss its meaning. We begin with the concept of an 'unbiased' estimator. Let us consider any method of estimation (i.e. least squares, least absolute differences, least fourth power differences, etc.) and denote the arbitrary estimator $\breve{\beta}$. Such an estimator will, in repeated samples, generate the set of values $_\tau\breve{\beta}$ (where τ is the experiment number). An unbiased estimator is one for which the expected value averaged over all repeated samples (i.e. all sets of error terms) is equal to the true value: that is, if

$$E_\tau(_\tau\breve{\beta}) = \beta \qquad (3.13)$$

then $\breve{\beta}$ is unbiased. This property must also hold whatever the true value of β. If $E(\breve{\beta}) \neq \beta$ then the estimator is said to be *biased* and $[\beta - E(\breve{\beta})]$ is the measure of the bias. It is easy to construct biased estimators, e.g. the estimator $\breve{\beta} = \frac{1}{2}$ (all τ) is in general biased (although it would have the correct value if $\beta = \frac{1}{2}$), and we can see that there must be infinitely many biased techniques of estimating the parameters of a model. What is not perhaps so obvious is that there are also infinitely many unbiased estimators—we shall demonstrate some later. Clearly, given the choice between a method of estimation which on average gives the correct answer and one which systematically overstates (or understates), we may well prefer an unbiased estimator.

The existence of a multiplicity of unbiased estimators means that in choosing between them we can use a second criterion of performance. However the use of the natural criterion of 'closeness' to the true value (i.e. from sample to sample), which would be the average squared difference between the estimator and the true value, does not yield a useful solution if applied to the set of all unbiased estimators. Hence, it has been found necessary to restrict consideration to estimators which are also linear in the Y_t. These are defined as those estimators which can be written in the form:

$$\beta^\circ = \Sigma w_t Y_t \qquad (3.14)$$

where w_t are weights which are fixed in repeated sampling, that is, for a given value of X_t the repeated outcomes $_\tau Y_t$ are always assigned the weight w_t. This formulation rules out estimators which are non-linear (e.g. in which powers of the dependent variable enter),[1] and also estimators which are inhomogeneous with respect to Y_t, i.e. which also include a constant term.

The set of all linear estimators (i.e. all possible sets of w_t) is also infinitely large and includes both biased and unbiased estimators (but not all of either set). Hence we restrict ourselves to the intersection of the two sets of estimators and consider solely the set of all linear unbiased estimators. Within this set, which still contains an infinite number of estimators, we can apply a second criterion of optimality and look for that estimator which has the smallest average squared difference (error) from the true value; that is,

$$\min E_\tau[(_\tau\beta^\circ - \beta)^2]. \tag{3.15}$$

Since β° is restricted to be unbiased we can equivalently write (3.15) as

$$\min E_\tau[\{_\tau\beta^\circ - E_\tau(_\tau\beta^\circ)\}^2] \tag{3.16}$$

The average squared deviation of any variable from its own average value is known as the *variance* so that we are looking for the linear unbiased estimator with the smallest variance.[2] Hence to prove the Gauss–Markov theorem for the model $Y_t = \alpha + \beta X_t + U_t$, we must show that if the assumptions (i)–(v) hold then the OLS estimators can be written as linear functions of Y, are unbiased, and have the smallest variance of any such estimators.

We recall that application of OLS gives

$$\hat{\beta} = \Sigma x_t y_t / \Sigma x_t^2 \tag{3.17}$$

$$\hat{\alpha} = \bar{Y} - \beta \bar{X}. \tag{3.18}$$

It is immediately apparent that these are both linear estimators:

$$\hat{\beta} = \Sigma w_t Y_t \tag{3.19}$$

where

$$w_t = x_t / \Sigma x_t^2 \tag{3.20}$$

noting that

$$\Sigma x_t y_t = \Sigma x_t Y_t$$

and

$$\hat{\alpha} = \Sigma v_t Y_t \tag{3.21}$$

---

[1] It is important to be clear what is meant by 'linear'. The restriction is in the functional form by which the dependent variable Y (whatever it is in economic terms) enters into the estimating formula. Estimators of the forms $\beta^\circ = \Sigma W_t \log Y_t$ or $\Sigma W_t Y_t^2$ are ruled out. However the economic concept described by Y may well be a non-linear function of some more basic variable, e.g. in an equation explaining wages the functional form chosen might relate the *rate of change* of wages (the Y variable) to the level of unemployment (the X variable) in a linear fashion: $Y = \alpha + \beta X$. The class of linear estimators in this situation are all weighted sums of the rates of change of wages.

[2] For biased estimators (3.15) and (3.16) are not the same—the former criterion is known as the *mean square error*.

where
$$v_t = \left(\frac{1}{T} - \frac{x_t \bar{X}}{\Sigma x_t^2}\right) \qquad (3.22)$$

With X_t fixed in repeated samples these weights are also fixed in repeated samples.

It is next shown that the estimators are unbiased. In order to show this we always adopt these same tactic. Since we wish to relate the estimator (3.17) or (3.18) to the true value, it is necessary to introduce the true parameter into these equations by replacing Y_t with its representation in the true equation (3.1). For example:

$$\hat{\beta} = \Sigma x_t (\beta x_t + u_t)/\Sigma x_t^2 \qquad (3.23)$$

$$\hat{\beta} = \beta + \Sigma x_t u_t / \Sigma x_t^2. \qquad (3.24)$$

Now consider the typical term within the summation: $x_t u_t / \Sigma x_t^2$. In repeated samples for a particular t the x_t and the Σx_t^2 will remain constant and only U_t would vary. Hence the average overall of this term is determined by the average value of U_t using (3.10):

i.e.
$$E(x_t u_t / \Sigma x_t^2) = \{x_t E(u_t)/\Sigma x_t^2\}. \qquad (3.25)$$

However, by assumption (i) this is zero since $E(U_t) = 0$ and hence

$$E(u_t) = E(U_t - \bar{U}) = 0.[1]$$

Since the average of a sum is the sum of the averages we can take expectations to (3.24) to derive

$$E(\hat{\beta}) = \beta + E(\Sigma x_t u_t)/\Sigma x_t^2 = \beta \qquad (3.26)$$

so that OLS is unbiased for β. It is left to the reader to show that $\hat{\alpha}$ is unbiased (problem 3.5).

Hence we have shown that OLS is a linear unbiased estimator whatever the true values of α and β and whatever the values of the X_t (in the fixed sample scheme). We must next show that it has the smallest variance. There are two routes to proving this—we can set out an expression for a general linear unbiased estimator and choose the weights to minimize the variance, finding the resultant estimator to be OLS; or we can derive the variance of OLS and show that it is smaller than that of any other linear unbiased estimator. We follow the second route but problem 3.2 is addressed to the first line of proof.

We require now the variance of the OLS estimators. Since from (3.24)

$$\hat{\beta} = \beta + \Sigma x_t u_t / \Sigma x_t^2$$

---

[1] Note that $\bar{U}$ is not identically equal to zero for a particular sample but rather that on average it is zero.

and $E(\hat{\beta}) = \beta$

∴ $\text{Var } \beta = E\{(\hat{\beta}-\beta)^2\} = E\{(\Sigma x_t u_t)^2/(\Sigma x_t^2)^2\}$ (3.27)

In order to evaluate this we must first expand the squared terms *before* taking averages (expectations) over all samples.[1] This can be written:

$$\Sigma(x_t u_t)^2/(\Sigma x_t^2)^2 = \Sigma x_t^2 u_t^2/(\Sigma x_t^2)^2 + \sum_{t}\sum_{s}^{t \ne s} x_t x_s u_t u_s/(\Sigma x_t^2)^2.$$ (3.28)

The second double sum in effect takes a given value for $s(x_s u_s)$ and multiplies it by all 'cross-product' terms $(t \ne s)$ and adds them up. This is repeated over all s. The same operation is sometimes written

$$2\sum_{s}\sum_{t}^{s<t} x_t x_s u_t u_s$$

since $x_s x_t u_s u_t = x_t x_s u_t u_s.$

We can now take expectations, and using the assumptions that the x_t are fixed, that the U_t are pairwise independent (so that the expectations of the second item is zero), and that all U_t have the same variance we obtain, using (3.10):

$$\text{Var } \hat{\beta} = \frac{\sigma^2}{\Sigma x_t^2}.$$ (3.29)

It can be similarly shown (problem 3.5) that

$$\text{Var } \hat{\alpha} = \sigma^2 \left(\frac{1}{T} + \frac{\bar{X}^2}{\Sigma x_t^2} \right)$$ (3.30)

$$= \frac{\sigma^2 \Sigma X_t^2}{T \Sigma x_t^2}.$$ (3.31)

Finally we can introduce the concept of the covariance between the estimated parameters, which is defined by:

$$\text{Cov}(\hat{\alpha}, \hat{\beta}) = E\{(\hat{\alpha}-\alpha)(\hat{\beta}-\beta)\}.$$ (3.32)

For OLS the value of the covariance of the estimated parameters can be shown to be

$$\text{Cov}(\hat{\alpha}, \hat{\beta}) = -\frac{\bar{X}\sigma^2}{\Sigma X_t^2}.$$ (3.33)

Once the variance of the OLS estimator has been derived the proof needs to show that no other linear unbiased estimators have smaller variances than (3.29) and (3.31). Let us denote an arbitrary linear unbiased estimator by:

$$\beta^+ = \Sigma k_t Y_t$$ (3.34)

---

[1] A numerical example would show that in general the average of squared terms is not equal to the square of the average so that the order of operation is important.

where k_t are fixed in repeated samples. This can be rewritten

$$\beta^+ = \Sigma(w_t + c_t) Y_t \qquad (3.35)$$

where
$$k_t = w_t + c_t$$

and w_t are the ordinary least squares weights. Now if β^+ is unbiased we have

$$E(\beta^+) = \beta \qquad (3.36)$$

and hence
$$E\{\Sigma(w_t + c_t) Y_t\} = \beta. \qquad (3.37)$$

However, $E(\Sigma w_t Y_t) = \beta$ because this is the unbiased estimator OLS. Hence we require $E(\Sigma c_t Y_t) = 0$ to generate unbiasedness. Thus the set of all linear unbiased estimators can be defined relative to the OLS estimator by (3.35) subject to $E(\Sigma c_t Y_t) = 0$. If we substitute for Y_t this condition is expressed as:

$$E\{\Sigma c_t(\alpha + \beta X_t + U_t)\} = 0 \qquad (3.38)$$

Now since the c_t are fixed in repeated samples the term $E(\Sigma c_t u_t)$ is zero for all c_t, and since X_t and c_t are fixed in repeated samples the restriction is:

$$\alpha \Sigma c_t + \beta \Sigma c_t X_t = 0. \qquad (3.39)$$

Now the only values of c_t which can be *guaranteed to satisfy this condition whatever the values of* α, β *and the* X_t (i.e. irrespective of the actual model being considered) are those obeying

$$\Sigma c_t = 0 \qquad (3.40)$$

$$\Sigma c_t X_t = 0 \qquad (3.41)$$

These two restrictions allow us to generate the set of all linear unbiased estimators for the parameter β, which have that property whatever the numerical values of the true parameters and independent variables.[1] We can next derive the variance for the arbitrary linear unbiased estimator β^+.

$$\text{Var } \beta^+ = E\{(\Sigma k_t Y_t - \beta)^2\}. \qquad (3.42)$$

Substituting in for k_t and Y_t we have

$$\text{Var } \beta^+ = E[\Sigma\{(w_t + c_t)(\alpha + \beta X_t + U_t) - \beta\}^2]. \qquad (3.43)$$

Now we have from OLS (3.20) that

$$w_t = x_t / \Sigma x_t^2$$

and using (3.40) and (3.41)

$$\text{Var } \beta^+ = E[\{\Sigma(w_t U_t + c_t U_t)\}^2] \qquad (3.44)$$
$$= E\{(\Sigma w_t U_t)^2 + (\Sigma c_t U_t)^2$$

---

[1] As we have pointed out before it is possible to find other linear unbiased estimators for particular models (e.g. $\hat{\beta} = \frac{1}{2}$ when $\beta = \frac{1}{2}$) but they are not universally unbiased and since we do not know the true parameters then we require an estimator that always has the desired properties.

$$+2(\Sigma c_t U_t)(\Sigma w_t U_t)\} \qquad (3.45)$$
$$=\sigma^2 \Sigma w_t^2 + \sigma^2 \Sigma c_t^2 + 2\sigma^2 \Sigma c_t w_t \qquad (3.46)$$

using assumptions (ii) and (iii). But

$$\Sigma c_t w_t = \Sigma c_t x_t / \Sigma x_t^2 \qquad (3.47)$$

and this is zero by (3.40) and (3.41). Hence we have proved that for any linear unbiased estimator the variance is

$$\text{Var } \beta^+ = \sigma^2 \Sigma w_t^2 + \sigma^2 \Sigma c_t^2 \qquad (3.48)$$

where w_t are the weights of the OLS estimator. Finally we note that from (3.4)

$$\Sigma w_t^2 = 1/\Sigma x_t^2 \qquad (3.49)$$

so that
$$\text{Var } \beta^+ = \text{Var } \hat{\beta} + \sigma^2 \Sigma c_t^2. \qquad (3.50)$$

Since the second term on the right-hand side is strictly positive (unless all the c_t are zero) it follows that the variance of any other linear unbiased estimator is greater than that of OLS. Moreover, we have a direct measure of the increase in variance: it is given by the second term on the right-hand side, which is a function of the c_t and these are the differences between the OLS weights and the weights for the alternative estimator. A similar line of argument establishes that the OLS estimator $\hat{\alpha}$ also has the smallest variance of all linear unbiased estimators of α (problem 3.5). The results are summarized in the following statement of the Gauss–Markov theorem.

The Gauss–Markov Theorem

For the model $\quad Y_t = \alpha + \beta X_t + U_t \qquad t = 1 \ldots T$

If
- (i) $\quad E(U_t) = 0 \quad$ all t
- (ii) $\quad E(U_t^2) = \sigma^2 \quad$ all t
- (iii) $\quad E(U_t U_s) = 0 \quad$ all $t, s \quad$ with $s \neq t$
- (iv) $\quad X_t$ fixed in repeated samples
- (v) $\quad \Sigma x_t^2 > 0$

then the OLS estimators

$$\hat{\beta} = \Sigma x_t y_t / \Sigma x_t^2$$
$$\hat{\alpha} = \bar{Y} - \hat{\beta} \bar{X}$$

have the smallest variances of all linear unbiased estimators for the parameters out of the class of estimators that are linear unbiased whatever the actual values of α and β and whatever the values of the X_t.

The properties of the OLS estimators under these conditions are often described by the acronym BLUE (best linear unbiased estimators) where 'best' is the technical term for the smallest variance among estimators.

Some final points should be made about the status of the theorem:

1. It provides a set of *sufficient* conditions for OLS to be BLUE *whatever* the model. These conditions are not necessary however. It can be shown that there is a set of *necessary and sufficient* conditions on the assumptions about the error properties for OLS to be BLUE whatever the values of the parameters or exogenous variables. These conditions allow only a slight relaxation on the assumptions that we have used and in practice virtually all departures from the set of assumptions (i) to (v) will have the effect that OLS is no longer BLUE and hence by implication some other estimator will be BLUE. A substantial part of the rest of the book is devoted to this situation.

2. It is important to notice that for any given set of data or true parameters there exist estimators which may be superior to OLS. However since these estimators in fact require knowledge of the true parameters they are unusable in practice.

3. The restriction to estimators which are (homogeneous) linear functions of the Y_t can be seen to have a very desirable result in that we can prove OLS to be optimal whatever the true data and parameter values; and it is for this reason that we primarily concentrate on such estimators. It must be recognized that in some circumstances non-linear or inhomogeneous linear estimators will be required and indeed may improve upon OLS.

4. The condition (v), that the set of X_t should not all have the same value and hence have zero deviation from means, in effect allows a unique solution for ordinary least squares. If this condition failed—so that the X_t were constant—then the estimator could be seen to be undefined and its variance would be infinite (all values of $\hat{\beta}$, including infinitely large ones, would fit the data equally well).

5. We have established that the method of OLS when applied to a set of data has a certain optimality property over repeated samples. It is very important to realize that it says nothing about any single application of the technique—we can merely claim that the technique we use to estimate, with a particular set of data, is one which if repeated over all conceivable error situations would have the BLUE property.

6. In practical terms we can see that three of the conditions guaranteeing the optimality of OLS are unobservable. By the logic of the model the error terms (U_t) are not knowable (without knowing the parameters) and so we cannot hope *to know for certain* whether conditions (i), (ii), and (iii) do hold. Hence our use of OLS and assertion of its optimality are conditional on factors that we cannot know to be correct. This means that, in practice, tests for the plausibility of these assumptions are required as well as alternative estimation techniques when it is known or likely that the assumptions are not true.

7. For different models, i.e. with different numbers of dependent variables or without a constant, it is necessary, using the simple scalar algebra of this book, to reprove the Gauss–Markov theorem for each case. Only with matrix algebra can we construct a proof that holds for any model linear in the parameters.

8. The comparison between OLS and all other linear unbiased estimators can be presented in a slightly different way, which is very useful when we come to study situations where estimators of this general class are needed. To broaden our discussion we sketch the approach for a model without an intercept. Consider

$$Y_t = \beta X_t + U_t \tag{3.51}$$

where the U_t obey the standard conditions (are 'well behaved'). The OLS estimator is now:

$$\hat{\beta} = \frac{\Sigma X_t Y_t}{\Sigma X_t^2} \tag{3.52}$$

and is BLUE (see exercise 3.1). The linear weights are

$$w_t = X_t / \Sigma X_t^2. \tag{3.53}$$

Consider the general linear estimator

$$\beta^+ = \Sigma k_t Y_t \tag{3.54}$$

where for unbiasedness we also require

$$\Sigma k_t X_t = 1. \tag{3.55}$$

Hence *any set of weights Z_t*, which are fixed in repeated samples, lead to a set of weights k_t^* suitable for unbiased estimator by the equation

$$k_t^* = Z_t / \Sigma Z_t X_t \tag{3.56}$$

since this obeys (3.55). Hence substituting (3.56) into (3.54) we see that the complete class of linear unbiased estimators is described by

$$\beta^+ = \Sigma Z_t Y_t / \Sigma Z_t X_t \tag{3.57}$$

defined over all sets of Z_t (fixed in repeated samples). The variance of the general linear unbiased estimator, under our assumptions, can be shown to be

$$\text{Var } \beta^+ = \sigma^2 \Sigma Z_t^2 / (\Sigma Z X)^2 \tag{3.58}$$

and hence the efficiency (relative to OLS) is

$$\frac{\text{Var } \hat{\beta}}{\text{Var } \beta^+} = \frac{(\Sigma Z X)^2}{\Sigma Z^2 \Sigma X^2} \tag{3.59}$$

Now by the Cauchy–Schwarz inequality this is strictly less than unity unless

$$Z_t = M X_t \quad \text{for all} \quad t, \tag{3.60}$$

hence OLS is BLUE. The measure (3.59), which is the squared correlation between X and Z (in raw form), is a useful indicator of when the use of an alternative LUE will result in a large degree of inefficiency (increase in variance). The weaker the correlation between X and Z the greater the variance relative to that of OLS.

3.2 Estimation of the variances of OLS estimators

Our discussion of the Gauss–Markov theorem has concentrated upon showing that, under certain conditions, this estimator has the smallest variance of all linear unbiased estimators. Given that we know the weights we indeed know the formulae for the true variances of $\hat{\beta}$ and $\hat{\alpha}$ (i.e. how much they would vary between repeated samples on average). However these formulae involve the error variance σ^2. Since the true errors are unknown this true variance is also unknown so that, although we know OLS is best, we do not know its variance in practice. Thus we are led to consider how to estimate the variance of the OLS estimator of $\hat{\beta}$. This requires a method of estimating σ^2. Just as with the parameters α and β, there are many ways of estimating σ^2 and hence Var $\hat{\beta}$ and so on. Since the definition of σ^2 is given by assumption (ii), $E(U_t^2) = \sigma^2$, it is natural to attempt to use the residuals to replace the unknown true errors. The mean squared residual may seem a natural choice (since the mean residual is zero) but the estimator

$$\check{\sigma}^2 = \frac{1}{T}\Sigma \hat{U}_t^2 \qquad (3.61)$$

is in fact biased since:

$$E(\check{\sigma}^2) = \frac{T-2}{T}\sigma^2. \qquad (3.62)$$

The prior estimation of α and β has used up two 'degrees of freedom' and the set of T values of $\hat{U}_t$ are not T independent estimates of the U_t. Hence we have less variation than if the $\hat{U}_t$ were all independent of each other. The use of (3.61) produces a downwards bias in the estimation of the error variance. This bias would be substantial for very small values of T, but clearly the larger the sample size the less important the bias would become. It is now obvious how to convert (3.61) into an unbiased estimator (given that assumptions (i) to (v) are correct):

$$\hat{\sigma}^2 = \frac{1}{T-2}\Sigma \hat{U}_t^2. \qquad (3.63)$$

The square root of this estimator is the *standard error of estimate* referred to in Chapter 2. The estimator (3.63) is not the only unbiased estimator of the error variance (there are clearly infinitely many, based on weighted samples and sub-samples of the squared residuals). It can however be shown to have analogous properties to OLS estimators.

Once an estimate of the error variance is obtained we can then estimate the variances of the least squares parameters. It is clear that unbiased estimators are given by

$$\widehat{\text{Var } \hat{\beta}} \sim \frac{\hat{\sigma}^2}{\Sigma x_t^2} \tag{3.64}$$

$$\widehat{\text{Var } \hat{\alpha}} \sim \frac{\hat{\sigma}^2 \Sigma X_t^2}{T \Sigma x_t^2} \tag{3.65}$$

$$\widehat{\text{Cov}(\hat{\alpha}, \hat{\beta})} = \frac{-\bar{X} \hat{\sigma}^2}{\Sigma x_t^2}. \tag{3.66}$$

These formulae can be calculated solely from the X_t data (although in practice they are often calculated sequentially via $\hat{\alpha}$ and $\hat{\beta}$). Hence we can obtain a measure of how much on average a least squares estimator would fluctuate around the true value in repeated samples. If this value is small then intuitively we are likely to have more confidence that the value actually estimated for the parameter—our single sample—will be near the true value. A large value of the error variance makes us less confident that our estimates are near to the true values. These ideas, which are formalized in the techniques of statistical inference, and which we develop in Chapter 5, make it clear that the size of the variance is of considerable interest to us in allowing us to assess the degree of reliability of our results. We see that three factors determine the variance of the estimator of $\hat{\beta}$:

(i) the value of the error variance;
(ii) the degree of spread in the X_t variable;
(iii) the number of observations.

This decomposition is made clearer by rewriting (3.29)

$$\text{Var } \hat{\beta} = \frac{\sigma^2}{T \cdot S_{xx}} \tag{3.67}$$

where:

$$S_{xx} = \frac{1}{T} \Sigma (X_t - \bar{X})^2 \tag{3.68}$$

(i.e. the sample variance of X_t). Thus for a given model we can hope to obtain a more precise estimate if there are more observations and/or if the observations are more spread out. We can see the plausibility of the latter by considering figures 3.2a and 3.2b. The shaded boxes represent the areas in which the data might lie—the second case clearly is over a greater variation in X. Both boxes show the same variation in the error term (vertical deviations from the true line). The range of lines, fitted by eye, is going to be much more variable in the first case than in the second. In fact this predictable property of OLS could be exploited in cases where a survey was to be used to collect the data which was then to be analysed. A larger survey would tend to give more precise answers than a small survey, and a survey with a wider range of values of the independent variable would be more helpful than a survey with a narrower range of values (providing that all values were indeed generated by the same model).

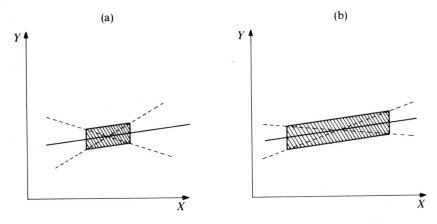

FIG. 3.2 The influence of the Spread of X on the Variance of the Estimator

---

Result 3.1

In the model
$$Y_t = \alpha + \beta X_t + U_t$$
where assumptions (i) to (v) hold, the OLS estimators
$$\hat{\beta} = \Sigma x_t y_t / \Sigma x_t^2$$
$$\hat{\alpha} = \bar{Y} - \hat{\beta} \bar{X}$$
have variances and covariance:
$$\text{Var } \hat{\beta} \sim \sigma^2 / \Sigma x_t^2$$
$$\text{Var } \hat{\alpha} = \sigma^2 \Sigma X_t^2 / T \Sigma x_t^2$$
$$\text{Cov}(\hat{\alpha}, \hat{\beta}) = \frac{-\bar{X} \sigma^2}{\Sigma x_t^2}.$$

An unbiased estimator of the error variance is given by:
$$\hat{\sigma}^2 = = \frac{1}{T-2} \Sigma \hat{u}_t^2$$

and unbiased estimators of the parameter variances are given by:
$$\widehat{\text{Var } \hat{\beta}} = \hat{\sigma}^2 / \Sigma x_t^2$$
$$\widehat{\text{Var } \hat{\alpha}} = \hat{\sigma}^2 \Sigma X_t^2 / T \Sigma x_t^2$$
$$\widehat{\text{Cov}}(\hat{\alpha}, \hat{\beta}) = -\bar{X} \hat{\sigma}^2 / \Sigma x_t^2$$

The use of these formulae is illustrated in example 3.1.

Example 3.1: Variances and covariance of least squares

From the data of Chapter 2 (in £ million) we can calculate the estimated error variance from the residual sum of squares. In fact it is conventional to give the square root of this term and the square root of the estimated parameter variances (the standard errors):

$$\hat{\sigma} = 1544.176$$
$$\text{SE } \hat{\beta} = 0.01650$$
$$\text{SE } \hat{\alpha} = 2163.379$$
$$\text{Cov}(\hat{\alpha}\hat{\beta}) = -35.373\ldots$$

The whole regression is often written with the standard errors in brackets below the coefficients:

$$Y = 15750 + 0.775X$$
$$(2163)\quad(0.016)$$

3.3 Least squares and forecasting

An apparently different problem, but in actuality very close to parameter estimation, is that of forecasting. We consider a situation where there is data available on both Y and X for periods 1 to T and then data only on X for periods $T+1$ to $T+F$. Not only can we estimate the relationship between Y and X but we can then use it to *forecast* or *predict* the value of the variable Y for the F periods for which we have no data on it.

Let the model be
$$Y_t = \beta X_t + U_t \qquad (3.69)$$

where $E(U_t) = 0$, $E(U_t^2) = \sigma^2$, $E(U_t U_{t-s}) = 0$

for $t = 1 \ldots T + F$.

It is important to note that for our analysis we shall require that the equation should remain the same into the forecasting period (there is no structural shift) and that the error properties should also be unchanging. Under this set up we know that the BLUE of β (linear in the T values of Y) is the OLS estimator:

$$\hat{\beta} = \sum_{t}^{T} X_t Y_t \Big/ \sum_{t}^{T} X_t^2 \qquad (3.70)$$

The new problem is that of predicting the actual $Y_{T+1} \ldots Y_{T+F}$, which either are unknown to us or have not even happened yet. These values are determined by a systematic component βX_{T+m} and a random component

U_{T+m}. By assumption we cannot know the random component so that we can hope only to predict the systematic component of Y (that which is determined by factors which we have been able to identify). Intuitively it seems that a sensible estimator would be:

$$\hat{Y}_{T+m} = \hat{\beta} X_{T+m} \qquad m = 1 \ldots F \qquad (3.71)$$

In fact we can prove that in a special sense this *predictor* is optimal, using an extension of the Gauss–Markov theorem. There is an important difference between the estimation of a parameter and of the outcome of a variable. The variable is itself *random in repeated samples* (as will be the value of the prediction) so that we cannot ask that the prediction should be unbiased in the sense that the average of all predictions is equal to the true outcome (there will be no single true outcome over all the repeated samples). Instead we require that the average prediction over repeated samples be equal to the average outcome in repeated samples:

$$E(\hat{Y}_t) = E(Y_t) \qquad (3.72)$$

If a predictor has this property we will say that it is unbiased. Clearly it is natural to wonder whether the least-squares-based predictor has the smallest variance among all unbiased linear predictors. It is clear that $\hat{Y}$ is unbiased and is linear (in the values $Y_1 \ldots Y_T$). The proof of optimality follows exactly the same route as the proof of the Gauss–Markov theorem given earlier. Instead of proving the optimality of $\hat{\beta}$, we are proving the optimality for a constant multiple of it—$\hat{\beta} X_{T+m}$ (for each value of m). The variance of the predictor, around the expected values of the predictor is

$$\operatorname{Var} \hat{Y}_{T+m} = E\{(\hat{\beta} X_{T+m} - \beta X_{T+m})^2\} \qquad (3.73)$$

and this can be shown to be

$$= \bar{X}_{T+m}^2 \operatorname{Var} \hat{\beta}. \qquad (3.74)$$

Any other predictor (Y^*_{T+m}) can be written

$$Y^*_{T+m} = \sum_1^T (w_t + c_t) Y_t \qquad (3.75)$$

where the w_t are the OLS weights:

$$w_t = X_{T+m} X_t / \Sigma X_t^2. \qquad (3.76)$$

If this predictor is to be unbiased it also requires

$$\Sigma c_t w_t = 0 \qquad (3.77)$$

and with this condition it can be shown that

$$\operatorname{Var} Y^*_{T+m} = \operatorname{Var} \hat{Y}_{T+m} + K \qquad (3.78)$$

where K is a strictly postive number unless all the c_t are zero. Hence the OLS

Properties of Least Squares Estimation 55

predictor is BLUE, with variance $X_{T+m}^2 \text{Var } \hat{\beta}$. This can be estimated, if required, by using the OLS residuals to estimate $\text{Var } \hat{\beta}$ in the usual way. It can be seen that the forecast variance depends not only on the variance of the parameter estimator (and hence in the factors which determine that) but also on the value of X_{T+m}. The larger the value of the independent variable the greater the range of values of the dependent variable we shall predict just because of the uncertainty as to the position of the true relationship—figure 3.3 illustrates this. The larger X_{T+m} is (the further from the origin) the more is the uncertainty about β reflected in the range of predictions that would be made in repeated samples. We can rewrite the formula:

$$\text{Var } \hat{Y} = \sigma^2 X_{T+m}^2 \Big/ \sum_1^T X_t^2 \tag{3.79}$$

—this shows that, particularly when the value of the independent variable in the prediction period is large relative to the values in the estimation period, the forecast variance will be high. The typical economic time series, where X values are growing ever larger over time, will tend to need to forecast in just such a situation (that is, where the value of X_{T+m} is greater than most of all of the values in the estimation period) and so the variance will tend to be high and become greater (at a more than proportional rate because of squared term) the further into the future we have to predict.

So far we have been predicting just the deterministic portion of the dependent variable (i.e the position of the regression line itself)—however we recognize that in repeated samples the outcome of the dependent variable is itself a random variable (Y does not always lie on the true line) because of the equation error in the forecast period. In constructing the variance of the

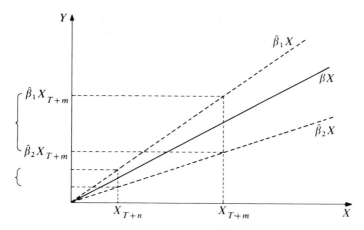

FIG. 3.3 The Effect of the Size of the Independent Variable on the Variance of the Predictor

predictor it is sensible to take this error variance also into account because we know that for this cause alone the dependent variable has a variance in repeated samples. We can allow for this most easily by constructing the 'variance' around the actual value of the dependent variable,[1] and not its systematic part as in (3.55); we use a separate symbol to denote this:

$$\text{Var } Y^*_{T+m} = E(\hat{\beta} X_{T+m} - Y_{T+m})^2. \tag{3.80}$$

Using the fact that

$$Y_{T+m} = \beta X_{T+m} + U_{T+m} \tag{3.81}$$

and the assumptions that U_{T+m} is independent of all errors in the estimation period (and hence of $\hat{\beta}$) and that it has the same variance as those errors, we obtain

$$\text{Var } Y^* = \left(\sigma^2 X^2_{T+m} \bigg/ \sum_1^T X_t^2 \right) + \sigma^2. \tag{3.82}$$

This is, not surprisingly, larger than the variance of the forecast of the systematic part of the dependent variable—we are now trying to forecast something which is itself random (i.e. has an unpredictable component). The same value of forecast ($\hat{\beta} X_{T+m}$) will vary more with respect to this variable target—given that the randomness in the forecast is not correlated with the randomness in the actual outcome in repeated samples. The variance of the systematic part of the dependent variable (3.79) and the variance of the forecast around the actual outcome (3.82) can be used as measures of how close our forecasts are likely to be the actual outcome. Again, since the true error variance is unknown we have to estimate the error variance by s^2:

$$\widehat{\text{Var } \hat{Y}_{T+F}} = s^2 X^2_{T+m} / \Sigma X_t^2. \tag{3.83}$$

This formula allows us in practice to estimate how far, on average, our forecast will be from the systematic and actual values of the dependent variable. The extension of these results to the case where there is an intercept indicates the general pattern. The model is now:

$$Y_t = \alpha + \beta X_t + U_t$$

and we require to forecast for periods $T+1$ to $T+F$. In the case where the errors for the whole period, $t=1$ to $t=T+F$, follow the standard assumptions we assert that the predictor based on OLS parameter values from the sub-period 1 to T is optimal (see problem 3.14).

$$\hat{Y}_{T+m} = \hat{\alpha} + \hat{\beta} X_{T+m} \tag{3.84}$$

This is both linear (in Y_1 to Y_T) and unbiased. The variance can be shown to

---

[1] This is not strictly speaking a variance—there is no variable here which has this average error around a mean value—however it is useful to refer to it as a variance.

be built up out of the parameter estimators' variances and covariances:

$$\text{Var}(\hat{Y}_{T+m}) = \text{Var}\,\hat{\alpha} + X_{T+m}^2\,\text{Var}\,\hat{\beta} + 2X_{T+m}\,\text{Cov}(\hat{\alpha},\hat{\beta}) \qquad (3.85)$$

(Substituting in from (3.29), (3.30), and (3.33) we can derive the alternative expression:

$$\text{Var}(\hat{Y}_{T+m}) = \sigma^2 \left\{ \frac{1}{T} + (x_{T+m})^2 / \Sigma x_t^2 \right\}. \qquad (3.86)$$

The variance of the regression line now depends rather on the distance of the exogenous variable in the forecasting period from its mean value in the estimation period as well as the error variance, the number of observations, and the general spread of the data in the estimation period. The variance of the regression line can be estimated by replacing the actual error variance by its value as estimated by residual variance (s^2). The 'variance' of the actual outcome (allowing for the error term in the forecast period) is then

$$\text{Var}\,Y_{T+m}^* = \sigma^2 \left\{ 1 + \frac{1}{T} + x_{T+m}^2 / \Sigma x_t^2 \right\}. \qquad (3.87)$$

The use of these formulae is illustrated for the data and model of the consumption function in example 3.2.

Example 3.2. Forecasting from least squares

We take three values of income in order to construct forecasts for points which lie respectively at the centre, the edge, and outside the experience of the data set used for the estimation of the consumption function.

$$X_{T+1} = £120{,}000 \text{ million}$$
$$X_{T+2} = £170{,}000 \text{ million}$$
$$X_{T+3} = £200{,}000 \text{ million}$$

The estimated regression line was

$$Y = 15750 + 0.775\,X$$

(in £ million) with

$$S.E(\hat{\alpha}) = 2163$$
$$S.E(\hat{\beta}) = 0.0165$$
$$\text{Cov}(\hat{\alpha},\hat{\beta}) = -35.27$$
$$\hat{\sigma} = 1544.$$

Using either (3.85) or (3.86) and (3.87) we obtain the results shown in Table 3.1.

TABLE 3.1 Forecasts From A Least Squares Estimation

X_F	$\hat{Y}_F$	$(\text{Var } \hat{Y}_F)^{1/2}$	$(\text{Var}^* \hat{Y}_F)^{1/2}$
120,000[a]	108,750	361	1585
170,000	147,500	740	1712
200,000	170,750	1204	1958

[a] All figures are in £ million.

The effect of the increasing distance from the mean of the independent variable in the variance of the forecasted regression line is shown clearly in column 3, while the effect is damped down by the addition of the variance of the error term in column 4 (since this is relatively large and does not alter with the location of the data).

Problems 3

3.1 (a) Prove the following statements where a and b are constants and U and V are random variables defined over a definite set of values 1 to T.

(i) $$E(aU) = a\bar{U}$$
where $$\bar{U} = E(U);$$

(ii) $$E(aU + bV) = a\bar{U} + b\bar{V};$$

(iii) $$\sigma_u^2 = \text{Var } U = E[\{U - E(U)\}^2]$$

(iv) $$\text{Var}(aU + bV) = a^2\sigma_u^2 + b^2\sigma_v^2 + 2ab\sigma_{uv}$$
where $$\sigma_{uv} = \text{Covar } UV.$$

(b) Construct a simple example to show that it is not generally true that $E(ab) = E(a)E(b)$.

3.2 For the model $Y = \beta X_t + U_t$ where the five (sufficient) conditions hold, prove that the OLS estimator of β is BLUE. (Note that (v) requires $\Sigma X_t^2 > 0$.)

3.3 (Alternative approach to proving the Gauss–Markov theorem). In the model $Y_t = \beta X_t + U_t$, consider a linear estimator $\beta^+ = \Sigma k_t Y_t$. Find the value of β^+ that minimizes the variance of β^+ subject to the restriction that β^+ is unbiased; hint: construct the Lagrangian of the form

$$L = \text{Var}(\beta^+) + \lambda[E(\beta^+) - \beta]$$

and solve for

$$\frac{\partial L}{\partial \beta^+} = \frac{\partial L}{\partial \lambda} = 0.$$

3.4 In the model $Y_t = \beta X_t + U_t$, where the errors obey the conditions sufficient for OLS to be BLUE, the following alternative estimators are considered:

(i) $\quad\quad\quad\quad\quad\quad\quad\quad \beta^+ = \bar{Y}/\bar{X};$

(ii) $\quad\quad\quad\quad\quad\quad\quad\quad \tilde{\beta} = (Y\max - Y\min)/(X\max - X\min)$

where $X\max$, $X\min$ are the largest and smallest values of X, and $Y\max$, $Y\min$ are their associated Y values;

(iii) $\quad\quad\quad\quad\quad\quad\quad\quad \beta^\circ = \Sigma Y_t X_t^{\gamma-1}/\Sigma X_t^\gamma$

(iv) $\quad\quad\quad\quad\quad\quad\quad\quad \beta^- = \Sigma t Y_t / \Sigma t X_t \quad\quad t = 1 \ldots T$

(where t is an index of time when the observations are in temporal sequence).

(v) $\quad\quad\quad\quad\quad\quad\quad\quad \beta^* = \Sigma r_t Y_t / \Sigma r_t X_t$

(where r_t is the rank (by size) or X_t, i.e. largest X_t has $r_t = 1$ etc).
Show that each estimator is linear (give the weights) and is unbiased, and derive its variances. Discuss the circumstances in which the differences in variance from OLS (the loss of efficiency) are likely to be large or small.

3.5 In the model $Y_t = \alpha + \beta X_t + U_t$, where the errors obey conditions (i) to (v) show that the OLS estimator of α (i.e. $\hat{\alpha} = \bar{Y} - \hat{\beta}\bar{X}$) is BLUE.

3.6 In the model $Y_t = \alpha + \beta X_t + U_t$, where the conditions sufficient for OLS to be optimal hold, show that the best linear unbiased estimator of a linear combination of the parameters $(c_1 \alpha + c_2 \beta)$ is the least squares equivalent $(c_1 \hat{\alpha} + c_2 \hat{\beta})$.

3.7 (a) In the model $Y_t = \beta X_t + U_t$ where the conditions sufficient for OLS to be BLUE hold, the residuals $\hat{U}_t$ are defined $\hat{U}_t = Y_t - \hat{\beta} X_t$. Show that

(i) $\quad\quad\quad\quad\quad\quad \text{Var}(\hat{U}_t^2) = \sigma^2(1 - X_t^2/\Sigma X_t^2)$

(ii) $\quad\quad\quad\quad\quad\quad \text{Cov}(\hat{U}_t, \hat{U}_s) = -\sigma^2 X_t X_s / \Sigma X_t^2 \quad\quad s \neq t.$

(b) Show that $\Sigma \hat{U}_t \neq 0$ (in general) and compare this to the case of the model which includes an intercept.

(c) Show that

$$E\left(\frac{1}{T-1} \Sigma \hat{U}_t^2\right) = \sigma^2$$

3.8 For the model $Y_t = \alpha + U_t$, where the U obey assumptions (i), (ii), and (iii), what is the OLS estimator of α? Prove that OLS is BLUE. What is the variance of the OLS estimator. Why do we not need assumptions (iv) and (v) for this problem?

3.9 Defining the squared correlation between estimators $\hat{\alpha}$ and $\hat{\beta}$ by the function $r^2(\hat{\alpha}, \hat{\beta}) = \{\text{Cov}(\hat{\alpha}, \hat{\beta})\}^2 / \text{Var } \hat{\alpha} \times \text{Var } \hat{\beta}$, show that in the case of the model of chapter 3 it has the value

$$\frac{T\bar{X}^2}{\Sigma X_t^2}$$

and that this is equal to $\tilde{r}^2$ between X_{0t} and $X_t (X_{0t} = 1$ all $t)$ (see (2.60)). When will this attain its upper and lower bounds?

3.10 Consider the model $Y_t = \alpha + \beta X_t + U_t$, where the conditions sufficient for OLS to be BLUE hold. Denote these OLS estimators $\hat{\alpha}$, $\hat{\beta}$ and that from the (incorrect) regression of Y on X (without an intercept) by $\hat{\hat{\beta}}$. Prove that if $\bar{X} = 0$:

(i) $\hat{\beta} = \hat{\hat{\beta}}$

(ii) $\text{Var } \hat{\beta} = \sigma^2 / \Sigma X_t^2$.

Does this mean that it does not matter which regression is actually carried out?

3.11 Consider the model $Y_t = \beta X_t + U_t$ where the conditions sufficient for OLS to be BLUE hold. What are the properties of the estimator $\beta^+ = \Sigma x_t y_t / \Sigma x_t^2$? Under what circumstances will the loss in efficiency (increase in variance) be small?

3.12 (a) Show that for any estimator the Mean Square Error: MSE = Variance + Bias^2.

(b) In the model $Y_t = \beta X_t + U_t$ where the conditions sufficient for OLS to be BLUE hold, consider the estimators:

(i) $\beta^\circ = \bar{X} \Sigma Y_t / \Sigma X_t^2$

(ii) $\beta^* = \Sigma X_t Y_t / \Sigma X_t^2 + \frac{1}{T}$.

Are these estimators (1) unbiased (2) linear? Derive the value of their Mean Square Errors and compare them with that of OLS. NB $\text{MSE}(\beta^+) = E\{(\beta^+ - \beta)^2\}$.

3.13 For the model $Y_t = \beta X_t + U_t$, where the conditions sufficient for OLS to be BLUE hold, show that the minimum mean square error estimator (linear in Y_t) and for all values of β and X_t is given by

$$\check{\beta} = \Sigma X_t Y_t \bigg/ \left(\frac{\sigma^2}{\beta^2} + \Sigma X_t^2\right)$$

with
$$\text{MSE}(\check{\beta}) = \frac{\sigma^2}{\beta^2} + \Sigma X_t^2.$$

(Hint: The structure of the Gauss–Markov theorem proof can be utilized to show that any other linear estimator will have a larger MSE). Discuss the importance of these results.

Properties of Least Squares Estimation

3.14 In the model $Y_t = \alpha + \beta X_t + U_t$ ($t = 1 \ldots T+1$), where there is data on both variables from 1 to T but only on X from $T+1$ to $T+F$, and the conditions (i) to (iv) hold and the condition (v) is $\frac{1}{T}\Sigma X_t^2 > 0$, show that the best linear (in Y_1 to Y_T) unbiased predictor of the regression line is

$$\hat{Y}_{T+m} = \hat{\alpha} + \hat{\beta} X_{T+m}.$$

3.15 (a) Prove the result that for two independent random variables P and Q:

$$\text{Var}(PQ) = \bar{Q}^2 \text{Var } P + \bar{P}^2 \text{Var } Q + \text{Var } P \text{ Var } Q$$

where $\bar{P} = E(P)$ etc.

(b) There is a model

$$Y_t = \beta X_t + U_t \qquad T = 1 \ldots T+F$$

where the errors obey conditions (i) to (v). There is data on both X and Y for the period 1 to T but for $T+1$ to $T+F$ only estimates of X are available ($\hat{X}_{T+m}$). These estimates are random (in repeated samples) with the properties that:

$$E(\hat{X}_{T+m}) = X_{T+m}$$
$$\text{Var}(\hat{X}_{T+m}) = \Delta^2$$
$$E(\hat{X}_{T+m} U_t) = 0 \qquad t = 1 \ldots T+F.$$

It is proposed to forecast Y (the systematic part of the equation) with the predictor:

$$Y^*_{T+m} = \hat{\beta} \hat{X}_{T+m}$$

(where $\hat{\beta}$ is the OLS estimator based on the data from 1 to T). Show that this predictor is unbiased and has a variance (around the regression line)

$$\text{Var } Y^*_{T+m} = \beta^2 \Delta^2 + X^2_{T+m} \text{Var } \hat{\beta} + \Delta^2 \text{Var } \hat{\beta}.$$

Discuss the effects of uncertainty about future levels of the exogenous variable on the accuracy of forecasts.

3.16 Show that for the model $Y_t = \alpha + \beta X_t + U_t$, where U_t is 'well behaved', the general linear unbiased estimator is of the form: $\beta^* = \Sigma z_t y_t / \Sigma z_t x_t$, where z_t is a variable fixed in repeated samples and $z_t = Z_t - \bar{Z}$. Derive the variance of the general linear unbiased estimator and show that the efficiency (relative to OLS) is proportional to the correlation (in deviation form) between X_t and Z_t.

3.17 For the model

$$Y_t = \beta X_t + U_t \qquad t = 1 \ldots T+F$$

where the errors obey the assumptions (i) to (v), there is data on both variables for periods 1 to T, and data on X for $T+1$ to $T+F$, there is a predictor

$$\hat{Y}_{T+m} = \hat{\beta} X_{T+m}$$

where $\hat{\beta}$ is the OLS estimator based on values 1 to T. Show that the 'variance' of the predictor around the actual outcome is equal to the variance around the expected outcome plus the error variance: i.e.

$$E\{(\hat{Y}_{T+F} - Y_{T+F})^2\} = E[\{\hat{Y}_{T+F} - E(Y_{T+F})\}^2] + \sigma^2$$

Answers 3

3.1. (a) (i) Let the probability of the ith outcome be p(i) i.e. in repeated sampling a proportion p(i) of all trials take the ith value.

$$E(U) \triangleq \Sigma p(i) U(i) = \bar{U}$$

$\therefore$
$$E(aU) = \Sigma a p(i) U(i)$$
$$= a \Sigma p(i) U(i) = a \bar{U}$$

(ii)
$$E(aU + bV) = \sum_i \sum_j (aU(i) + bV(j)) p(i,j)$$

where p(i, j) is the probability that the outcome $U(i)$ and $V(j)$ occurs.

$$= \Sigma \Sigma a U(i) p(i,j) + \Sigma \Sigma b V(j) p(i,j)$$

$$= a \sum_i U(i) \left\{ \sum_j p(i,j) \right\} + b \sum_j V(j) \left\{ \sum_i p(i,j) \right\}$$

$$= a\bar{U} + b\bar{V}$$

(in the final step of the proof we have added all the chances of $U(i)$ occurring to weight each $U(i)$ by a probability, etc.).

(iii)
$$\text{Var } U = E[\{U - E(U)\}^2]$$
$$= \Sigma p(i) \{U(i) - \bar{U}\}^2$$

$\therefore$
$$\text{Var } aU = \Sigma p(i) \{aU(i) - a\bar{U}\}^2$$
$$= a^2 \text{ Var } U.$$

(iv)
$$\text{Var}(aU + bV) = E\{(aU + bV - a\bar{U} - b\bar{V})^2\}$$
$$= E\{(aU - a\bar{U})^2\} + E\{(bV - b\bar{V})^2\}$$
$$\quad + 2E\{a(U - \bar{U})b(V - \bar{V})\}$$
$$= a^2 \sigma_u^2 + b^2 \sigma_v^2 + 2ab \sigma_{uv}$$

where $E\{(U - \bar{U})(V - \bar{V})\} = \text{Cov}(UV) = \sigma_{uv}$.

(b) Consider table 3.2 of probabilities and values, where

TABLE 3.2 The Probabilities of Values of a Bivariate Distribution

		a	
		1	3
b	1	5/12	2/12
	2	2/12	3/12

e.g. the combination $a=1$, $b=1$ occurs in 5/12 of all samples. Now the average (expected) value of the product ab weighted by probabilities is

$$E(ab) = (1 \times 1 \times 5/12) + (1 \times 2 \times 2/12) + (3 \times 1 \times 2/12) + (3 \times 2 \times 3/12)$$
$$= 33/12 = 2.75$$
$$E(a) = 1 \times [5/12 + 2/12] + 3[2/12 + 3/12] = 22/12$$
$$E(b) = 1 \times [5/12 + 2/12] + 2[2/12 + 3/12] = 17/12$$
$$\therefore E(a)E(b) = 2.60$$

so the variables are not independent.

3.2 Using OLS the estimator is

$$\hat{\beta} = \frac{\Sigma XY}{\Sigma X^2}$$

so that $\hat{\beta} = \Sigma w_t Y_t$ where $w_t = X_t / \Sigma X_t^2$. Now replacing Y by the equation we obtain:

$$\hat{\beta} = \beta + \frac{\Sigma UX}{\Sigma X^2}$$

Given a fixed X, with $E(U) = 0$ and $\Sigma X_t^2 > 0$ we see that $E(\hat{\beta}) = \beta$. From (*) the variance of $\hat{\beta}$, using assumptions (ii) and (iii) is given by

$$\text{Var } \hat{\beta} = \frac{\sigma^2}{\Sigma X^2} = \sigma^2 \Sigma w_t^2.$$

Consider the general linear estimator

$$\beta^* = \Sigma k_t Y_t = \Sigma (w_t + c_t) Y_t.$$

For unbiasedness we can see that we require additionally $\Sigma c_t X_t = 0$ (i.e. $\Sigma c_t w_t = 0$). But

$$\text{Var } \beta^* = \sigma^2 \Sigma w_t^2 + \sigma^2 \Sigma c_t^2 + 2\sigma^2 \Sigma c_t w_t$$

(using assumptions (ii) and (iii)) and since the last term must be zero, via the condition for unbiasedness, the variance of any other linear unbiased estimator is strictly greater than that of OLS. The difference squared of the weights is a measure of inefficiency.

3.3 Form the Lagrangian:

$$L = E[\{\Sigma k_t Y_t - E(\Sigma k_t Y_t)\}^2] - \lambda\{E(\Sigma k_t Y_t) - \beta\}$$

$$\therefore \quad L = \sigma^2 \Sigma k_t^2 + \lambda\{\beta(\Sigma k_t X_t - 1)\}$$

using (i), to (v)). Find a minimum:

$$\therefore \quad \frac{\partial L}{\partial k_t} = 2\sigma^2 k_t + \lambda \beta X_t = 0 \tag{1}$$

$$\frac{\partial L}{\partial \lambda} = \beta(\Sigma k_t X_t - 1) = 0 \tag{2}$$

$$\therefore \quad \Sigma k_t X_t = 1.$$

Multiply (i) by X_t and sum over t.

Hence
$$\frac{-2\sigma^2}{\beta \Sigma X_t^2} = \lambda.$$

Substituting back into (1) $X_t / \Sigma X_t^2 = k_t$, which is the OLS weight, so that the minimum variance linear unbiased estimator is indeed OLS. Note that the constraint (of unbiasedness) is binding ($\lambda \neq 0$) unless $\sigma^2 = 0$ or one of β or ΣX_t^2 is infinite. Only in such cases would OLS coincide with the minimum mean square error estimator (see exercise 3.12).

3.4 (i) $\quad\quad\quad w_t = 1/(T\bar{X})$ all t

$$\therefore \quad \beta^+ = \beta + \bar{U}/\bar{X} \quad \text{and so } E(\beta^+) = \beta.$$

$$\therefore \quad \text{Var } \beta^+ = \sigma^2/T\bar{X}^2$$

Efficiency
$$= \frac{T\bar{X}^2}{\Sigma X_t^2}.$$

Since this is a linear estimator with $Z_t = 1$ (all t) we can use the general result (3.59)

$$\text{Efficiency} = (\Sigma 1.X)^2 / \Sigma 1.\Sigma X^2.$$

Hence the stronger the (raw) correlation between the unit variable and X the less will be the loss in efficiency.

(ii)
$$W_{MAX} = \frac{1}{X_{MAX} - X_{MIN}},$$

$$W_{MIN} = \frac{-1}{X_{MAX} - X_{MIN}}$$

$$W_t = 0 \quad \text{other} \quad t.$$

Now
$$\tilde{\beta} = \beta + \frac{U_{MAX} - U_{MIN}}{X_{MAX} - X_{MIN}}$$

∴
$$E(\tilde{\beta}) = \beta$$

and
$$\text{Var } \tilde{\beta} = 2\sigma^2/(X_{MAX} - X_{MIN})^2$$

$$\text{Eff} = (X_{MAX} - X_{MIN})^2 / 2\Sigma X_t^2.$$

The denominator can be rewritten:

$$(X_{MAX} - X_{MIN})^2 + (X_{MAX} + X_{MIN})^2 + 2\overset{\neq MAX, MIN}{\Sigma X_t^2}$$

Hence efficiency of $\tilde{\beta}$ is low if (1) $(X_{MAX} - X_{MIN})^2$ is small relatively, (2) $(X_{MAX} + X_{MIN})^2$ is large, (3) the number of observations increases, (4) the extra observations are far from zero. If the other X_t tend towards zero and if $X_{MAX} \approx -X_{MIN}$ then $\text{Var } \hat{\beta} \to \text{Var } \tilde{\beta}$ and $\tilde{\beta} \to \hat{\beta}$). This is an example *for a particular data set* where apparently different estimation techniques converge to the same value of the estimator.

(iii)
$$w_t = X_t^{\gamma - 1} / \Sigma X_t^\gamma$$

$$\beta' = \beta + \Sigma U_t w_t \quad \therefore \quad E(\beta') = \beta.$$

$$\text{Var } \beta' = \sigma^2 \Sigma X_t^{(2\gamma - 2)} / (\Sigma X_t^\gamma)^2$$

$$\text{Eff} = (\Sigma X_t^\gamma)^2 / \Sigma X_t^2 \Sigma X_t^{(2\gamma - 2)}$$

(which is the 'raw' correlation between X and $X^{\gamma - 1}$). Clearly the nearer γ is to 2 the stronger the correlation and the less the loss of efficiency in using β'. This is clearly an infinite class of estimators. We can see that $\gamma = 2$ is OLS; $\gamma = 1$ is β^+, $\gamma = 0$ is $(1/T)\Sigma(Y/X)$.

(iv)
$$w_t = t/\Sigma t X_t$$

$$\beta^\circ = \beta + \Sigma U_t w_t \quad \therefore \quad E(\beta^\circ) = \beta.$$

$$\text{Var } \beta^\circ = \sigma^2 \Sigma t^2 / \Sigma (t X_t)^2$$

$$\text{Eff} = \Sigma (t X_t)^2 / \Sigma t^2 \Sigma X_t^2.$$

The stronger the correlation of X with a linear trend the less the loss in efficiency.

(v)
$$w_t = r_t / \Sigma r_t X_t$$

$$E(\beta^*) = \beta$$

$$\text{Var } \beta^* = \sigma^2 \Sigma r_t^2 / (\Sigma r_t X_t)^2$$

$$\text{Eff} = (\Sigma r X)^2 / \Sigma r^2 \Sigma X^2.$$

As the correlation between X and its rank increases (Xs tend to be arranged by t in strictly increasing or decreasing size) then the loss of efficiency is small.

NB These linear unbiased estimators are important because they offer useful alternatives in situations where OLS is biased (because there is a correlation between U and X). They are in fact examples of the technique of instrumental variables (cf. Chapter 8).

3.5 The OLS estimator is

$$\hat{\alpha} = \bar{Y} - \hat{\beta}\bar{X}$$
$$= \Sigma v_t Y_t$$

where

$$v_t = \left[\frac{1}{T} - x_t \bar{X}/\Sigma x_t^2\right]$$

so that OLS is linear. Substituting in for Y_t

$$\hat{\alpha} = (\alpha + \beta\bar{X} + \bar{U} - \hat{\beta}\bar{X})$$

where $\bar{U}$ is the *sample* mean error. Therefore, $E(\hat{\alpha}) = \alpha$ (using unbiasedness of $\hat{\beta}$). Now,

$$\text{Var } \hat{\alpha} = E[\{\Sigma v_t Y_t - E(\Sigma v_t Y_t)\}^2]$$
$$= \sigma^2 \Sigma v_t^2 = \sigma^2 \Sigma X_t^2 / T \Sigma x_t^2$$

(using assumptions ii and iii). Consider any other linear estimator

$$\alpha^* = \Sigma k_t Y_t = \Sigma(v_t + w_t) Y_t$$

For unbiasedness we require:

$$E(\Sigma w_t Y_t) = E[\Sigma\{w_t(\alpha + \beta X_t + U_t)\}] = 0.$$

This can be satisfied for all values of α, β and X_t only if we choose $\Sigma w_t = 0$, $\Sigma w_t X_t = 0$. The variance of this general linear unbiased estimator is

$$\text{Var } \alpha^* = E[\{\Sigma(v_t + w_t) Y_t - E\Sigma(v_t + w_t) Y_t\}^2]$$
$$= E[\{\Sigma(v_t + w_t) U_t\}^2]$$
$$= \sigma^2 \Sigma v_t^2 + \sigma^2 \Sigma w_t^2 + 2\sigma^2 \Sigma v_t w_t$$

(using ii and iii). Now,

$$\Sigma v_t w_t = \Sigma\left(\frac{1}{T} - x_t X/\Sigma x_t^2\right) w_t.$$

However by the criteria required for α^* to be unbiased this is identically zero, so that any other linear unbiased estimator for α has a greater variance than OLS (and the loss of efficiency can be measured by the squared deviations (w_t) of the linear weights from those of OLS).

3.6 The least squares estimators are

$$\hat{\alpha} = \Sigma v_t Y_t \qquad v_t = \frac{1}{T} - x_t \bar{X}/\Sigma x_t^2$$

$$\hat{\beta} = \Sigma w_t Y_t \qquad w_t = x_t/\Sigma x_t^2.$$

To obtain unbiasedness for the sum of parameters we require

$$c_1 \alpha + c_2 \beta = c_1 (\alpha \Sigma v_t + \beta \Sigma v_t X_t) + c_2 (\alpha \Sigma w_t + \beta \Sigma w_t X_t).$$

Clearly the only weights that will satisfy this equation for *all* values of α, β, c_1, c_2, and the X_t must satisfy $\Sigma v_t = 1$, $\Sigma v_t X_t = 0$, $\Sigma w_t = 0$, $\Sigma w_t X_t = 1$. The variance of the estimator is then

$$\text{Var}(c_1 \hat{\alpha} + c_2 \hat{\beta}) = c_1^2 \sigma^2 \Sigma v_t^2 + c_2^2 \sigma^2 \Sigma w_t^2 + 2 c_1 c_2 \sigma^2 \Sigma v_t w_t$$

$$= \{c_1^2 \text{ Var } \hat{\alpha} + c_2^2 \text{ Var } \hat{\beta} + 2 c_1 c_2 \text{ Cov}(\hat{\alpha}, \hat{\beta})\}.$$

Consider any other linear estimator

$$c_1 \alpha^* + c_2 \beta^* = c_1 \Sigma (v_t + d_t) Y_t + c_2 \Sigma (w_t + e_t) Y_t.$$

For unbiasedness the weights must satisfy: $\Sigma d_t = 0$, $\Sigma d_t X_t = 0$, $\Sigma e_t = 0$, $\Sigma e_t X_t = 0$. Now the variance of this alternative estimator is

$$= \text{Var}(c_1 \alpha^* + c_2 \beta^*)$$

$$= E\{(c_1 \Sigma v_t U_t + c_2 \Sigma w_t U_t + c_1 \Sigma d_t U_t + c_2 \Sigma e_t U_t)^2\}$$

$$= c_1^2 \sigma^2 \Sigma v_t^2 + c_2^2 \sigma^2 \Sigma w_t^2 + 2 c_1 c_2 \sigma^2 \Sigma v_t w_t$$

$$+ E\{(c_1 \Sigma d_t U_t + c_2 \Sigma e_t U_t)^2\}$$

$$+ 2E\{(c_1 \Sigma v_t U_t + c_2 \Sigma w_t U_t)(c_1 \Sigma d_t U_t + c_2 \Sigma e_t U_t)\}$$

However all the cross products in the final term are zero e.g. $2 c_1 c_2 \sigma^2 \Sigma w_t d_t = 0$ (using the value of w_t and the fact that $\Sigma d_t X_t$ must equal 0). Hence the variance of the alternative estimator of the linear combination is equal to the variance of the OLS estimator of the combination plus a squared term. Hence OLS is BLUE for this more general situation.

This proof includes proofs for the optimality of $\hat{\alpha}$ and of $\hat{\beta}$ taken separately. It also leads to the optimal predictor of Y_F by letting $c_1 = 1$ and $c_2 = X_F$.

3.7 (a) Since $\hat{U}_t = Y_t - \hat{\beta} X_t$

$\therefore \qquad \hat{U}_t = \beta X_t + U_t - X_t \Sigma (\beta X_t + U_t) X_t / \Sigma X_t^2$

$\qquad\qquad = U_t - X_t \Sigma U_t X_t / \Sigma X_t^2$

$\therefore \qquad E(\hat{U}_t^2) = E\left\{U_t^2 + \frac{X_t^2 (\Sigma U_t X_t)^2}{(\Sigma X_t^2)^2} - \frac{2 X_t U_t \Sigma U_t X_t}{\Sigma X_t^2}\right\}$

$\qquad\qquad = \sigma^2 - \frac{\sigma^2 X_t^2}{\Sigma X_t^2}$

and
$$E(\hat{U}_t\hat{U}_s) = E\left\{U_tU_s - \frac{X_tU_s\Sigma X_tU_t}{\Sigma X_t^2} - \frac{X_sU_t\Sigma X_tU_t}{\Sigma X_t^2} + \frac{X_tX_s(\Sigma U_tX_t)^2}{(\Sigma X_t^2)^2}\right\}$$
$$= -\sigma^2 X_t X_s/\Sigma X_t^2.$$

Hence these estimators are biased, in the sense that $E(\hat{U}_t\hat{U}_s) \neq E(U_tU_s)$ all t, s, but the bias will disappear as the sample size increases.

(b)
$$\Sigma\hat{U}_t = \Sigma(Y_t - \hat{\beta}X_t)$$
$$= \Sigma\{Y_t - X_t(\Sigma X_tY_t/\Sigma X_t^2)\}.$$

In the case of the model with an intercept the parallel equation is
$$\Sigma\hat{U}_t = \Sigma(Y_t - \hat{\alpha} - \hat{\beta}X_t)$$

but this is the 'normal equation' associated with the partial derivative with respect to α, and hence is zero at the least squares value. Since this derivative is not used when there is no intercept the condition does not automatically hold in the latter case. Substituting in for Y we have as usual
$$\Sigma\hat{U}_t = \Sigma(U_t - X_t\Sigma X_tU_t/\Sigma X_t^2)$$

which clearly is equal to zero only in special circumstances. However, its expected value is zero.

(c) Using assumptions (i) or (ii),
$$E\left(\frac{\Sigma\hat{U}_t^2}{T-1}\right) = \frac{\Sigma(\sigma^2 - \sigma^2 X_t^2/\Sigma X_t^2)}{T-1} = \sigma^2.$$

3.8 Forming the sum of squares function and minimizing we have
$$S(\hat{\alpha}) = \Sigma(Y_t - \hat{\alpha})^2$$
$$\frac{\partial S}{\partial \hat{\alpha}} = -2\Sigma(Y_t - \hat{\alpha}) = 0$$
$$\therefore \quad \hat{\alpha} = \bar{Y}$$

(The linear weights are $w_t = 1/T$ all t.)
$$\therefore \quad E(\hat{\alpha}) = \alpha$$

(using assumption (i)). Also,
$$\text{Var } \hat{\alpha} = E\{(\bar{Y} - \alpha)^2\}$$
$$= \frac{\sigma^2}{T} = \sigma^2\Sigma w_t^2.$$

Consider any other linear estimator α^*,

$$\alpha^* = \Sigma(w_t + c_t) Y_t.$$

For this to be unbiased we must have $\Sigma c_t = 0$. But

$$\text{Var } \alpha^* = E[\{\Sigma(w_t + c_t)U_t\}^2]$$
$$= \sigma^2 \Sigma w_t^2 + \sigma^2 \Sigma c_t^2 + 2\sigma^2 \Sigma w_t c_t$$

(using assumptions (ii) and (iii)). But from the value for w_t and the restriction on Σc_t we see that the third term is zero, so that OLS is BLUE.

The second-order condition for a minimum is satisfied automatically:

$$\frac{\partial^2 S}{\partial \alpha^2} = 2T > 0$$

so that there is no need to restrict the data (assumption (v)). By construction the independent variable implicit in the model, $X_{0t} = 1$ for all t, is constant in repeated samples so that we do not need to add assumption (iv). To estimate the variance we use the value of $\hat{\alpha}$ to give:

$$\hat{U}_t = U_t - \frac{1}{T}\Sigma U_t$$

$$\therefore \quad E(\Sigma \hat{U}_t^2) = (T-1)\sigma^2$$

$$\therefore \quad \frac{1}{T(T-1)}\Sigma \hat{U}_t^2 \text{ is an unbiased estimator for Var } \hat{\alpha}.$$

3.9 From the standard formulae for a model with single variable plus intercept (3.29), (3.31), and (3.33):

$$r^2(\hat{\alpha}, \hat{\beta}) = T \bar{X}^2 / \Sigma X_t^2.$$

But the correlation between X_0 and X is

$$\tilde{r}^2 = (\Sigma 1 \cdot X)^2 / \Sigma 1^2 \Sigma X^2$$
$$= T^2 \bar{X}^2 / T\Sigma X_t^2$$

(the sign of the unsquared correlation between coefficients is the opposite of that between the variables). This result shows that as the variable X_t tends toward a constant set of values the coefficients become more poorly determined—it is harder to separate out the intercept and slope effects. The upper bound is when $X_t = K$ for all t and in this case there is no minimum to the sum of squares function since the second-order condition fails. The lower bound is at zero, which would happen if the mean value of the X variable was zero.

70 Properties of Least Squares Estimation

3.10 Now by standard results:
$$\hat{\beta} = \Sigma xy/\Sigma x^2$$
$$\hat{\alpha} = \bar{Y} - \hat{\beta}\bar{X}$$
$$\hat{\hat{\beta}} = \Sigma YX/\Sigma X^2$$
$$\text{Var } \hat{\beta} = \sigma^2/\Sigma x_t^2$$

(i) Expanding $\hat{\beta} = \Sigma(X - \bar{X})(Y - \bar{Y})/\Sigma(X - \bar{X})^2$, using the property that $\Sigma xy = \Sigma xY$ and the restriction that $\bar{X} = 0$, we see that here
$$\hat{\hat{\beta}} = \Sigma YX/\Sigma X^2 = \hat{\beta}.$$

(ii) $$\text{Var } \hat{\beta} = \sigma^2/\Sigma x_t^2$$

which with $\bar{X} = 0$ gives
$$\text{Var } \hat{\beta} = \sigma^2/\Sigma X_t^2$$

which is the formula for the variance of the slope when the *true* model is: $Y_t = \beta X_t + U_t$. In our case, the variance of the estimator $\hat{\hat{\beta}} = \Sigma YX/\Sigma X^2$ can be seen to be $\sigma^2/\Sigma X^2$ (substituting in for Y and using $\Sigma X = 0$). The theoretical formulae for estimates of β and their true variances are indeed identical in this case. However in practice we would need to estimate the variance and the failure to allow for the presence of α (which is estimated to be equal to $\bar{Y}$) does make an important difference.

We know from standard results that the estimated error variance is unbiased if the true model is used, i.e.
$$E(s^2) = E(\Sigma \hat{U}_t^2/T - 2) = \sigma^2.$$

Consider the residuals from the incorrect estimator
$$\hat{\hat{U}}_t = Y_t - \hat{\hat{\beta}} X_t$$

and the associated estimated variance:
$$E(\hat{\hat{s}}^2) = E\left(\frac{1}{T-1} \Sigma \hat{\hat{U}}_t^2\right)$$
$$= \frac{1}{T-1} E\Sigma (Y_t - \hat{\hat{\beta}} X_t)^2$$

replacing Y_t by $(\alpha + \beta X_t + U_t)$ and $\hat{\hat{\beta}}$ by
$$\frac{\Sigma X_t(\alpha + \beta X_t + U_t)}{\Sigma X_t^2}$$

and using the restriction that $\Sigma X_t = 0$, we obtain
$$E(\hat{\hat{S}}^2) = \frac{T\alpha^2}{T-1} + \sigma^2$$

(and this shows that the estimated variance is too large. We would attribute less accuracy to the estimation of β than if we had used the correct model).

3.11 Now $\beta^+ = \Sigma xy/\Sigma x^2 = \Sigma xY/\Sigma x^2$ so β^+ is linear. Substituting in for y we have $E(\beta^+) = E\Sigma x\{\beta x + (U - \bar{U})\}/\Sigma x^2 = \beta$ so that the estimator is still unbiased. The variance is

$$\text{Var } \beta^+ = E[\{\Sigma x(U - \bar{U})\}^2/(\Sigma x_t^2)^2].$$

Following the usual steps we obtain:

$$\text{Var } \beta^+ = \sigma^2/\Sigma x_t^2.$$

(Since β^+ is the linear unbiased estimator with weights $x_t/\Sigma x_t^2$ the variance could also be obtained using the standard result that it is equal to the error variance times the sum of the squared weights).

We can immediately see the efficiency is given by the ratio:

$$\text{Var } \beta^+/\text{Var } \hat{\hat{\beta}} = \Sigma X_t^2/\Sigma x_t^2.$$

Now expanding the denominator we obtain

$$\text{Eff} = \Sigma X_t^2/(\Sigma X_t^2 - T\bar{X}^2).$$

Only if $\bar{X}$ is zero will there be no loss in efficiency by treating the model as if there were no intercept.

3.12 (a) $\quad E\{(\beta^+ - \beta)^2\} = E([\{\beta^+ - E(\beta^+)\} + \{E(\beta^+) - \beta\}]^2)$
$\quad\quad\quad\quad = \text{Var } \beta^+ + (\text{Bias } \beta^+)^2$

since the cross-product, on expanding and taking expectations, is identically zero.

(b) (i) $\quad\quad\quad\quad\quad\quad\quad\quad \beta^\circ = \Sigma w_t Y_t$

where $\quad\quad\quad\quad\quad\quad w_t = \bar{X}/\Sigma X_t^2$ for all t

Now $\quad\quad\quad\quad\quad\quad E(\beta^\circ) = \bar{X}\Sigma(\beta X_t)/\Sigma X_t^2$

using assumption (i),
$\quad\quad\quad\quad\quad\quad\quad\quad\quad = \beta K$

where $\quad\quad\quad\quad K = \bar{X} \Sigma X_t/\Sigma X_t^2 = (\Sigma X_t)^2/T\Sigma X_t^2.$

Now $\quad\quad\quad\quad\quad\quad T\Sigma X_t^2 \geq (\Sigma X_t)^2$

(Cauchy–Schwarz inequality—see problem 2.1—where $a_t = 1$, $b_t = X_t$ with equality only if $a_t = M b_t$ for all t, i.e. X_t constant.) Hence in general $K < 1$ and β° is biased downwards.

The mean square error of β° also requires its variance. Substituting again for Y_t into the basic formula we have

$$\text{Var } \beta^\circ = \sigma^2 \bar{X}^2 T/(\Sigma X_t^2)^2 \text{ (i.e. } \sigma^2 \Sigma w_t^2)$$

∴ $\quad\quad\quad \text{MSE}(\beta^\circ) = \sigma^2 \bar{X}^2 T/(\Sigma X_t^2)^2 + \beta^2(K - 1)^2.$

72 Properties of Least Squares Estimation

For OLS there is no bias so
$$\text{MSE}(\hat{\beta}) = \sigma^2 / \Sigma X_t^2.$$

Comparing these two expression we can see that neither MSE is always smaller. For example if our data has $\bar{X} = 0$ (in repeated samples) then
$$\frac{\text{MSE}(\hat{\beta})}{\text{MSE}(\beta^\circ)} = \frac{\sigma^2 / \Sigma X_t^2}{\beta^2}.$$

For some values of β, σ^2, and ΣX^2 one estimator dominates, and at other values the other has a smaller MSE. However since to know which is better we need to know the value of β and σ^2, this information is not very useful.

(ii)
$$\beta^* = \Sigma X_t Y_t / \Sigma X_t^2 + \frac{1}{T}$$

This estimator is *not* linear in Y_t (remembering that linearity does not allow for a constant term in the expression $\Sigma w_t Y_t$).

$$E(\beta^*) = \beta + \frac{1}{T}$$

using the standard substitution. Hence the estimator is biased, but the bias will be small for large T.

$$\text{MSE}(\beta^*) = \sigma^2 / \Sigma X_t^2 + \frac{1}{T^2}$$

(using assumption (i)). Hence the MSE of this estimator is always greater than that of OLS.

3.13 To find the MSE of $\breve{\beta}$ we substitute in for Y and use the general expression for an MSE:
$$\text{MSE}(\breve{\beta}) = E\left[\{\Sigma X(\beta X + U)/(\sigma^2/\beta^2 + \Sigma X^2) - \beta\}^2\right]$$

$$= E\left\{\frac{\left(\Sigma XU - \frac{\sigma^2}{\beta}\right)^2}{K}\right\}.$$

Squaring and using assumptions (ii) and (iii)
$$= \frac{\sigma^2(K)}{K^2} = \frac{\sigma^2}{K}$$

where
$$K = \sigma^2/\beta^2 + \Sigma X^2.$$

The estimator $\breve{\beta}$ is linear with $W_t = X_t/K$. Consider any other linear estimator
$$\beta^* = \Sigma(W_t + C_t) Y_t.$$

then
$$\text{MSE}(\beta^*) = E[\{\Sigma(W_t + C_t)(\beta X_t + U_t) - \beta\}^2]$$
$$= E[\{\Sigma W_t(\beta X_t + U_t) - \beta + \Sigma C_t(\beta X_t + U_t)\}^2]$$
$$= \text{MSE } \tilde{\beta} + \sigma^2 \Sigma C_t^2 + \beta^2 (\Sigma C_t X_t)^2$$
$$+ 2\beta^2 \Sigma C_t X_t \Sigma W_t X_t - 2\beta^2 \Sigma C_t X_t + 2\sigma^2 \Sigma C_t W_t$$

But substituting in for W_t and using the value of K we can see that the sum of the last three terms (the cross-product $\sigma^2 \Sigma W_t C_t$) is identically zero. Hence the MSE of any other linear estimator is strictly greater than that of $\tilde{\beta}$. This shows that OLS has a greater MSE than $\tilde{\beta}$ (since its MSE is $\sigma^2/\Sigma X_t^2$) but that this result cannot be utilized because it requires knowledge of the ratio β^2/σ^2.

Finally it should be noticed that this estimator has minimum MSE for all estimators that are linear in Y and whatever the values of β, σ^2, and the X_t. Clearly in certain circumstances other estimators can do better still but their algebraic form will either not be linear in Y (e.g. the estimator $\beta^* = \beta$) or else will change depending on circumstances.

3.14 Consider the OLS equation as a linear predictor
$$\hat{Y}_{T+m} = \Sigma W_t Y_t$$
when
$$W_t = \frac{1}{T} - \frac{-x_t \bar{X}}{\Sigma x_t^2} + x_t X_{T+m}/\Sigma x_t^2$$

(from least squares formulae for $\hat{\alpha}$ and $\hat{\beta}$). Using properties of errors we can show that

$$E(\hat{Y}_{T+m}) = E(Y_{T+m}) = \alpha + \beta X_{T+m}$$

so that in this special sense OLS is an unbiased predictor. The variance around the regression line is

$$\text{Var } \hat{Y}_{T+m} = \sigma^2 \Sigma W_t^2$$

(using Var $Y_t = \sigma^2 W_t^2$ and Cov $Y_t Y_s = 0$) because of error properties (ii) and (iii). Consider any other linear unbiased predictor,

$$Y^*_{T+m} = \Sigma(W_t + C_t) Y_t.$$

For unbiasedness we will restrict the choice of C_t to values such that $E(\Sigma C_t Y_t) = 0$, which is satisfied in general only for $\Sigma C_t = \Sigma C_t X_t = 0$. The variance of the general linear unbiased predictor is

$$\text{Var } Y^*_{T+m} = \sigma^2 \Sigma W_t^2 + \sigma^2 \Sigma C_t^2 + 2\sigma^2 \Sigma C_t W_t$$

(using assumptions (ii) and (iii)). However the third term is identically zero by the properties of the C_t and the value of W_t. Hence the OLS predictor has a smaller variance than that of any other linear unbiased predictor.

3.15 (a) $\quad\quad\quad \text{Var } PQ = E[\{PQ - E(PQ)\}^2]$

Using $\quad\quad\quad E(PQ) = \bar{P}\bar{Q}$

$\quad\quad\quad\quad\quad \text{Var } PQ = E\{(PQ)^2\} - \bar{P}^2\bar{Q}^2.$ $\quad\quad\quad$ (1)

But $\quad\quad \bar{P}^2 \text{ Var } Q + \bar{Q}^2 \text{ Var } P + \text{Var } Q \text{ Var } P$

$\quad\quad\quad\quad = \bar{P}^2 E\{(Q-\bar{Q})^2\} + \bar{Q}^2 E\{(P-\bar{P})^2\}$
$\quad\quad\quad\quad\quad + E\{(Q-\bar{Q})^2\}E\{(P-\bar{P})^2\}$

$\quad\quad\quad\quad = \bar{P}^2\{E(Q^2) - \bar{Q}^2\} + \bar{Q}^2\{E(P^2) - \bar{P}^2\}$
$\quad\quad\quad\quad\quad + \{E(Q^2) - \bar{Q}^2\}\cdot\{E(P^2) - \bar{P}^2\}$ $\quad\quad\quad$ (2)

$\quad\quad\quad\quad = E(P^2)E(Q^2) - \bar{P}^2\bar{Q}^2$ $\quad\quad\quad$ (1)

since from independence:

$$E(P^2 Q^2) = E(P^2) E(Q^2)$$

so that the result holds.

(b) The least squares predictor is now

$$Y^*_{T+m} = \hat{\beta} \hat{X}_{T+m}.$$

But $\quad\quad\quad E(Y^*_{T+m}) = E(\hat{\beta})E(\hat{X}_{T+m})$

(since $\hat{\beta}$ is a function of the U_t, which are independent of $\hat{X}_{T+m}$):

$\therefore \quad\quad\quad E(Y^*_{T+m}) = \beta X_{T+m} = E(Y_{T+m})$

and OLS is unbiased. The variance of the predictor is

$$\text{Var}(\hat{\beta} \hat{X}_{T+m}) = \beta^2 \text{ Var } \hat{X}_{T+m} + X^2_{T+m} \text{ Var } \hat{\beta} + \text{Var } \hat{\beta} \text{ Var } \hat{X}_{T+m}$$

using the first part of the question and the independence of $\hat{\beta}$ and $\hat{X}_{T+m}$.

Comparing this result to the case where $\text{Var } \hat{X}_{T+m}$ is zero (the standard forecast with certain exogenous values) we see that the variance increases by the percentage factor:

$$\frac{\beta^2 \text{ Var } \hat{X}_{T+m} + \text{Var } \hat{\beta} \text{ Var } \hat{X}_{T+m}}{X^2_{T+m} \text{ Var } \hat{\beta}} = [\{CV(\hat{\beta})\}^2 + 1]/\{CV(X_{T+m})\}^2$$

where CV is the coefficient of variation, $\beta/SE(\beta)$ etc., which measures the variability of a variable relative to its mean. As $\hat{X}_{T+m}$ becomes relatively more uncertain the variance of the forecast grows.

3.16 Consider the general linear estimator $\beta^* = \Sigma W_t Y_t$. For unbiasedness we must restrict the set of W_t:

$$E(\beta^*) = E[\Sigma W_t(\alpha + \beta X_t + U_t)]$$
$$= \alpha \Sigma W_t + \beta \Sigma W_t X_t$$
$$\therefore \quad \Sigma W_t = 0$$
$$\Sigma W_t X_t = 1.$$

Any variable in deviation form will satisfy the first criterion while any variable P_t derived as $P_t = R_t/\Sigma R_t X_t$ satisfies the latter type of condition ($\Sigma P_t X_t = 1$) (whatever R_t). Hence choosing $W_t = z_t/\Sigma z_t x_t = z_t/\Sigma z_t X_t$ over the set of all possible Z_t generates the set of all linear unbiased estimators, i.e. $\beta^* = \Sigma z_t y_t/\Sigma z_t x_t = \Sigma z_t Y_t/\Sigma z_t x_t$ is the general linear unbiased estimator. Substituting in for Y_t or using the properties of the errors to show that Var $Y_t = \sigma^2 W_t^2$ we have

$$\text{Var } \beta^* = \sigma^2 \Sigma w_t^2$$
$$= \sigma^2 \Sigma z^2/(\Sigma zx)^2.$$

The efficiency is then measured by

$$\text{Var } \hat{\beta}/\text{Var } \beta^* = (\Sigma zx)^2/\Sigma z^2 \Sigma x^2$$

which lies between zero and unity (Cauchy–Schwarz inequality) with the upper bound attained for $z_t = Kx_t$ all t. In this case the general linear estimator is OLS.

3.17 $E[(\hat{Y}_{T+F} - Y_{T+F})^2] = E([\{\hat{Y}_{T+F} - E(Y_{T+F})\} + \{E(Y_{T+F}) - Y_{T+F}\}]^2)$
$$= \text{Var}^* Y_{T+F} + \sigma^2 + 2E[\{\hat{Y}_{T+F} - E(Y_{T+F})\}$$
$$\times \{E(Y_{T+F}) - Y_{T+F}\}]$$

Expanding the cross-product and using the result that for the OLS predictor $E(\hat{Y}_{T+F}) = E(Y_{T+F})$ we have the cross-product zero and hence the desired result.

4
Multiple Regression

So far we have considered a basic model with one explanatory variable (with or without an intercept). We have seen how, under certain assumptions, OLS is an optimal estimator. The next step is clearly to generalize to the case when there is more than one explanatory variable. In this chapter we analyse fully the case of two explanatory variables—the cases of three or more variables are so cumbersome with scalar algebra that we do not attempt them but merely report the results that can be derived in the most general case by use of matrix algebra. As well as being interested in whether the Gauss–Markov theorem is generalizable we shall also focus on the question of whether the two-variable case raises any problems not encountered in the single-variable case.

4.1 The model and assumptions

We begin with a two-variable model which includes an intercept:

$$Y_t = \alpha + \beta_1 X_{1t} + \beta_2 X_{2t} + U_t \qquad (4.1)$$

for $t = 1 \ldots T$.

We assume the following:

(i) $\quad E(U_t) = 0$ all t

(ii) $\quad E(U_t^2) = \sigma^2$ all t

(iii) $\quad E(U_t U_s) = 0$ all $s, t; s \neq t$

(iv) $\quad X_{1t}, X_{2t}$ are fixed in repeated samples.

(v) $\quad \Sigma x_{1t}^2 \Sigma x_{2t}^2 - (\Sigma x_{1t} x_{2t})^2 > 0$.

These assumptions are exactly the same as for the single-variable model except that assumption (v) is generalized for a reason that will become clear.

Applying the OLS principle we wish to find estimators which minimize the residual sum of squares. Denoting arbitrary estimators by $\check{\alpha}, \check{\beta}_1, \check{\beta}_2$ we have the sum of squares S as a function of these estimators:

$$S(\check{\alpha}, \check{\beta}_1, \check{\beta}_2) = \Sigma(Y_t - \check{\alpha} - \check{\beta}_1 X_{1t} - \check{\beta}_2 X_{2t})^2 \qquad (4.2)$$

Differentiating with respect to the parameters and equating to zero, we obtain the three normal equations:

$$\Sigma(Y_t - \hat{\alpha} - \hat{\beta}_1 X_{1t} - \hat{\beta}_2 X_{2t}) = 0 \qquad (4.3)$$

$$\Sigma X_{1t}(Y_t - \hat{\alpha} - \hat{\beta}_1 X_{1t} - \hat{\beta}_2 X_{2t}) = 0 \qquad (4.4)$$

$$\Sigma X_{2t}(Y_t - \hat{\alpha} - \hat{\beta}_1 X_{1t} - \hat{\beta}_2 X_{2t}) = 0 \qquad (4.5)$$

(where ^ is used to denote an optimal value of the function).

The first equation yields the standard result that, for a model with an intercept, OLS passes through the point of means:

$$\bar{Y} = \hat{\alpha} + \hat{\beta}_1 \bar{X}_1 + \hat{\beta}_2 \bar{X}_2. \qquad (4.6)$$

This, as in the single-version case, can be substituted into (4.4) and (4.5) to eliminate one unknown and to put the data into deviation form:

$$\Sigma x_{1t}(y_t - \hat{\beta}_1 x_{1t} - \hat{\beta}_2 x_{2t}) = 0 \qquad (4.7)$$

$$\Sigma x_{2t}(y_t - \hat{\beta}_1 x_{1t} - \hat{\beta}_2 x_{2t}) = 0. \qquad (4.8)$$

Solving for $\hat{\beta}_1$ we obtain

$$\hat{\beta}_1 = (\Sigma x_1 y \, \Sigma x_2^2 - \Sigma x_2 y \, \Sigma x_1 x_2) / \{\Sigma x_1^2 \Sigma x_2^2 - (\Sigma x_1 x_2)^2\} \qquad (4.9)$$

(where we drop the summation subscript). We can now solve for $\hat{\beta}_2$ or, by noticing that in the two-variable model X_1 and X_2 are entirely symmetric, we can obtain $\hat{\beta}_2$ by exchanging X_1 and X_2 in (4.9):

$$\hat{\beta}_2 = (\Sigma x_2 y \, \Sigma x_1^2 - \Sigma x_1 y \, \Sigma x_1 x_2) / \{\Sigma x_1^2 \Sigma x_2^2 - (\Sigma x_1 x_2)^2\} \qquad (4.10)$$

Knowing $\hat{\beta}_1$ and $\hat{\beta}_2$ we can next substitute them into (4.6) to obtain $\hat{\alpha}$ expressed solely in terms of the data. We see that the estimators will not exist if the denominator is zero (i.e. OLS has no unique minimum)—hence the new form of assumption (v) (see problem 4.1).

4.2 The Gauss–Markov theorem

Next we turn to the question of the optimality of these estimators. They are all, by inspection, linear in Y_t. For instance

$$\hat{\beta}_1 = \Sigma w_t y_t \qquad (4.11)$$

where:

$$w_t = (x_{1t} \Sigma x_{2t}^2 - x_{2t} \Sigma x_{1t} x_{2t}) / \{\Sigma x_{1t}^2 \Sigma x_{2t}^2 - (\Sigma x_{1t} x_{2t})^2\} \qquad (4.12)$$

It is straightforward also to prove unbiasedness. We again illustrate for $\hat{\beta}_1$. As in the single-variable case we need to substitute in the true equation to replace the dependent variable but expressed in deviation form:

$$\hat{\beta}_1 = \frac{\Sigma x_1(\beta_1 x_1 + \beta_2 x + u)\Sigma x_2^2 - \Sigma x_2(\beta x_1 + \beta_2 x_2 + u)\Sigma x_1 x_2}{\Sigma x_1^2 \Sigma x_2^2 - (\Sigma x_1 x_2)^2} \qquad (4.13)$$

78 Multiple Regression

(where u_t is $U_t - \bar{U}$).

$$\therefore \quad \hat{\beta}_1 = \beta_1 + (\Sigma x_1 U \Sigma x_2^2 - \Sigma x_2 U \Sigma x_1 x_2)/D \tag{4.14}$$

where:
$$D = \Sigma x_1^2 \Sigma x_2^2 - (\Sigma x_1 x_2)^2. \tag{4.15}$$

Taking expectations and using the fact that the X variables are fixed in repeated samples we have:

$$E(\hat{\beta}_1) = \beta_1 \tag{4.16}$$

Similar arguments show the other estimators to be unbiased also. As before, unbiasedness does not require assumptions (ii) and (iii) to hold, since we have not yet used them.

We can next turn to the question of variance—our strategy as before will be to obtain the OLS variance and then show that no other linear unbiased estimator can have a smaller variance. Since OLS is unbiased we see immediately from (4.14) that

$$\operatorname{Var} \hat{\beta}_1 = E\left\{\left(\frac{\Sigma x_1 u \Sigma x_2^2 - \Sigma x_2 u \Sigma x_1 x_2}{D}\right)^2\right\} \tag{4.17}$$

We next square and then take expectations—using assumptions (ii) and (iii):

$$\operatorname{Var} \hat{\beta}_1 = \sigma^2 \Sigma x_2^2 / \{\Sigma x_1^2 \Sigma x_2^2 - (\Sigma x_1 x_2)^2\} \tag{4.18}$$

Similarly

$$\operatorname{Var} \hat{\beta}_2 = \sigma^2 \Sigma x_1^2 / \{\Sigma x_1^2 \Sigma x_2^2 - (\Sigma x_1 x_2)^2\} \tag{4.19}$$

$$\operatorname{Var} \hat{\alpha} = \sigma^2 \left(\frac{1}{T} + \frac{\bar{X}_1^2 \Sigma x_2^2 + \bar{X}_2^2 \Sigma x_1^2 - 2\bar{X}_1 \bar{X}_2 \Sigma x_1 x_2}{\Sigma x_1^2 \Sigma x_2^2 - (\Sigma x_1 x_2)^2}\right)$$

and
$$\operatorname{Cov}(\hat{\beta}_1, \hat{\beta}_2) = -\sigma^2 \Sigma x_1 x_2 / \{\Sigma x_1^2 \Sigma x_2^2 - (\Sigma x_1 x_2)^2\} \tag{4.20}$$

By inspection it can be seen that from (4.12)

$$\operatorname{Var} \hat{\beta}_1 = \sigma^2 \Sigma w_t^2. \tag{4.21}$$

Consider the general linear estimator for β_1:

$$\beta_1^+ = \Sigma k_t Y_t \tag{4.22}$$

which can be written as

$$\beta_1^+ = \Sigma (w_t + c_t) Y_t \tag{4.23}$$

(where the c_t are fixed in repeated samples). For this to be an unbiased estimator we require

$$E(\beta_1^+) = \Sigma E\{(w_t + c_t) Y_t\} = \beta. \tag{4.24}$$

Since $E(\Sigma w_t Y_t) = \beta_1$, this implies
$$\Sigma E(c_t Y_t) = 0 \tag{4.25}$$

which in turn requires
$$\Sigma c_t(\alpha + \beta_1 X_{1t} + \beta_2 X_{2t}) = 0 \tag{4.26}$$

In order for this to hold, whatever the values of the true parameters and whatever the values the X variables take in the sample, this can be expressed as:

$$\Sigma c_t = 0 \qquad (4.27)$$

$$\Sigma c_t X_{1t} = 0 = \Sigma c_t x_{1t} \qquad (4.28)$$

$$\Sigma c_t X_{2t} = 0 = \Sigma c_t x_{2t}. \qquad (4.29)$$

Thus the set of all linear unbiased estimators is generated by considering all sets of c_t which obey conditions (4.27), (4.28), and (4.29).

The final step is to generate the variance of the general linear unbiased estimator. As before this can be expressed as

$$\text{Var } \beta_1^+ = E\{(k_t Y_t - \beta_1)^2\} \qquad (4.30)$$

$$= E[\{\Sigma(w_t + c_t)(\alpha + \beta_1 X_{1t} + \beta_2 X_{2t} + U_t) - \beta_1\}^2] \qquad (4.31)$$

Using (4.12) and (4.27), (4.28), (4.29):

$$\text{Var } \beta_1^+ = E\{(\Sigma w_t U_t + \Sigma c_t U_t)^2\} \qquad (4.32)$$

$$= \sigma^2 \Sigma w_t^2 + \sigma^2 \Sigma c_t^2 + 2\sigma^2 \Sigma c_t w_t. \qquad (4.33)$$

However, using the restrictions on c_t and the value of w_t we see that the final term is identically zero:

$$\text{Var } \beta_1^+ = \sigma^2 \Sigma w_t^2 + \sigma^2 \Sigma c_t^2 \qquad (4.34)$$

And using (4.21)

$$\text{Var } \beta_1^+ = \text{Var } \hat{\beta}_1 + \sigma^2 \Sigma c_t^2. \qquad (4.35)$$

Hence the OLS estimator has a smaller variance for β_1 than any other linear unbiased estimator where assumptions (i) to (v) hold (see problem 4.1). As before the proofs can be repeated to show that all the OLS estimators are BLUE.

This result is generalizable to the many-variable case—if assumptions (i) to (iv) hold, plus the extension of (v), then OLS is BLUE (in the sense that in the class of all linear homogeneous (in Y) estimators, which are unbiased whatever the actual values of the true parameters and the fixed variables, it has the smallest variance). We must say a brief word here about condition (v), although it will be necessary to return to it later on. The key to understanding the role of this condition lies in the 'normal' equations. These are the equations which are solved for the optimal values of the estimators. Let us first consider the one-variable model with a constant—the equations can be written:

$$\Sigma(Y_t - \hat{\alpha} - \hat{\beta} X_t) = 0 \qquad (2.10)$$

$$\Sigma X_t (Y_t - \hat{\alpha} - \hat{\beta} X_t) = 0. \qquad (2.11)$$

If the variable X_t were a constant (K) for all t then the two equations would in fact be identical (divide the second by K) and we cannot solve two *identical* equations for two unknowns—infinitely many pairs of $\hat{\alpha}$ and $\hat{\beta}$ that would satisfy (2.10) would then automatically satisfy (2.11). Hence the condition for a unique solution has to be that X_t is not constant for all t, which can be expressed: $\Sigma(X_t - \bar{X})^2 > 0$. If X_t were constant it would always equal its own mean and the deviations would all be zero. We are in fact assuming that there is no exact relationship between the weight on the intercept $\alpha(1)$ and on the slope $\beta(X_t)$, i.e. that they are not perfectly *multicollinear*. As we said before, if the two factors are the same then we cannot distinguish between them.

A similar argument applies to the two variable model. The normal equations here are:

$$\Sigma(Y_t - \hat{\alpha} - \hat{\beta}_1 X_{1t} - \hat{\beta}_2 X_{2t}) = 0 \tag{4.3}$$

$$\Sigma X_{1t}(Y_t - \hat{\alpha} - \hat{\beta}_1 X_{1t} - \hat{\beta}_2 X_{2t}) = 0 \tag{4.4}$$

$$\Sigma X_{2t}(Y_t - \hat{\alpha} - \hat{\beta}_1 X_{1t} - \hat{\beta}_2 X_{2t}) = 0. \tag{4.5}$$

Any exact interrelations between the normal equations would render solution impossible, i.e. we cannot have

$$X_{1t} = K_1 \quad \text{all } t$$

or
$$X_{2t} = K_2 \quad \text{all } t$$

or
$$X_{1t} = K_3 X_{2t} \quad \text{all } t$$

or
$$X_{1t} = K_4 + K_5 X_{2t} \quad \text{all } t. \tag{4.36}$$

These three conditions are summarized effectively by requiring that

$$\Sigma x_{1t}^2 \Sigma x_{2t}^2 - (\Sigma x_{1t} x_{2t})^2 > 0.$$

If either of the first two fail the inequality fails because both terms are zero; while if the third or fourth fail the inequality fails because the two terms (although non-zero) are equal and of opposite sign. In this model there are effectively three variables, the intercept (X_{0t}) with unit-value X_{1t} and X_{2t}. For there to be a unique solution to the OLS minimization problem we require that there should not exist any weights K_0, K_1, and K_2 which have the property

$$K_0 X_{0t} + K_1 X_{1t} + K_2 X_{2t} = 0 \quad \text{all } t \tag{4.37}$$

(there is no 'linear dependence' between the variables). This condition has an exact analogue in every model. For example the model of Chapter 2 with an intercept and a single variable requires that there exist no K_0 and K_1 for which

$$K_0 X_{0t} + K_1 X_{1t} = 0 \quad \text{all } t \tag{4.38}$$

while the two-variable model without intercept requires that the following does not hold (see problem 4.1):

$$K_1 X_{1t} + K_2 X_{2t} = 0 \text{ all } t. \tag{4.39}$$

4.3 Multiple regression and goodness of fit

The principles of measuring the goodness of fit for a multiple regression are a straightforward extension of those for single-variable regression. Once we have estimated the equation by ordinary least squares we can obtain the fitted values and the residuals:

$$\hat{Y}_t = \hat{\alpha} + \hat{\beta}_1 X_{1t} + \hat{\beta}_2 X_{2t} \tag{4.40}$$

$$\hat{U}_t = Y_t - \hat{Y}_t. \tag{4.41}$$

We see from the 'normal' equations that

$$\Sigma \hat{U}_t = 0 = \Sigma \hat{U}_t X_{1t} = \Sigma \hat{U}_t X_{2t} = 0. \tag{4.42}$$

The residual sum of squares (RSSQ), which is the function minimized by OLS, is as before:

$$\text{RSSQ} = \Sigma \hat{U}_t^2. \tag{4.43}$$

The unbiased estimator of the error variance is used to generate the standard error of estimate:

$$\text{SEE} = \sqrt{\frac{1}{T-3} \Sigma \hat{U}_t^2} \tag{4.44}$$

The divisor becomes $(T-3)$ because that is the number of degrees of freedom between the $\hat{U}_t$ after estimating the three parameters. In the many-variable case, with K independent variables and an intercept, the SEE (as an unbiased estimator of the error variance) is given by:

$$\text{SEE} = \sqrt{\frac{1}{T-K-1} \Sigma \hat{U}_t^2}. \tag{4.45}$$

The correlation coefficient can now be derived only in two ways. It is either the correlation between actual and fitted values:

$$R^2 = (\Sigma y_t \hat{y}_t)^2 / \Sigma y_t^2 \Sigma \hat{y}_t^2 \tag{4.46}$$

or the 'explained' percentage of the total variance

$$R^2 = \frac{\text{ESSQ}}{\text{TSSQ}} = 1 - \frac{\text{RSSQ}}{\text{TSSQ}} \tag{4.47}$$

where
$$\text{TSSQ} = \Sigma y_t^2 \tag{4.48}$$
$$\text{ESSQ} = \Sigma y_t^2 - \Sigma \hat{y}_t^2 \tag{4.49}$$
since
$$\Sigma y_t^2 = \Sigma \hat{y}_t^2 + \Sigma \hat{U}_t^2.$$

Obviously this coefficient, which is known as the multiple correlation coefficient, cannot be derived from the relation between Y and a single X variable since it summarizes the effect of both independent variables.

The generalization of the multiple correlation to the many-variable case is by the use of (4.42) or (4.43). In all cases the range of R^2 is given by the Cauchy–Schwarz inequality so that:

$$0 \leq R^2 \leq 1. \tag{4.50}$$

If R^2 is zero then RSSQ = TSSQ, which means that the regression line is the mean of the Y_t—the X variables are given exactly zero coefficients by OLS. If R^2 is unity then the line fits the data exactly and the residuals are all zero.

As we shall discuss later, there is in fact a problem of deciding how many variables should enter a regression in order to explain a given variable Y_t. If the choice is made on the grounds of goodness of fit then it appears that R^2 will favour models with larger numbers of variables. Consider the choice of a model with just X_1 or a model with X_1 and X_2. In estimating the latter by OLS the parameter of X_2 could be set equal to zero if that did minimize RSSQ, that is, it includes all parameter combinations available to the first model as a subset of cases. Hence the RSSQ can never be lower for model one than for model two, and, since the TSSQ is the same in both cases, the R^2 must be at least as great for model two as for model one. In order to make an adjustment for this effect it is often suggested that the multiple correlation should be 'corrected for degrees of freedom'. To effect this we first rewrite R^2

$$R^2 = 1 - (\Sigma \hat{U}_t^2 / T)/(\Sigma y_t^2 / T). \tag{4.51}$$

This version uses estimates of the variance of the residuals and the variance of the dependent variable, but both are biased as estimates of population variances. We replace them by unbiased estimates and obtain the corrected $\bar{R}^2$ (R bar)

$$\bar{R}^2 = 1 - \frac{\left(\dfrac{1}{T-K-1} \Sigma \hat{U}_t^2\right)}{\left(\dfrac{1}{T-1} \Sigma y_t^2\right)} \tag{4.52}$$

where there are K explanatory variables as well as the constant. The relationship between the two measures is found by substituting (4.51) into (4.52).

$$\bar{R}^2 = 1 - \left(\frac{T-1}{T-K-1}\right)(1 - R^2) \tag{4.53}$$

so that $\bar{R}^2 \leq R^2$ for $K \leq 1$. Since this is a purely derived statistic, and does not measure the correlation between any two variables, it is not governed by the Cauchy–Schwartz inequality. Its upper limit is still unity (for very large T and low K) where R^2 tends to unity, but its lower limit is $-K/(T-K-1)$ as R^2 tends to zero. This is unbounded in the lower limit. A large value of K with T only just greater than K can produce a negative corrected correlation coefficient even when R^2 is actually positive.

As well as being interested in the overall goodness of fit of an equation we may want to analyse the contributions of the individual variables to the total goodness of fit and to assess their relative importance.

We obviously cannot assess the relative importance of different variables simply by looking at the regression coefficients (the parameter estimates). First, a change in the units of a variable changes the parameter by the inverse factor, so that the size of coefficients in general depends on the measurement dimension of the associated variable. Second, some variables have a wider range of values than others so that this tends to lead to small variance of their estimates even though the *marginal* impact of the variable may be small. A scaling of the regression which puts all variables onto an equal footing is to divide every independent variable X by its standard deviation and take deviations from its means:

$$X_{it}^* = (X_{it} - \bar{X}_i) \bigg/ \frac{1}{T} \Sigma (X_{it} - \bar{X}_i)^2. \qquad (4.54)$$

The dependent variable is similarly transformed. The new equation, in the two-variable case, becomes

$$Y_t^* = A + \beta_1^* X_{1t}^* + \beta_2^* X_{2t}^* + U_t \qquad (4.55)$$

(where the intercept has been augmented by functions of the means of all the variables). These resulting coefficients of the regression of the 'standardized' variables are known as 'beta' coefficients and are related to the original parameters by the scaling factors. For example,

$$\beta_1^* = \beta_1 \cdot \left\{ \frac{\Sigma (X_1 - \bar{X}_1)^2}{\Sigma (Y - \bar{Y})^2} \right\}^{\frac{1}{2}} \qquad (4.56)$$

The relative sizes of the $\hat{\beta}_1^*$ give a measure of the relative importance of marginal impact of a change in the X_i. They measure the effect of a change in X_i^* by one unit (in X_i by one standardized unit on the dependent variable also measured in standardized units) (see problem 4.4).

If we wish to assess, not the size of the marginal influences on the dependent variable, but rather the marginal contributions to the explanation of its variation then we must focus on a correlation type measure. The approach to this is in essence to compare the multiple correlation coefficients obtained with and without the variable in question. Consider the two-variable model with a constant. The multiple correlation of the full model we denote by R^2, while that of the model with variable $i (i = 1, 2)$ deleted we denote R_i^2. The 'incremental' contributions are then $(R^2 - R_i^2)$. In practice, rather than work with these incremental differences (which must be less than R^2 and so usually have a range smaller than that of a correlation coefficient) it is usual to convert these into percentages. After the other variable has made its contribution there is a fraction $(1 - R_i^2)$ of the TSSQ to be explained. The additional variable then explains a fraction (r_i^2) of this: i.e.

$$r_i^2 = (R^2 - R_i^2)/(1 - R_i^2). \tag{4.57}$$

This concept (which varies between zero and unity) is known as a partial correlation coefficient and is usually written in the form $r_{y1.2}^2$ to denote the extra correlation between the dependent variable Y and variable X_1 once X_2 has first been used to explain Y. It could be reached by an equivalent route of regressing Y on X_2, obtaining the residuals, and regressing these on X_1 to yield a correlation the same as that derived from the regressions of Y on X_1 and X_2, and of Y on X_2 via the use of (4.51) (see problem 4.3). The important aspect of this concept is that the partial correlation between Y and X_1 (say) can be very different from the simple (total) correlation between those variables. If X_1 and X_2 themselves are strongly correlated, then once Y is regressed on X_2 much of the influence of X_1 will have been picked up indirectly so that any extra contribution it has to make in explaining Y is relatively smaller than its simple correlation with Y would suggest. A series of complex relationships tie all these concepts together. For example in the two-variable (plus constant) case it can be shown that (see problem 4.2):

$$R_{y.12}^2 = \frac{r_{y1}^2 + r_{y2}^2 - 2r_{y1} r_{y2} r_{12}}{1 - r_{12}^2} \tag{4.58}$$

$$r_{y1.2}^2 = \frac{(r_{y1} - r_{y2} r_{12})^2}{(1 - r_{12}^2)(1 - r_{y2}^2)} \tag{4.59}$$

where:

r_{yi}^2 is squared (total) correlation between Y and $X_i (i = 1, 2)$;
r_{12}^2 is squared (total) correlation between X_1 and X_2;
$r_{y1.2}^2$ is squared partial correlation between Y and X_1 given that X_2 has already been related to Y;
$R_{y.12}^2$ is the squared multiple correlation of Y on variables X_1 and X_2.

Example 4.1. A multiple regression

To illustrate the technique of multiple regression we hypothesize that as well as income, unemployment also affects consumption. High levels of unemployment (U) are expected to create a greater precautionary demand for savings so that the relationship is supposed to be negative between U and C. The data for total unemployment is given is table 4.1.

TABLE 4.1 UK Unemployment 1961–1982

Year	U in 000s (excluding school leavers)
1961	339.0
1962	453.9
1963	539.4
1964	393.7
1965	338.2
1966	353.2
1967	547.2
1968	574.4
1969	566.3
1970	602.0
1971	775.8
1972	826.1
1973	591.2
1974	590.9
1975	902.3
1976	1229.4
1977	1313.0
1978	1299.1
1979	1277.4
1980	1560.8
1981	2419.8
1982	2793.4

Source: Economic Trends (1984).

The actual plotting of the new series can be managed as in Chapter 2, in that it can be plotted against time, against income, or against consumption. However the three-dimensional scattergram, which would reveal the simultaneous actions of income and unemployment on consumption, is clearly impractical (the true relationship and the fitted relationship would both be planes in this space).

We present the results of the regression of C on Y and U (measuring money values in millions of pounds and unemployment in thousands).

$$C = 17880 + 0.752\ Y + 0.930\ U$$
$$(2817)\quad (0.026)\quad (0.798)$$
$$R^2 = 0.9916,\ \bar{R}^2 = 0.9907,\ \text{SEE} = 1531,\ \text{RSSQ} = 4.45 \times 10^7.$$

We next turn to the estimates and their standard errors. Here it is helpful to look back at the single-variable regression on page 53. We see that the constant is larger (but with a larger standard error) in the single-variable regression. The coefficient on disposable income is very similar (but again the variance is larger). The coefficient on unemployment is positive, contrary to expectations. The overall goodness of fit as measured by R^2 is higher (as it must be) in the multiple regression but only by a very small amount and correspondingly the standard error of estimate is smaller. The corrected $\bar{R}^2$ is virtually unchanged.

We can obtain the other correlation measures by the formulae given in the text. Firstly we have the matrix of simple pairwise unsquared correlations (Table 4.2).

TABLE 4.2. Matrix of Pairwise correlations

	C	Y	U
C	1.0000	0.9955	0.7828
Y		1.0000	0.7707
U			1.0000

The entries below the diagonal are not given since they are the same as those above the diagonal. From these correlations we see that the strongest correlation is between consumption and income. Consumption is positively correlated with unemployment, and income and unemployment are also positively correlated.

The partial correlations can be obtained from (4.57):

$$r_{cu.y} = 0.067$$
$$r_{cy.u} = 0.978.$$

These make it clear that if we first include income, then unemployment has a very low extra explanatory power; but if we first include unemployment, income despite its correlation with unemployment, can still explain a large fraction of that part of consumption unexplained by unemployment. Finally we give the estimated covariances between parameters: $\text{Cov}(\hat{\alpha}, \hat{\beta}_1) = -70.58$, $\text{Cov}(\hat{\alpha}, \hat{\beta}_2) = 1458$, $\text{Cov}(\hat{\beta}_1, \hat{\beta}_2) = -0.016$.

4.4 Multiple regression and the problem of multicollinearity

In all the models we have so far considered it has been necessary to restrict ourselves to consideration of cases where certain conditions on the independent variables held. For example in the single-variable model with a constant the condition was $\Sigma(X_t - \bar{X})^2 > 0$, while in the two-variable model with a constant the condition became

$$D = \{\Sigma x_{1t}^2 \, \Sigma x_{2t}^2 - (\Sigma x_{1t} x_{2t})^2\} > 0.$$

We explained that these conditions had to hold in order that there should be as many independent 'normal' equations as there were parameters to be estimated, that is, that the second-order conditions for a minimum be satisfied. If the conditions failed then no unique solution to OLS could be found. In both cases this could be seen in the variance of the parameters. Equations (4.18) and (4.19) would yield infinitely large variances (indicating that all values of the parameter are possible). This case, where one variable is an exact linear function of one (or more) other variable is known as 'perfect multicollinearity'. In practice it seems that only logical mistakes of putting the same variable twice into a regression could produce the exact linear dependence that we must avoid. Hence the problem of perfect multicollinearity is seen not as a practical problem but rather as a theoretical curiosity. However a much more important issue is raised when the conditions do not fail exactly, but in some sense nearly fail. We have already pointed out that in the single-variable model as Σx_t^2 decreases the variance of OLS will increase; in this case, when all the X data are close to their own mean, OLS will yield a very imprecise estimate of the parameter. It will still of course be the best (minimum variance) estimator but it will be a rather poor 'best'. Similarly as the condition 4 (v) is only just satisfied the variances of $\hat{\beta}_1$ and $\hat{\beta}_2$ will tend to increase and OLS to become more imprecise. In the single-variable model we can see immediately that the problem is that the X variable is too clustered round its mean. In the two-variable model we can obtain a better insight by rewriting (4.18):

$$\text{Var } \hat{\beta}_1 = \sigma^2 / \Sigma x_1^2 (1 - r_{12}^2) \tag{4.60}$$

where r_{12}^2 is the (total) squared correlation between X_1 and X_2. Hence we can see that, for a given variance of X_1 and a given error variance, the variance of ordinary least squares depends also on the correlation between X_1 and X_2. The stronger the correlation between these variables (with either a negative or positive association) the higher the variance of $\hat{\beta}_1$. As the two variables become more alike it becomes harder to distinguish their separate influences on Y and hence the OLS parameter estimates become more variable in repeated sampling. A similar result holds for Var $\hat{\beta}_2$—for a given variation in X_2 a stronger correlation between X_1 and X_2 will raise the variance of $\hat{\beta}_2$.

Finally the covariance between $\hat{\beta}_1$ and $\hat{\beta}_2$ can be written

$$\text{Cov}(\hat{\beta}_1, \hat{\beta}_2) = -\frac{r}{1-r^2}\frac{\sigma^2}{\sqrt{D}}. \tag{4.61}$$

This shows that for a set of data with a strong positive correlation the coefficients will have a high negative covariance. In repeated sampling there will be a tendency not only for a larger variation of the estimates around the true values of the parameters but for overshooting by one to be matched by undershooting on the other.

Now that we can see that in the two-variable model a key factor is the size of the correlation between the two independent variables we can return to the effect of this on goodness of fit and partial correlation. When there is zero intercorrelation between the two variables we can simplify (4.59) and (4.57).

$$R^2_{y.12} = r^2_{y1} + r^2_{y2} \tag{4.62}$$

$$r^2_{y1.2} = r^2_{y1}/(1-r^2_{y2}) \tag{4.63}$$

The first shows that in the special case where the two independent variables are uncorrelated the multiple correlation is the sum of the individual squared correlations while the second therefore is the fraction explained by $X_1(r^2_{y1})$ relative to that unexplained by $X_2(1-r^2_{y2})$—the incremental correlation of course reduces to $R^2_i = r^2_{yi}$. Finally we can examine the regression coefficients in the case where the two independent variables are uncorrelated. From (4.9) we have

$$\hat{\beta}_1 = \Sigma x_1 y/\Sigma x_1^2 \tag{4.64}$$

and from (4.60) we have

$$\text{Var } \hat{\beta}_1 = \sigma^2/\Sigma x_1^2.$$

Hence the regression coefficients reduce to what we would have obtained had we carried out a regression of Y on just X_1. This is an important result since it shows that had we forgotten to include X_2 in the regression where it genuinely should have been included, the coefficient of X_1 could only have been the same if the variables had been uncorrelated; further as we shall show in a later chapter the estimator from the single variable regression would have been biased unless the variables had been uncorrelated.

The results of this section are generalizable to the many-variable case. If there is any exact linear relation between the K variables (and the intercept) then there will be no unique solution to the normal equations and OLS will not have a solution. If there is some strong correlation between a set or subset of the independent variables then the variance of OLS may increase. It is no longer the case that this effect can be related simply to a single measure (the correlation between the two independent variables)—each pair may have only a modest correlation but nevertheless a nearly perfect relation may hold between some combination of the independent variables which makes the variance of some OLS parameters very large.

In summary, we see that perfect multicollinearity would make OLS (or other methods of estimation) unusable since no estimator could distinguish between the exactly related variables. If instead the independent variables are strongly, but not perfectly, correlated, then OLS is BLUE but the variances of the OLS estimators of the parameters become larger, the greater is the degree of intercorrelation. When there is zero intercorrelation the OLS estimates and their variances are those which would have been obtained in separate single-variable regression, and the multiple correlation is the sum of the correlations that would have been obtained from the separate regressions.

4.5 Estimation with parameter restrictions

In all our discussions on the single- and multiple-variable models we have so far assumed that there is no information available on the parameters. These estimated values are dictated solely by the data. In certain circumstances we do have information on the parameters about which we are certain, or with which we wish to experiment. For example we may be certain that the marginal propensity to consume in the long run steady state is unity, and we would then wish to estimate our two-variable equation relating consumption to income and unemployment subject to this restriction. Or in estimating a Cobb–Douglas production function:

$$\text{Log } O = \log A + \alpha \text{ Log } L + \beta \log K$$

we may wish to constrain it to have constant returns to scale and hence to impose the restriction $\alpha + \beta = 1$.

We might even wish to impose more than one restriction on the equation. In general it is straightforward to impose restrictions linear in the parameters on an equation[1] by an extension of the least squares method. We give a treatment of a single parameter restriction and then indicate how this can be generalized to a multi-parameter restriction and to the many-restriction case.

Consider the model

$$Y_t = \alpha + \beta_1 X_{1t} + \beta_2 X_{2t} + U_t \tag{4.65}$$

---

[1] As many restrictions as parameters can be imposed provided that the restrictions are linearly independent.

90 Multiple Regression

which has to be estimated subject to the restriction

$$\beta_1 = r_1 \tag{4.66}$$

where r_1 is a known number. Our least squares problem is then to choose $\hat{\hat{\alpha}}$, $\hat{\hat{\beta}}_1$, and $\hat{\hat{\beta}}_2$ to minimize the sum of squares S, subject to the restriction that $\hat{\hat{\beta}}_1 = r_1$. Since the restriction gives us the true value of β_1 it clearly should be utilized since we require a minimum variance unbiased estimator.

Hence we need
$$\text{Min } S(\hat{\hat{\alpha}}_1, \hat{\hat{\beta}}_1, \hat{\hat{\beta}}_2) \tag{4.67}$$

Subject to
$$\hat{\hat{\beta}}_1 = r_1. \tag{4.68}$$

This can either be solved by Lagrangian techniques or else by substituting (4.68) into the minimand and reducing its dimension by one (see problems 4.7 and 4.8). We follow the latter approach. The restricted sum of squares, i.e. the one obeying (4.68) can be written:

$$S = \Sigma(Y_t - \hat{\hat{\alpha}} - r_1 X_{1t} - \hat{\hat{\beta}}_2 X_{2t})^2. \tag{4.69}$$

There are two normal equations to solve for the unknowns:

$$\frac{\partial S}{\partial \hat{\hat{\alpha}}} = -2 \Sigma(Y_t - \hat{\hat{\alpha}} - r_1 X_{1t} - \hat{\hat{\beta}}_2 X_{2t}) = 0 \tag{4.70}$$

$$\frac{\partial S}{\partial \hat{\hat{\beta}}_2} = -2 \Sigma(Y_t - \hat{\hat{\alpha}} - r_1 X_{1t} - \hat{\hat{\beta}}_2 X_{2t}) X_{2t} = 0. \tag{4.71}$$

Clearly if we collect all variables with known coefficients as a new dependent variable—

$$Y_t^* = Y_t - r_1 X_{1t} \tag{4.72}$$

—these are the normal equations for the single-variable regression of Y^* on X_2 with an intercept, and hence the solutions are:

$$\hat{\hat{\beta}}_2 = \Sigma y_t^* x_{2t} / \Sigma x_{2t}^2 \tag{4.73}$$

$$\hat{\hat{\alpha}} = \bar{Y}^* - \hat{\hat{\beta}}_2 \bar{X}_2 \tag{4.74}$$

with variances:
$$\text{Var } \hat{\hat{\beta}}_2 = \sigma^2 / \Sigma x_{2t}^2$$
$$\text{Var } \hat{\hat{\alpha}} = \sigma^2 \Sigma X_{2t}^2 / T \Sigma x_{2t}^2. \tag{4.75}$$

These are the restricted least squares (RLS) estimators and their variances. The variances can be estimated in the usual fashion from the residuals of the restricted least squares fit, remembering that only two degrees of freedom have been utilized in their estimation:

$$\hat{\hat{S}}^2 = \frac{1}{T-2} \Sigma \hat{\hat{U}}_t^2 = \frac{1}{T-2} \Sigma (Y_t^* - \hat{\hat{\alpha}} - \hat{\hat{\beta}}_2 X_{2t})^2. \tag{4.76}$$

These estimators are those which 'fit' the data best subject to the restriction that the estimated value of β_1 is equal to the known value r_1. It can also be seen that the variance of the restricted estimator is in general smaller that of OLS using (4.60):

$$\text{Var } \hat{\hat{\beta}}_2 = \frac{\sigma^2}{\Sigma x_{2t}^2} \leq \frac{\sigma^2}{\Sigma x_{2t}^2 (1 - r_{12}^2)} = \text{Var } \hat{\beta}_2. \tag{4.77}$$

Clearly the restriction appears to improve on OLS and is most valuable when X_1 and X_2 are highly correlated (r_{12} is large). We need to explain this result in the light of the fact that OLS is known to be optimal (in a certain sense). This is best done by directly considering the optimality of the restricted least squares estimators.

In considering the relation of constraints to the equation for estimation there are two separate situations. One situation makes the constraint homogeneous in some (or all) of the parameters of the equation, including that on the intercepts, the other is not expressible in homogeneous form. Some examples will make the situation clear.

(a) Models with homogeneous restrictions

$$Y_t = \beta_1 X_{1t} + \beta_2 X_{2t} + U_t \tag{4.78}$$

(This could have $X_{1t} = 1$ to allow for the intercept term.) The restrictions are of the form

$$\sum_{i=1}^{2} \theta_i \beta_i = 0 \tag{4.79}$$

where the θ_i are known. For example,

$$\beta_1 + \beta_2 = 0. \tag{4.80}$$

Any restriction of this form can be substituted into the equation to yield a new right-hand side with fewer parameters to be estimated (the number reduced by the number of restrictions). Our example gives:

$$Y_t = \beta_1 (X_{1t} - X_{2t}) + U_t \tag{4.81}$$

Ordinary least squares can be applied to these new equations relating Y to the composite X variables. Since the error terms continue to be well behaved (if the restrictions are correct) the estimators are the Best Linear (in Y) Unbiased by the Gauss-Markov theorem. However the model they relate to is not the same as the unrestricted model in that *fewer parameters* are involved. This means that the set of weights, linear in Y, that are consistent with unbiasedness will be greater. Consider the set of all linear (in Y) estimators of β_1 in model (4.81):

$$\beta_1^* = \Sigma W_t Y_t. \qquad (4.82)$$

For unbiasedness we have the single restriction via (4.81) on the weights:

$$\Sigma W_t (X_{1t} - X_{2t}) = 1 \qquad (4.83)$$

Had we considered the unrestricted model (4.78) the class of all linear estimators is

$$\beta_1^+ = \Sigma V_t Y_t \qquad (4.84)$$

and substituting in (4.78) we require *for β_1^+ to be unbiased whatever the true β_1 and β_2* the *two* restrictions:

$$\Sigma V_t X_{1t} = 1; \quad \Sigma V_t X_{2t} = 0. \qquad (4.85)$$

However we see that all weights V_t which obey (4.85) will automatically obey (4.83) but that the reverse is not necessarily true, that is that in the restricted model we can consider all the weights allowable for unrestricted unbiased estimators and also other weights which would not be available—the class of W_t includes all the class of V_t but not vice-versa. Hence the best estimator in the larger class can indeed be better than the best estimator in the subclass. The straight reduction in the number of parameters to be estimated in fact gives more choice to the class of linear unbiased estimators for these parameters remaining and hence the change in model can reduce variance of the estimated parameter. We have indeed already seen (4.60) that the two-variable (multiple) regression model has a larger variance for the estimated parameters than it would as a single-variable model when the exclusion restriction ($\beta_2 = 0$) is imposed by just carrying out the single-variable model when the exclusion restriction ($\beta_2 = 0$) is imposed by just carrying out the single-variable regression (see problem 4.12).

(b) Models with inhomogeneous restrictions

With the same equation as before—(4.78)—we consider restrictions of the form

$$\sum_{1}^{2} \theta_i \beta_i = \theta_0 \qquad (4.86)$$

where all the θ are known and $\theta_0 \neq 0$. An example would be:

$$\beta_1 + \beta_2 = 1. \tag{4.87}$$

Substituting into the equation and collecting all *known* terms on the left-hand side of the equation we have

$$(Y_t - X_{1t}) = \beta_2(X_{2t} - X_{1t}) + U_t. \tag{4.88}$$

The dimension of the right-hand side of the equation has again been reduced but the left-hand side has also been changed. Considering this equation as one in Y_t^* ($Y_t^* = Y_t - X_{1t}$) we see that OLS is best linear (in Y^*) unbiased for model (4.88). The set of all linear estimators is

$$\beta_2^* = \Sigma W_t Y_t^* \tag{4.89}$$

and for unbiasedness we require again

$$\Sigma W_t(X_{2t} - X_{1t}) = 1 \tag{4.90}$$

whereas for OLS on the unrestricted equation we considered (4.84) subject to (4.85). As we have already seen the set of W_t is larger than the set of V_t (and includes all of it). However the sets are applied to different variables—Y^* and Y. The final step is to see that the set of all Y_t^* can be written

$$\Sigma W_t Y_t^* = \Sigma W_t Y_t - \Sigma W_t X_{1t}. \tag{4.91}$$

In the unrestricted case the second term has to be zero (4.85) but in the restricted case it does not, so we are indeed considering a wider class of estimators. This class is sometimes said to be 'linear in the Y_t and the restriction', i.e. $\Sigma W_t Y_t + W_0 \theta_0$.[1] For both cases the key to understanding how the restriction can reduce the variance is in observing that the restriction has the effect of allowing a wider choice of linear unbiased estimators for the *subset* of parameters remaining to be estimated. Finally, we notice RLS can offer a solution to the problem of perfect multicollinearity. If, for example, X_1 and X_2 were perfectly correlated ($X_{1t} = KX_{2t}$ all t) then there would exist no minimum for the OLS technique. However, applying a restriction such as (4.87) leads to an estimating equation (4.88) where there is a single explanatory variable which will now obey the second-order conditions unless for all t we had $X_{1t} = X_{2t}$ ($K = 1$). Thus the equation is estimable by RLS. This result is an extension of the variance-reducing properties of RLS (see problem 4.14).

---

[1] Clearly such a characterization is not useful in the homogeneous case where the restriction (θ_0) is zero.

Multiple Regression

> ### Result 4.2. Restricted least squares
>
> For a model $Y_t = \alpha + \beta_1 X_{1t} + \beta_2 X_{2t} + U_t$ where the error terms are well behaved and the true parameters are known to obey the linear restriction
>
> $$\theta_0 = \theta_1 \beta_1 + \theta_2 \beta_2.$$
>
> Substitute in the restriction, take all variables with known coefficients to the left-hand side and gather all variables with common coefficients on the right-hand side: i.e.
>
> $$Y_t - \left(\frac{\theta_0}{\theta_1}\right) X_{1t} = \alpha + \beta_2 \left\{ X_{2t} - \left(\frac{\theta_2}{\theta_1}\right) X_{1t} \right\} + U_t.$$
>
> Regress the new variables on each other. The variances of the remaining parameters, obtained from standard OLS formulae, will be no greater than those from OLS estimation of the unrestricted model.
>
> $$\hat{\hat{\beta}}_2 = \Sigma y_t^* x^* / \Sigma x^{*2}$$
> $$\hat{\hat{\alpha}}_2 = \bar{Y}^* - \hat{\hat{\beta}}_2 \bar{X}^*$$
> $$\text{Var } \hat{\hat{\beta}}_2 = \sigma^2 / \Sigma x^{*2}$$
> $$\text{Var } \hat{\hat{\alpha}}_2 = \sigma^2 \Sigma X^{*2} / \Sigma x^{*2}$$
>
> where
> $$Y_t^* = Y_t - \frac{\theta_0}{\theta_1} X_{1t}$$
> $$X_t^* = X_{2t} - \frac{\theta_2}{\theta_1} X_{1t}$$

Having established the theoretical benefits of the use of restricted least squares there are a number of practical issues which must be considered. We have already seen that the variance of the restricted estimator can itself be estimated by using the estimated variance of the residuals (4.76). There is then the question of how to estimate parameters which have been substituted out of the equation in its estimating form. For example in the model

with the restriction
$$Y_t = \alpha + \beta_1 X_{1t} + \beta_2 X_{2t} + U_t \tag{4.92}$$

$$\beta_1 + \beta_2 = r \tag{4.93}$$

we obtain by substituting out β_2 and applying OLS to the transformed equation

$$\hat{\hat{\beta}}_1 = \Sigma y^* x^* / \Sigma x^{*2} \tag{4.94}$$

where
$$Y^* = Y - r X_1, \quad X^* = X_2 - X_1 \tag{4.95}$$

and
$$\operatorname{Var} \hat{\hat{\beta}}_1 = \sigma^2/\Sigma x^{*2}. \tag{4.96}$$

However the estimated parameters also obey the restriction so that

$$\hat{\hat{\beta}}_2 = r - \hat{\hat{\beta}}_1. \tag{4.97}$$

(which can be seen to be linear and unbiased)

$$\operatorname{Var} \hat{\hat{\beta}}_2 = E[\{(r - \hat{\hat{\beta}}_1) - \beta_2\}^2] \tag{4.98}$$
$$= \operatorname{Var} \hat{\hat{\beta}}_1,$$

A similar treatment will allow us to derive the variance of any such indirectly estimated parameter.

Finally we consider the case of more than one restriction. We might have the restrictions:

$$\alpha = r_1 \tag{4.99}$$

$$\beta_1 + 2\beta_2 = r_2. \tag{4.100}$$

The restricted least squares equation would be:

$$Y_t - r_1 - r_2 X_1 = \beta_2(X_2 - 2X_1) + U.$$

This could be estimated by OLS between $Y^*(Y - r_1 - r_2 X_1)$ and $X^*(X_2 - 2X_1)$ in the form of a single-variable model without an intercept. As before $\hat{\hat{\beta}}_1$ would be recovered from (4.100). The error variance can again be estimated from the residuals of the restricted least squares equation. However the degrees of freedom will be greater than with OLS since fewer parameters are actually estimated. In the case above there are $(T-1)$ degrees of freedom for RLS and $(T-3)$ degrees of freedom for OLS.

All models with linear restrictions can be handled in the same way and providing the restrictions are correct (and independent) the resulting estimators will be unbiased and have a smaller variance than the (unrestricted) OLS estimators (see problems 4.10 and 4.15).

We see that providing the restrictions are correct the error term is not disturbed by the imposition of the restriction. This in turn implies that the expected value of the error variance obtained from OLS and that obtained from RLS would be equal. A large divergence between the two would suggest that perhaps the restriction imposed was not correct. This idea is utilized extensively in hypothesis testing, to which we turn in the next chapter.

Although the expected value of the standard errors of estimate (corrected by the appropriate degrees of freedom) will indeed be the same if the restrictions are correct it is obvious that for any given set of data the value of the RSSQ (i.e. not corrected for degrees of freedom) for the restricted equation can be no smaller (and is usually greater) than for that of OLS applied to the unrestricted equation. The RSSQ is the criterion minimized by OLS with a free choice of the value of every parameter—if we restrict some of this freedom of choice then we are considering only a subset of the numerical values over which OLS optimized. Hence RLS usually has a larger residual variance than OLS. The fact that the (true) variances of the parameters are smaller for RLS than for OLS is not paradoxical although it may be puzzling at first sight. These variances are averages over repeated samples and are measures of the variation of a parameter estimated by different methods. The RSSQ is of course a single-sample concept, is not corrected for degrees of freedom, and, even if it were, focuses on the parameter itself (σ^2) and not on the variation of the estimates of that parameter between samples. To summarize, we see that RLS will have a larger RSSQ than OLS but smaller parameter variances. If in addition we are imposing homogenous restrictions so that the dependent variable is unchanged,[1] then the TSSQ will be the same for both the OLS and RLS equations used for estimating; the correlation coefficient will then also be lower for RLS since the RSSQ is higher.

Example 4.2. Multiple regression subject to a linear restriction

We next estimate the consumption function subject to the restriction that the m.p.c. is unity. We assume also that this restriction is correct so that the errors have the properties necessary for RLS to be BLUE (and for the various formulae to be valid). We transform the equation by creating a new dependent variable—consumption less income—and we regress this on unemployment by OLS. The resulting equation is

$$C - Y = -8769 - 5.012\,U$$
$$(1354) \quad (1.205)$$
$$\text{SEE} = 3627,\ \text{RSSQ} = 26.3 \times 10^7.$$

Comparing this result with the unrestricted equation given in example 4.1 we see that the sign of the coefficient on unemployment has changed dramatically from positive to negative, while the intercept has also changed from positive to negative. The goodness of fit as measured by the standard error of estimate has worsened—the 'average' error in the restricted equation is twice as large as that for the unrestricted equation. Clearly there must be some doubt as to whether the restriction is valid since its imposition affects the other parameters so greatly. Moreover the estimated variance of the parameter on U is larger than that for OLS. This was of course obtained from the conventional formula ($\sigma^2/\Sigma x_{2t}^2$) with residuals calculated via fitted values. This formula is correct only if the restriction is correct so that the increase in calculated variance, rather than the expected decrease, further points to the incorrectness of the restriction.

Finally we can calculate a derived correlation between the fitted values of consumption (from RLS) and the actual data—this is 0.953 which is course lower than that of the unrestricted model at 0.992.

---

[1] If inhomogeneous restrictions are imposed we can still calculate an equivalent correlation coefficient by computing the fitted values of Y predicted by the restricted model and correlating these with the actual Y values.

4.6 Forecasting with multiple regressions

Just as in the case of the single-variable model of Chapters 2 and 3, an important use of an econometrically estimated equation is to construct a forecast of the dependent variable for some observation outside the estimation period. In order to do this we need the values of the exogenous variables for the forecast period. Consider the two-variable equation without an intercept:

$$Y_t = \beta_1 X_{1t} + \beta_2 X_{2t} + U_t \qquad (4.101)$$

where there is data on all variables for periods 1 to T and also data on the exogenous variables for periods $T+1$ to $T+F$. The errors over the whole period (1 to $T+F$) are 'well behaved'. As before the optimal predictor is based on the least squares parameters:

$$\hat{Y}_{T+F} = \hat{\beta}_1 X_{1,T+F} + \hat{\beta}_2 X_{2,T+F}. \qquad (4.102)$$

This can be shown to be unbiased linear in the Y (1 to T) and has a variance

$$\operatorname{Var}(\hat{Y}_{T+F}) = X_{1,T+F}^2 \operatorname{Var} \hat{\beta}_1 + X_{2,T+F}^2 \operatorname{Var} \hat{\beta}_2$$
$$+ 2 X_{1,T+F} X_{2,T+F} \operatorname{Cov}(\hat{\beta}_1, \hat{\beta}_2). \qquad (4.103)$$

This variance is estimated in the usual way by using the value of the estimated variances and covariance of the least squares estimators. As in the case of the variances of the estimators of the parameters the effect of stronger multicollinearity between the series will be to increase the variance of the forecast.

This observation leads us finally to consider the possibility of forecasting when there is a restriction on the parameters of the equation in the hope that restrictions can reduce variance. In exactly analogous fashion to our previous results the best linear unbiased predictor subject to the restriction will be based on the RLS estimators. Suppose that in the case of equation (4.101) we also have available the restriction:

$$\beta_1 + \beta_2 = r. \qquad (4.104)$$

Then the optimal predictor is

$$\hat{\hat{Y}}_{T+F} = \hat{\hat{\beta}}_1 X_{1,T+F} + \hat{\hat{\beta}}_2 X_{2,T+F} \qquad (4.105)$$

$$\operatorname{Var}(\hat{\hat{Y}}_{T+F}) = X_{1,T+F}^2 \operatorname{Var} \hat{\hat{\beta}}_1 + X_{2,T+F}^2 \operatorname{Var} \hat{\hat{\beta}}_2$$
$$+ 2 X_{1,T+F} \cdot X_{2,T+F} \operatorname{Cov}(\hat{\hat{\beta}}_1 \hat{\hat{\beta}}_2) \qquad (4.106)$$

But it can easily be shown, following (4.98), that in this case

$$\operatorname{Cov}(\hat{\hat{\beta}}_1 \hat{\hat{\beta}}_2) = -\operatorname{Var} \hat{\hat{\beta}}_1 = -\operatorname{Var} \hat{\hat{\beta}}_2 \qquad (4.107)$$

so that
$$\operatorname{Var} \hat{\hat{Y}}_{T+F} = (X_{1,T+F} - X_{2,T+F})^2 \operatorname{Var} \hat{\hat{\beta}}_1. \qquad (4.108)$$

Again the imposition of the restriction can be shown to reduce the variance of the predictor relative to that of the unconstrained case (problem 4.16).

Example 4.3. Forecasting from a multiple regression

We consider making a forecast from the unrestricted equation as given in example 4.1. The value of income is assumed to be £170,000 million in the forecast period and the value of unemployment 2,900 (thousands). The formula for an expected predictor of an equation of the form

$$Y_t = \beta_0 X_{0t} + \beta_1 X_{1t} + \beta_2 X_{2t}$$

where $X_{0t} = 1$ all t

is $\hat{Y}_F = \Sigma \hat{\beta}_i X_{iF}$

with variance around the expected outcome:

$$\operatorname{Var} \hat{Y}_F = \sum_i \sum_j X_{iF} X_{jF} \widehat{\operatorname{Cov}(\hat{\beta}_i, \hat{\beta}_j)}$$

where $\operatorname{Cov}(\hat{\beta}_i, \hat{\beta}_i) = \operatorname{Var} \hat{\beta}_i.$

Using the estimates and the estimated variances and covariances given in example 4.1 we have

$$\hat{Y}_F = £148{,}417 \text{ million}$$

$$(\operatorname{Var} \hat{Y}_F)^{1/2} = £1{,}228 \text{ million.}$$

If we include also the variance of the error term we obtain

$$(\operatorname{Var} \hat{Y}_F^*)^{1/2} = £1{,}963 \text{ million.}$$

These values can be compared to those in example 3.2. With the same value of income for the forecast value, but with the restriction that unemployment has zero influence, the 'point' forecasts are very similar (differing by less than 1%). However the variance of the multiple regression forecast is greater—the imposition of the (zero) restriction has reduced the variance but at the possible cost of bias.

4.7 The many-variable generalization

We have proved that for any model linear in the parameters with an intercept and two explanatory variables, under the set of conditions indicated, ordinary least squares is the best linear (in Y) unbiased estimator. The Gauss–Markov theorem is indeed generalizable to any case. The proof of this general result requires matrix algebra and so is beyond the scope of this book. There are many books which provide both the general OLS formulae and the proof that OLS is BLUE. The only condition that needs to be generalized is (v)—this requires that no set of the X variables (including the unit variable if there is an intercept) should be an exact linear multiple of any other of the X variables.

The technique of restricted least squares can also be generalized to the many-variable many-restriction case. The substitution of the restrictions into the basic equation and subsequent application of OLS to the resulting equation of fewer parameters results in a decrease in the variance of the estimators.

The use of OLS estimators of the parameters to construct a forecast can also be generalized in a straightforward fashion.

Problems 4

4.1. For the model $Y_t = \beta_1 X_{1t} + \beta_2 X_{2t} + U_t$, where

(i) $\qquad E(U_t) = 0$ all t

(ii) $\qquad E(U_t^2) = \sigma^2$ all t

(iii) $\qquad E(U_t U_s) = 0$ all $s, t, s \neq t$

(iv) $\qquad X_1$ and X_2 are fixed in repeated samples.

(v) $\qquad \Sigma X_{1t}^2 \Sigma X_{2t}^2 - (\Sigma X_{1t} X_{2t})^2 > 0$

(a) Derive the OLS estimators for β_1 and β_2.
(b) Prove that OLS is BLUE.
(c) Discuss the case where $X_{1t} = 1$ all t.

4.2. For the model $Y_t = \alpha + \beta_1 X_{1t} + \beta_2 X_{2t} + U_t$ where the assumptions of Chapter 4 hold, show that:

(a) $\qquad R^2 = (\hat{\beta}_1^2 \Sigma x_1^2 + \hat{\beta}_2^2 \Sigma x_2^2 + 2\hat{\beta}_1 \hat{\beta}_2 \Sigma x_1 x_2)/\Sigma y^2$

(b) $\qquad R^2 = (\hat{\beta}_1 \Sigma x_1 y + \hat{\beta}_2 \Sigma x_2 y)/\Sigma y^2$

(c) $\qquad R^2 = (r_{y1}^2 + r_{y2}^2 - 2r_{y1} r_{y2} r_{12})/(1 - r_{12}^2)$

(d) $\qquad r_{y1.2}^2 = (r_{y1} - r_{y2} r_{12})^2/(1 - r_{12}^2)(1 - r_{y2}^2)$

Hence in the two-variable model show how, for given correlations between the X variables and the Y variable, the coefficient of multiple correlation and the partial correlation coefficients will be affected as the intercorrelation between the variables changes.

4.3. For the model $Y_t = \alpha + \beta_1 X_{1t} + \beta_2 X_{2t} + U_t$, show that the formula for the partial correlation of Y on X_1 given X_2

$$r_{y1.2}^2 = (R^2 - R_1)/(1 - R_1^2)$$

(where R_1 is the correlation of Y on X_2 alone and R^2 is the multiple correlation with both X variables) is identical to a correlation obtained by the alternative procedure:

(i) regress Y on X_2 and obtain the residuals $\hat{u}_t$;
(ii) regress X_1 on X_2 and obtain the residuals $\hat{v}_t$;
(iii) correlate the residuals $\hat{v}_t$ and $\hat{u}_t$.

4.4. In the model $y_t = \beta_1 X_{1t} + \beta_2 X_{2t} + U_t$, the data is rescaled;
$$Y_t^* = \lambda Y_t$$
$$X_{1t}^* = \mu_1 X_{1t}$$
$$X_{2t}^* = \mu_2 X_{2t}.$$

The assumptions of question 4.1 hold.

(a) What are the relations between the OLS estimators on the unscaled data $(\hat{\beta}_1, \hat{\beta}_2)$ and those from the scaled data $(\hat{\hat{\beta}}_1, \hat{\hat{\beta}}_2)$?
(b) How is the goodness of fit affected by the rescaling?

4.5. In the model $Y_t = \alpha + \beta_1 X_{1t} + \beta_2 X_{2t} + U_t$, which is estimated by OLS, what is the relationship between the correlation between X_1 and X_2 and the correlation between the estimated regression coefficients $\hat{\beta}_1$ and $\hat{\beta}_2$?

4.6. Consider the equation
$$Y_t = \beta_1 X_{1t} + \beta_2 X_{2t} + U_t$$

where the assumptions of question 4.1 hold. Consider the following transformations and show how the OLS estimators of coefficients in regressions using transformed variables are related to those which would be obtained from applying OLS to the original equation:

(i) $(Y_t - X_{1t}) = \gamma_1 X_{1t} + \gamma_2 X_{2t} + U_t$
(ii) $Y_t = \delta_1 (X_{1t} - X_{2t}) + \delta_2 X_{2t} + U_t$

Would the goodness of fit be altered by these transformations?

4.7. The model $Y_t = \alpha + \beta_1 X_{1t} + \beta_2 X_{2t} + U_t$ is to be estimated subject to the restriction $\beta_1 + \beta_2 = r$. Show that the estimators obtained by substituting in for β_1 or for β_2 are identical.

4.8. The model $Y_t = \beta_1 X_{1t} + \beta_2 X_{2t} + U_t$ is subject to the linear restriction $\beta_1 + \beta_2 = r$, where r is a known constant. The assumptions necessary for OLS to be BLUE hold. Derive the RLS estimators:

(i) substituting the restriction into the structural equation;
(ii) minimizing subject to the restriction (Lagrange technique).

Prove for one approach or the other that RLS in BLUE in the class of estimators linear in Y_t and r and which obey the restriction.

4.9. The model $Y_t = \beta_1 X_{1t} + \beta_2 X_{2t} + U_t$. is to be estimated subject to the restriction $\beta_1 = r_1$ where r_1 is a known constant. Compare the OLS and RLS estimators and their variances when

(i) $\tilde{r}_{12}^2 \to 1$

(ii) $\tilde{r}_{12}^2 \to 0$

where $\tilde{r}_{12}^2 = (\Sigma X_1 X_2)^2 / (\Sigma X_1^2 \cdot \Sigma X_2^2)$.

Comment on these results.

4.10. Discuss how the following models can be estimated (both parameters and standard errors).

$$Y_t = \beta_1 X_{1t} + \beta_2 X_{2t} + \beta_3 X_{3t} + U_t$$

subject to

(i) $\beta_1 + \beta_2 + \beta_3 = 0$.

(ii) $\beta_1 - \beta_2 = \beta_2 - \beta_3$

(iii) $\beta_2 = 2\beta_1$

and $\beta_3 = 3\beta_1$.

4.11 The model $Y_t = \beta_1 X_{1t} + \beta_2 X_{2t} + U_t$ is estimated subject to the *incorrect* restriction $\beta_1 = r_1$, where r_1 is a known number, and the error term U_t obeys the conditions necessary for OLS to be BLUE. Show that the bias in the estimator of β_1 derived from regressing $(Y_t - r_1 X_{1t})$ on X_{2t} is $(\beta_1 - r_1)\Sigma X_1 X_2 / \Sigma X_2^2$. Show that

$$E\left(\frac{\hat{\hat{\sigma}}^2}{\Sigma X_{2t}^2}\right) \neq \operatorname{Var} \hat{\hat{\beta}}_2$$

(where $\hat{\hat{\sigma}}$ is obtained from the RLS residuals) and hence show that RLS can, if incorrectly applied, increase the variance of the estimator relative to that of OLS.

4.12. The model $Y_t = \beta_1 X_{1t} + \beta_2 X_{2t} + U_t$ is to be estimated subject to the restriction that $\beta_1 + \beta_2 = 0$ (the assumptions of section 4.1 hold). Derive the variance of RLS and show under what condition this is (*a*) strictly less than that of OLS and (*b*) equal to that of OLS.

4.13. The model $Y_t = \beta_1 X_{1t} + \beta_2 X_{2t} + U_t$ is to be estimated subject to the restriction $\beta_2 = r$ (*r* known) (the assumptions of section 4.1 hold). Show that the restricted least squares estimator $\hat{\hat{\beta}}_1$ of β_1 (derived from substituting in for β_2) can be written:

$$\hat{\hat{\beta}}_1 = \hat{\beta}_1 - (\Sigma X_1 X_2 / \Sigma X_1^2)(r - \hat{\beta}_2), \quad \hat{\beta}_i = \text{OLS}.$$

4.14. In the model $Y_t = \beta_1 X_{1t} + \beta_2 X_{2t} + U_t$ the assumptions of section 4.1 hold except that $X_{1t} = K X_{2t}$ all t. Show that a unique estimator of β_1 (and β_2) can generally be obtained despite this perfect multicollinearity when we have a parameter restriction of the form: $\beta_1 + \beta_2 = r$ (*r* known). Would the presence of a single linear restriction always allow us to obtain unique estimates of parameters whatever the form of the data?

4.15. Consider the model $Y_t = \beta_1 X_{1t} + \beta_2 X_{2t} + U_t$, where the conditions for OLS to be BLUE hold. The experimenter imposes the incorrect restriction $\beta_1 = C$ and estimates the equation by RLS. What is the ratio of the MSE of RLS to the MSE of OLS? Under what circumstances will that of RLS be smaller?

4.16. There is a model $Y_t = \beta_1 X_{1t} + \beta_2 X_{2t} + U_t$, where the U_t are well behaved. There is also a restriction on the parameters, $\beta_1 + \beta_2 = 0$.

(a) Consider the restricted least squares predictor (i.e. based on RLS estimators of β_1 and β_2). Show that the predictor and its variance are identical whether we work with the restricted equation (i.e. $Y = \beta_1(X_1 - X_2) + U$) or we substitute the estimated values back into the original equation.

(b) Show that the variance of the RLS predictor is less than that of the OLS predictor.

Answers 4

4.1. (a) The sum of squares function is $S(\hat{\beta}_1, \hat{\beta}_2) = \Sigma(Y - \hat{\beta}_1 X_1 - \hat{\beta}_2 X_2)^2$ (dropping t subscripts). The first-order conditions for a minimum are given by the 'normal' equations:

$$\frac{\partial S}{\partial \hat{\beta}_1} = -2\Sigma(Y - \hat{\beta}_1 X_1 - \hat{\beta}_2 X_2) X_1 = 0$$

$$\frac{\partial S}{\partial \hat{\beta}_2} = -2\Sigma(Y - \hat{\beta}_1 X_1 - \hat{\beta}_2 X_2) X_2 = 0.$$

Solving for the two unknowns we obtain (using v)

$$\hat{\beta}_1 = (\Sigma Y X_1 \Sigma X_2^2 - \Sigma Y X_2 \Sigma X_1 X_2) / \{\Sigma X_1^2 \Sigma X_2^2 - (\Sigma X_1 X_2)^2\}$$

($\hat{\beta}_2$ by symmetry—interchange X_1 and X_2).

(b) Now $\hat{\beta}_1 = \Sigma w_{1t} Y_t$

where $w_{1t} = (X_{1t} \Sigma X_{2t}^2 - X_{2t} \Sigma X_{1t} X_{2t})/D$

(where D is the denominator of the expression for $\hat{\beta}_1$. It can be seen that

$$E(\hat{\beta}_1) = E(\Sigma w_{1t} Y_t)$$
$$= \Sigma w_{1t}(\beta_1 X_{1t} + \beta_2 X_{2t})$$

(using assumptions (i) and (iv))

$$= \beta_1$$

(from formula for w_{1t}).

Also
$$\text{Var } \hat{\beta}_1 = E(\Sigma w_{1t} Y_t - E\Sigma w_{1t} Y_t)^2$$
$$= E(\Sigma w_{1t} u_t)^2$$
$$= \sigma^2 \Sigma w_{1t}^2$$

using (ii) and (iii)

$$= \sigma^2 \Sigma X_2^2 / D.$$

Consider any other linear unbiased estimator β^+. The set of all linear estimators is

$$\beta^+ = \Sigma k_t Y_t = \Sigma (w_t + c_t) Y_t.$$

For unbiasedness we restrict this

$$E(\beta^+) = E(\Sigma w_t Y_t) + E(\Sigma c_t Y_t) = \beta$$

and since the first term is β_1 the second must be zero. This in turn requires $\Sigma c_t(\beta_1 X_{1t} + \beta_2 X_{2t}) = 0$. For this to hold, whatever the model (i.e. true values of β_1 and β_2) and whatever the data, we consider only c_t such that

$$\Sigma c_t X_{1t} = \Sigma c_t X_{2t} = 0.$$

Now variance of this alternative estimator is

$$\text{Var } \beta^+ = E[\{\Sigma(w_t + c_t) Y_t - \beta_1\}^2]$$
$$= E[\{\Sigma(w_t + c_t) u_t\}^2]$$

(from substituting in Y_t and using the value of w_t and the restrictions on c_t). Hence:

$$\text{Var } \beta^+ = \sigma^2 \Sigma w_t^2 + \sigma^2 \Sigma c_t^2 + 2\sigma^2 \Sigma c_t w_t.$$

Since all cross-products in error terms are zero by (iii). However, from the definition of w_t and the restrictions on c_t it follows that $\Sigma c_t w_t$ is zero. Therefore the variance of any other LUE is strictly greater than OLS. The sum of the squared deviations (c_t) relative to the weights for OLS gives a measure of the loss in efficiency from the use of alternative estimators.

(c) Substituting $X_{1t} = K$ into the formula for $\hat{\beta}_1$ we have

$$\hat{\beta}_1 = (\bar{Y}\Sigma X^2 - \bar{X}\Sigma YX) \Big/ \left\{ \Sigma X^2 - \frac{1}{T}(\Sigma X)^2 \right\}$$

(where X_t represents X_{2t}). But

$$\Sigma x_t^2 = \Sigma X_t^2 - T\bar{X}^2$$
$$\Sigma xy = \Sigma XY - T\bar{X}\bar{Y}$$
$$\therefore \quad \hat{\beta}_1 = (\bar{Y}\Sigma x_t^2 - \bar{X}\Sigma xy)/\Sigma x_t^2$$
$$= \bar{Y} - \hat{\beta}\bar{X}$$

(where $\hat{\beta}$ is the formula for the regression of Y on X when there is an intercept)—so that we obtain the standard formula for the estimator of the intercept in the single-variable case. In general the second-order conditions for a minimum are satisfied (see problem 2.2):

$$\frac{\partial^2 S}{\partial \beta_1^2} = 2\Sigma X_1^2 > 0$$

$$\frac{\partial^2 S}{\partial \beta_2^2} = 2\Sigma X_2^2 > 0$$

$$\frac{\partial^2 S}{\partial \beta_1 \partial \beta_2} = 2\Sigma X_1 X_2$$

so that provided (v) is satisfied the OLS formulae do correspond to a minimum. Even if X_{1t} is constant (v) will be satisfied provided that X_{2t} is not also constant.

4.2. (a) Using the definition of R^2

$$R^2 = \frac{ESSQ}{TSSQ} = \Sigma \hat{y}_t / \Sigma y_t^2$$

and replacing $\hat{y}_t$ by $(\hat{\beta}_1 x_{1t} + \hat{\beta}_2 x_{2t})$.

(b) Since $\hat{y}_t = y_t - \hat{u}_t$; we also have

$$R^2 = \Sigma(y - \hat{u})(\hat{\beta}_1 x_1 + \hat{\beta}_2 x_2)/\Sigma y^2$$

But $\Sigma x_1 \hat{u} = \Sigma x_2 \hat{u} = 0$ (these are 'normal' equations (4.7), (4.8), and illustrate the property that the residuals are by construction uncorrelated with each of the independent variables and hence also with the fitted value $\hat{y}$). Hence:

$$R^2 = (\hat{\beta}_1 \Sigma x_1 y + \hat{\beta}_2 \Sigma x_2 y)/\Sigma y^2$$

(c) Take the first term in (b) and substitute in for $\hat{\beta}_1$:

$$(\Sigma x_1 y \Sigma y x_1 \Sigma x_2^2 - \Sigma y x_2 \Sigma x_1 x_2 \Sigma x_1 y)\{\Sigma x_2^2 \Sigma y^2 \Sigma x_1^2 - \Sigma y^2 (\Sigma x_1 x_2)^2\}$$

Divide numerator and denominator by $\Sigma y^2 \Sigma x_1^2 \Sigma x_2^2$ and for the second term in the resulting numerator split Σy^2 into $\sqrt{\Sigma y^2} \cdot \sqrt{\Sigma y^2}$, etc. to give for the first term: $(r_{1y}^2 - r_{1y} r_{2y} r_{12})/(1 - r_{12}^2)$ $(r_{1y} = \Sigma x_1 y/\sqrt{\Sigma y^2}\sqrt{\Sigma x_1^2}$, etc.). The second term in the original expression can be obtained by interchanging X_1 and X_2 so that we obtain the formula for (c).

(d) Since by definition:

$$r_{y1.2}^2 = (R^2 - r_{y2}^2)/(1 - r_{y2}^2)$$

we substitute in the expression for R^2 from (c) and immediately obtain (d). We can see from 2(c) that as r_{12} tends to unity the values of r_{y1} and r_{y2} must tend to each other (whatever value this is) and so the numerator and denominator both tend to zero. To evaluate the limit we use L'Hôpital's rule and differentiate numerator and denominator by r_{12}:

$$\frac{f'}{g'} = r_{y1} r_{y2}/r_{12}$$

Hence the limit as r_{12} tends to unity is r_{y1}^2 (since $r_{y1} \to r_{y2}$) as would be expected. As r_{12} tends to zero we see that R^2 tends to $r_{y1}^2 + r_{y2}^2$ (this sum cannot of course be greater than unity when the intercorrelation is zero because otherwise R^2 would be greater than unity, which can be shown to be false from its definition and the use of the Cauchy–Schwarz inequality). Applying similar arguments we see that as r_{12} tends to unity the partial correlations tend to zero (as expected), while as r_{12} tends to zero the partial correlation tends to $r_{y1}^2/(1 - r_{y2}^2)$.

4.3. The coefficient derived in the alternative way is

$$\frac{(\Sigma \hat{u}\hat{v})^2}{\Sigma \hat{u}^2 \Sigma \hat{v}^2} = \frac{\left\{\Sigma\left(Y - X_2 \frac{\Sigma Y X_2}{\Sigma X_2^2}\right)\left(X_1 - X_2 \frac{\Sigma X_1 X_2}{\Sigma X_2^2}\right)\right\}^2}{\Sigma\left(Y - X_2 \frac{\Sigma Y X_2}{\Sigma X_2^2}\right)^2 \Sigma\left(X_1 - X_2 \frac{\Sigma X_1 X_2}{\Sigma X_2^2}\right)^2}$$

Expanding and dividing by ΣY^2 and ΣX_1^2 we obtain

$$(r_{y1} - r_{y2} r_{12})^2 / (1 - r_{y2}^2)(1 - r_{12}^2).$$

This is the expression for the partial correlation derived in 4.2(c).

4.4. (a) The model in scaled form can be written

$$Y^* = \frac{\lambda \beta_1 X_1^*}{\mu_1} + \frac{\lambda \beta_2}{\mu_2} X_2^* + U.$$

Using OLS on the transformed data we have:

$$\hat{\hat{\beta}}_1 = \frac{\Sigma Y^* X_1 \Sigma X_2^{*2} - \Sigma Y^* X_2^* \Sigma X_1^* X_2^*}{\Sigma X_1^{*2} \Sigma X_2^{*2} - (\Sigma X_1^* X_2^*)^2}.$$

Rewriting Y^* as λY etc.

$$\hat{\hat{\beta}}_1 = \frac{\lambda}{\mu} \hat{\beta}_1$$

and so on. Hence the estimated coefficients of scaled and unscaled data are in the same relation as the true coefficients.

(b) $$R^{*2} = (\Sigma Y^* \hat{Y}^*)^2 / (\Sigma Y^{*2} \Sigma \hat{Y}^{*2})$$

But $$\hat{Y}_t^* = \lambda \hat{Y}_t$$

using (a)

$$Y_t = \lambda Y_t$$

∴ $$R^{*2} = R^2$$

and $$\text{SEE}^* = \sqrt{\frac{1}{T-2} \Sigma U_t^{*2}}$$

$$= \lambda \, \text{SEE}$$

(using $\hat{U}_t^* = Y_t^* - \hat{Y}_t^*$ etc.)

4.5. Define

$$\text{Corr}^2(\hat{\beta}_1, \hat{\beta}_2) = \frac{[E\{(\hat{\beta}_1 - \beta_1)(\hat{\beta}_2 - \beta_2)\}]^2}{E\{(\hat{\beta}_1 - \beta_1)^2\}E\{(\hat{\beta}_2 - \beta_2)^2\}}$$

using standard formulae:

$$\text{Var } \hat{\beta}_1 = \sigma^2 \Sigma x_2^2 / D$$

$$\text{Var } \hat{\beta}_2 = \sigma^2 \Sigma x_1^2 / D$$

$$\text{Cov}(\hat{\beta}_1, \hat{\beta}_2) = -\sigma^2 \Sigma x_1 x_2 / D$$

$$\therefore \quad \text{Corr}^2(\hat{\beta}_1, \hat{\beta}_2) = \frac{(\Sigma x_1 x_2)^2}{\Sigma x_1^2 \Sigma x_2^2} = r_{12}^2$$

(where r_{12} is the correlation between X_1 and X_2). Note that the sign of the (unsquared) correlation between coefficients is opposite to that between the variables.

4.6(i) Using OLS on the transformed data,

$$\hat{\gamma}_1 = \frac{\Sigma(Y - X_1)X_1 \Sigma X_2^2 - \Sigma(Y - X_1)X_2 \Sigma X_1 X_2}{\Sigma X_1^2 \Sigma X_2^2 - (\Sigma X_1 X_2)^2}$$

$$= \hat{\beta}_1 - 1$$

(where $\hat{\beta}_1$ is the OLS estimator for the basic equation). Similarly $\hat{\gamma}_2 = \hat{\beta}_2$ (the estimated parameters from this linear transformation bear the same relation to those from the untransformed equation as do the true parameters: i.e.

$$\gamma_1 = \beta_1 - 1, \quad \gamma_2 = \beta_2.$$

$$R^{*2} = 1 - \frac{\text{RSSQ}^*}{\widetilde{\text{TSSQ}^*}}$$

(remembering that there is no intercept in the model) so that deviations from zero are taken.) The value of RSSQ* is the same as for the untransformed equation (RSSQ) but $\widetilde{\text{TSSQ}^*}$ is now different.

$$\widetilde{\text{TSSQ}^*} = \Sigma(Y - X_1)^2 = \Sigma Y^2 + \Sigma X_1^2 - 2\Sigma X_1 Y$$

$$\therefore \quad \widetilde{\text{TSSQ}^*} > \widetilde{\text{TSSQ}}$$

if

$$\Sigma X_1^2 > 2\Sigma X_1 Y$$

or

$$\tfrac{1}{2} > \Sigma X_1 Y / \Sigma X_1^2$$

so that if the coefficient of a regression of Y on the single variable X_1 is less than one half, then $\widetilde{\text{TSSQ}^*} > \widetilde{\text{TSSQ}}$ and $R^{*2} > R^2$.

(ii) $$\hat{\delta}_1 = \frac{\Sigma Y(X_1-X_2)\Sigma X_2^2 - \Sigma YX_2 \Sigma(X_1-X_2)X_2}{\Sigma(X_1-X_2)^2 \Sigma X_2^2 - \{\Sigma(X_1-X_2)X_2\}^2}$$

Similarly $\hat{\delta}_2 = \hat{\beta}_1 + \hat{\beta}_2$ (but $\delta_1 = \beta_1$ and $\delta_2 = \beta_1 + \beta_2$). $\tilde{R}^{*2} = \tilde{R}^2$ and SEE* = SEE (Since both the total sum of squares and the residual sum of squares are unaffected by the transformation).

4.7 Substituting in for β_2, as in the answer to 4.6, we have the regression of $(Y - rX_2)$ on an intercept and $(X_1 - X_2)$.

$$\hat{\hat{\beta}}_1 = \Sigma(y - rx_2)(x_1 - x_2)/\Sigma(x_1 - x_2)^2$$

and hence
$$\hat{\hat{\beta}}_2 = r - \hat{\hat{\beta}}_1 = r - \frac{\Sigma(y - rx_2)(x_1 - x_2)}{\Sigma(x_1 - x_2)^2}. \quad (a)$$

Substituting in for β_1 and regressing $(Y - rX_1)$ on an intercept and $(X_2 - X_1)$ we obtain

$$\hat{\hat{\beta}}_2 = \Sigma(y - rx_1)(x_2 - x_1)/\Sigma(x_2 - x_1)^2. \quad (b)$$

By putting all of (a) over the denominator-term and expanding (noting that the denominator is the same as that of (b)) it easily follows that the two expressions are identical.

4.8. Note that the restriction imposed is on the *estimated* values (not their expectations):

$$\hat{\hat{\beta}}_1 + \hat{\hat{\beta}}_2 = r$$

(all unbiased estimators will obey the condition

$$E[\beta_1^* + \beta_2^*] = r).$$

(i) $$Y_t^* = Y - rX_2 = \beta_1(X_1 - X_2) + U = \beta_1 X^* + U$$

∴ the estimator $$\hat{\hat{\beta}}_1 = \frac{\Sigma Y^* X^*}{\Sigma X^{*2}} = \frac{\Sigma(Y - rX_2)(X_1 - X_2)}{\Sigma(X_1 - X_2)^2}$$

is best linear in (Y^*) unbiased by the Gauss–Markov theorem since U is 'well behaved'

$$(\check{\beta}_2 = r - \check{\beta}_1).$$

This is a member of the class of linear estimators

$$\beta_1^* = \Sigma w_t Y_t^* = \Sigma w_t Y_t - r\Sigma X_{2t} w_t$$

where the condition for unbiasedness can be shown to be:

$$\Sigma w_t(X_{1t} - X_{2t}) = 1.$$

(Note that for OLS on the unrestricted form we consider the class of estimators $\beta_1^o = \Sigma v_t Y_t$ where the conditions for unbiasedness are $\Sigma v_t X_{1t} = 1$ and $\Sigma v_t X_{2t} = 0$. These two conditions on the set of all v_t are more restrictive

108 Multiple Regression

than the single condition on w_t (all acceptable sets of v_t satisfying the condition for the w_t but not the reverse) so that the set of $L(Y)$ unbiased estimators is a subset of the $L(Y^*)$ estimators. This wider class of estimators can then have a smaller minimum variance in terms of an estimator which is in $L(Y^*)$ but not in $L(Y)$.)

(ii) Define the function

$$\phi = \Sigma(Y - \hat{\beta}_1 X_1 - \hat{\beta}_2 X_2)^2 + \lambda(\hat{\beta}_1 + \hat{\beta}_2 - r)$$

For a minimum we have the first-order conditions:

$$\frac{\partial \phi}{\partial \hat{\beta}_1} = -2\Sigma(Y - \hat{\beta}_1 X_1 - \hat{\beta}_2 X_2)X_1 + \lambda = 0$$

$$\frac{\partial \phi}{\partial \hat{\beta}_2} = -2\Sigma(Y - \hat{\beta}_1 X - \hat{\beta}_2 X_2)X_2 + \lambda = 0$$

$$\frac{\partial \phi}{\partial \lambda} = \hat{\beta}_1 + \hat{\beta}_2 - r \qquad\qquad = 0$$

Subtracting the second condition from the first and substituting in the third we obtain:

$$\hat{\beta}_1 = \frac{\Sigma(Y - rX_2)(X_1 - X_2)}{\Sigma(X_1 - X_2)^2}, \quad \hat{\beta}_2 = r - \hat{\beta}_1$$

which are the same as that obtained from the regression using transformed variables. It can then be shown to be unbiased and minimum variance in the same way.

4.9. The model subject to the restriction is written:

$$(Y - rX_1) = \beta_2 X_2 + U$$

so that the RLS estimator of β_2 is:

$$_R\hat{\beta}_2 = \frac{\Sigma(Y - rX_1)X_2}{\Sigma X_2^2}$$

while OLS gives

$$\hat{\beta}_2 = \frac{\Sigma YX_2 \Sigma X_1^2 - \Sigma YX_1 \Sigma X_1 X_2}{\Sigma X_1^2 \Sigma X_2^2 - (\Sigma X_1 X_2)^2}$$

From these it follows that (substituting in for Y using the true restriction)

$$\text{Var } _R\hat{\beta}_2 = \sigma^2 / \Sigma X_2^2$$

$$\text{Var } \hat{\beta}_2 = \sigma^2 / \Sigma X_2^2 (1 - \tilde{r}_{12}^2)$$

(i) As $\tilde{r}_{12} \to 1$ we see that Var $\hat{\beta}_2$ tends to infinity. The second-order condition for the existence of a minimum fails, so that there is no unique solution ($\hat{\beta}_2$

would be undefined). However, Var $_R\hat{\beta}_2$ is unaffected by the intercorrelation between the variables. On average the value of $_R\hat{\beta}_2$ is unaffected since it exists and is an unbiased estimator even when the correlation is perfect.

(ii) As $\tilde{r}_{12} \to 0$ we see that Var $\hat{\beta}_2 \to$ Var $_R\hat{\beta}_2$ and $_R\hat{\beta}_2 \to \hat{\beta}_2$ (in each sample). The restriction eventually affects neither the value of the estimator nor its variance.

This illustrates the point that although RLS has a strictly smaller variance than OLS if we are comparing all possible data sets, there can be particular sets for which the two estimators coincide. It also shows that an equation, which at first sight is 'underidentified' because of the perfect correlation between regressors, can be identified (estimatable) if a restriction on the parameters is available. This idea is particularly important when we come to study systems of equations. The 'value' of the restriction (i.e. the reduction in variance it can achieve) is proportional to the correlation between the independent variables.

4.10. Substituting in the restriction and carrying out regressions with transformed variables (denoted by*) we have:

(i)
$$Y = \beta_1(X_1 - X_3) + \beta_2(X_2 - X_3) + U$$

$$\therefore \quad \hat{\beta}_1 = \frac{\Sigma YX_1^* \Sigma X_2^{*2} - \Sigma YX_2^* \Sigma X_1^* X_2^*}{\Sigma X_1^{*2} \Sigma X_2^{*2} - (\Sigma X_1^* X_2^*)^2}$$

$$\text{Var } \hat{\beta}_1 = \sigma^2 / \Sigma X_1^{*2}(1 - \tilde{r}^2)$$

where $\tilde{r}$ is the correlation between $(X_1 - X_3)$ and $(X_2 - X_3)$.

(ii)
$$Y = \beta_1(X_1 - X_3) + \beta_2(X_2 + 2X_3) + U$$

The same formula as for (i) except that:

$$X_1^* = X_1 - X_3;$$
$$X_2^* = X_2 + 2X_3;$$

(iii)
$$Y = \beta_1(X_1 + 2X_2 + 3X_3) + U$$
$$\hat{\beta}_1 = \Sigma YX_1^* / \Sigma X_1^{*2}$$
$$\text{Var } \hat{\beta}_1 = \sigma^2 / \Sigma X_1^{*2}$$

where
$$X_1^* = X_1 + 2X_2 + 3X_3$$

4.11 Proceeding as if the restriction were correct we have: (replacing β_1):

$$_r\hat{\beta}_2 = \frac{\Sigma(Y - rX_1)X_2}{\Sigma X_2^2}$$

$$E(_r\hat{\beta}_2) = \beta_2 + (\beta_1 - r)\frac{\Sigma X_1 X_2}{\Sigma X_2^2}$$

110 Multiple Regression

which indicates bias unless the X variables are uncorrelated.

$$\text{Var}_r\hat{\beta}_2 = \frac{\sigma^2}{\Sigma X_2^2}$$

(which is smaller than that from OLS). Rewrite the equation as

$$(Y - rX_1) = \beta_2 X_2 + V$$

where
$$V = (\beta - r)X_1 + U$$

$$\therefore \quad \hat{\sigma}^2 = \frac{\Sigma \hat{U}_t^2}{T-1} = \frac{\Sigma\left(V - X_2 \frac{\Sigma V X_2}{\Sigma X_2^2}\right)^2}{T-1}$$

(see answer 3.7). Substituting in for V we have

$$E(\hat{\sigma}^2) = \frac{1}{T-1} E\left(\Sigma(U + \{\beta_1 - r\}X_1)^2 - \frac{[\Sigma(U + \{\beta_1 - r\}X_1)X_2]^2}{\Sigma X_2^2}\right)$$

$$= \frac{1}{T-1}\left\{T\sigma^2 + (\beta_1 - r)^2 \Sigma X_1^2 - \sigma^2 - \frac{(\beta_1 - r)^2 (\Sigma X_1 X_2)^2}{\Sigma X_2^2}\right\}$$

$$\therefore \quad E(\hat{\sigma}^2) = \sigma^2 + \frac{(\beta_1 - r)^2}{T-1}\left\{\frac{\Sigma X_1^2 \Sigma X_2^2 - (\Sigma X_1 X_2)^2}{\Sigma X_2^2}\right\}$$

The second term is strictly positive by the Cauchy–Schwarz inequality (unless the independent variables are perfectly collinear) so that the true variance of the estimator will be overestimated if we base it on the estimated residuals. Clearly the more incorrect the restriction the greater this effect so that the expected value of the estimated variance of the (incorrect) restricted least squares estimator can be greater than the expected value of the variance of the OLS estimator.

These results have some importance because they provide reasons to be cautious in the use of restricted least squares. We can see that although the restriction produces an estimator whose *true* variance is less than that of OLS so that the bias incurred might be acceptable on any finite trade-off of bias against variance, the actual bias is unknown (since it involves the true parameter) so that the idea is non-operational. Further we cannot know the true variance of the restricted estimator and our attempt to estimate it would be systematically too large.

4.12 Substituting in the restriction and regressing with the transformed variables:

$$Y = \beta_1(X_1 - X_2) + U$$

$$\therefore \quad _R\hat{\beta}_1 = \frac{\Sigma Y(X_1 - X_2)}{\Sigma(X_1 - X_2)^2}$$

$$\text{Var}\,_R\hat{\beta}_1 = \frac{\sigma^2}{\Sigma(X_1 - X_2)^2}$$

while for OLS

$$\operatorname{Var} \hat{\beta}_1 = \frac{\sigma^2}{\Sigma X_1^2 (1 - \tilde{r}_{12}^2)}.$$

The difference between denominators is

$$\Sigma(X_1 - X_2)^2 - \Sigma X_1^2 + \frac{(\Sigma X_1 X_2)^2}{\Sigma X_2^2}$$

$$= \frac{\Sigma X_2^2 (\Sigma X_1^2 + \Sigma X_2^2 - 2\Sigma X_1 X_2) - \Sigma X_1^2 \Sigma X_2^2 + (\Sigma X_1 X_2)^2}{\Sigma X_2^2}$$

$$= \frac{(\Sigma X_1 X_2 - \Sigma X_2^2)^2}{\Sigma X_2^2}.$$

This is strictly positive unless $X_{1t} = X_{2t}$ for all t. Hence the restriction reduces the variance of the estimator relative to OLS not only when the data are not perfectly collinear, but even when there is perfect collinearity with one series a multiple (not equal to unity) of the other. If the two series are identically equal then even in the restricted model the second-order condition for the existence of a minimum fails (see Chapter 2 where, in the single-variable case, ΣX_t^2 must be strictly positive for a minimum to exist).

4.13 Substituting in the restriction and regressing using the transformed variables we obtain:

$$Y - rX_2 = \beta_1 X_1 + U$$

$$_R\hat{\beta}_1 = \frac{\Sigma(Y - rX_2)X_1}{\Sigma X_1^2} = \frac{\Sigma YX_1}{\Sigma X_1^2} - \frac{r\Sigma X_2 X_1}{\Sigma X_1^2}.$$

Now consider the expression

$$\hat{\beta}_1 - \frac{\Sigma X_1 X_2}{\Sigma X_1^2}(r - \hat{\beta}_2).$$

To show equality we must prove:

$$\frac{\Sigma YX_1}{\Sigma X_1^2} = \hat{\beta}_1 + \hat{\beta}_2 \frac{\Sigma X_1 X_2}{\Sigma X_1^2}$$

i.e.
$$\Sigma YX_1 = \frac{\Sigma X_1^2}{D}(\Sigma YX_1 \Sigma X_2^2 - \Sigma YX_2 \Sigma X_1 X_2)$$

$$+ \frac{(\Sigma YX_2 \Sigma X_1^2 - \Sigma YX_1 \Sigma X_1 X_2)\Sigma X_1 X_2}{D}$$

where $\quad D = \Sigma X_1^2 \Sigma X_2^2 - (\Sigma X_1 X_2)^2.$

This can be seen to be correct so that the expression is true. We see that the further the restricted value is from the OLS estimator, the greater is the effect on the restricted estimator of the other parameter (unless the data are uncorrelated).

4.14 We see immediately that *whatever* the value of K there is no unique minimum for the unrestricted problem because the second-order conditions are not satisfied (see problems 4.1 and 2.2). Now substituting in the restriction and carrying out the regression using the transformed variables we have:

$$Y - rX_2 = \beta_1(X_1 - X_2) + U$$

$$_R\hat{\beta}_1 = \frac{\Sigma(Y - rX_2)(X_1 - X_2)}{\Sigma(X_1 - X_2)^2}$$

$$\text{Var } _R\hat{\beta}_1 = \frac{\sigma^2}{\Sigma(X_1 - X_2)^2}$$

In the case of a single variable without an intercept model a unique minimum exists provided that $\Sigma X^2 > 0$ (*). Hence only if $X_1 = X_2$ for all t will this fail and no estimator exist. In general we can consider any linear restriction; $\beta_1 + M\beta_2 = r$, where r can be zero. Restricted least squares will exist (and have a smaller variance from OLS) unless the conditions (*) fails for the transformed variables. The general transformed equation is

$$(Y - rX_1) = \beta_2(X_2 - MX_1) + U$$

so that only if $X_{2t} = MX_{1t}$ for all t will no estimator exist. Hence, provided that the series do not bear the same (exact) relationship to each other as in the restriction, the technique of restricted least squares can be used even when there is perfect multicollinearity.

4.15 From the standard property that

$$\text{MSE} = \text{Bias}^2 + \text{Variance}$$

we have for OLS

$$\text{MSE}(\hat{\beta}_2) = \frac{\sigma^2}{\Sigma X_2^2(1 - \tilde{r}^2)}.$$

For RLS (see question 4.11)

$$\text{Bias }(_R\hat{\beta}_2) = \frac{(\beta_1 - c)\Sigma X_1 X_2}{\Sigma X_2^2}$$

and

$$\text{Variance }(_R\hat{\beta}_2) = \frac{\sigma^2}{\Sigma X_2^2}$$

$$\therefore \quad \frac{\text{MSE}(_R\hat{\beta}_2)}{\text{MSE}(\hat{\beta}_2)} = \frac{\frac{\sigma^2}{\Sigma X_2^2} + \frac{(\beta_1 - c)^2(\Sigma X_1 X_2)^2}{(\Sigma X_2^2)^2}}{\frac{\sigma^2}{\Sigma X_2^2(1 - r^2)}}$$

$$= (1 - \tilde{r}^2)\left\{1 + \frac{(\beta_1 - c)^2 \Sigma X_1^2 \tilde{r}^2}{\sigma^2}\right\}.$$

But
$$\frac{(1-\tilde{r}^2)\Sigma X_1^2}{\sigma^2} = (\text{Var } \hat{\beta}_1)^{-1}$$

∴
$$\frac{\text{MSE}(_R\hat{\beta}_2)}{\text{MSE}(\hat{\beta}_2)} = (1-\tilde{r}^2) + \frac{(\beta_1-c)^2 \tilde{r}^2}{\text{Var } \hat{\beta}_1}$$

$$= 1 + \tilde{r}^2 \left[\frac{(\beta_1-c)^2}{\text{Var } \hat{\beta}_1} - 1 \right]$$

Hence MSE $(_R\hat{\beta}_2)$ is the smaller if and only if

$$(\beta_1-c)^2 < \text{Var } \hat{\beta}_1$$

or

$$(\beta_1-c)^2 < \frac{\sigma^2}{\Sigma X_1^2(1-\tilde{r}^2)}.$$

This is more likely to be true
  (i) the more nearly correct is the restriction,
  (ii) the larger the residual variance,
  (iii) the larger the correlation between X_1 and X_2,
  (iv) the smaller are the X_1 values (rescaling would not affect this condition since β_1 and c would be inversely affected).

The ratio of MSE can be written as

$$1 + \tilde{r}^2(t^2 - 1)$$

where
$$t^2 = \frac{(\beta_1-c)^2}{\text{Var } \hat{\beta}_1}$$

and this is the *analogue* for a 't' test on whether $\beta_1 = c$ (see Chapter 5). However since t depends on true values and not their estimated values it cannot be used operationally to decide which estimator to use.

4.16 (a) The predictor based on the restricted equation is obtained using standard one-variable formulae;

$$_1\hat{\hat{Y}}_F = \hat{\hat{\beta}}_1(X_{1F} - X_{2F})$$

with variance $\text{Var }_1\hat{\hat{Y}}_F = (X_{1F} - X_{2F})^2 \text{ Var } \hat{\hat{\beta}}_1$; while that based on substituting estimated values back into the original equation is $_2\hat{\hat{Y}}_F = \hat{\hat{\beta}}_1 X_{1F} + \hat{\hat{\beta}}_2 X_{2F} = {_1}\hat{\hat{Y}}_F$ (because $\hat{\hat{\beta}}_2 = -\hat{\hat{\beta}}_1$ by construction).

$$\text{Var }_2\hat{\hat{Y}}_F = X_{1F}^2 \text{ Var } \hat{\hat{\beta}}_1 + X_{2F}^2 \text{ Var } \hat{\hat{\beta}}_1 + 2X_{1F}X_{2F} \text{ Cov } \hat{\hat{\beta}}_1\hat{\hat{\beta}}_2$$

But from (4.98) and (4.106) we have

$$\text{Var } \hat{\hat{\beta}}_2 = -\text{Cov } \hat{\hat{\beta}}_1\hat{\hat{\beta}}_2 = \text{Var } \hat{\hat{\beta}}_1$$

so
$$\text{Var }_2\hat{\hat{Y}}_F = \text{Var }_1\hat{\hat{Y}}_F.$$

(b)
$$\text{Var } \hat{\hat{Y}}_F = \frac{\sigma^2}{\Sigma(X_1 - X_2)^2} \cdot (X_{1F} - X_{2F})^2$$

while for OLS

$$\text{Var } \hat{Y}_F = X_{1F}^2 \text{ Var } \hat{\beta}_1 + X_{2F}^2 \text{ Var } \hat{\beta}_2 + 2 X_{1F} X_{2F} \text{ Cov } \hat{\beta}_1 \hat{\beta}_2.$$

Using the standard values for the variances and covariances of estimators we obtain

$$\text{Var } \hat{Y}_F = \frac{\sigma^2(X_{1F}^2 \Sigma X_2^2 + X_{2F}^2 \Sigma X_1^2 - 2 X_{1F} X_{2F} \Sigma X_1 X_2)}{D}$$

where
$$D = \Sigma X_1^2 \Sigma X_2^2 - (\Sigma X_1 X_2)^2.$$

Now
$$\text{Var } \hat{Y}_F - \text{Var } \hat{\hat{Y}}_F = \frac{\sigma^2 M}{D \Sigma (X_1 - X_2)^2}$$

where
$$M = \Sigma(X_1 - X_2)^2 (X_{1F}^2 \Sigma X_2^2 + X_{2F}^2 \Sigma X_1^2 - 2 X_{1F} X_{2F} \Sigma X_2 X_1\}$$
$$- (X_{1F} - X_{2F})^2 \{\Sigma X_1^2 \Sigma X_2^2 - (\Sigma X_1 X_2)^2\}].$$

Cancelling and collecting terms it can be shown that

$$M = \{X_{1F}(\Sigma X_2^2 - \Sigma X_1 X_2) + X_{2F}(\Sigma X_1^2 - \Sigma X_1 X_2)\}^2.$$

Hence the variance of OLS is strictly greater than that of RLS unless $\Sigma X_2^2 = \Sigma X_1 X_2$ and $\Sigma X_1^2 = \Sigma X_1 X_2$, which is attained only if $X_{1t} = X_{2t}$ all t. Hence the restriction reduces variance and the effect is clearly stronger the weaker the correlation in the basic data.

5
Hypothesis Testing

5.1 Introduction

In the preceding chapters we have been answering the question of how to estimate the parameters of an equation. The implication of this approach is that we accept whatever the data generates and therefore we have no prior conception of the value of the parameters. In many cases we may have some information about the parameters which we believe to be true and our main interest is to see whether the data support our hypothesis. For example, we might believe that the marginal propensity to consume is greater than zero but less than unity, or that the passing on of retail price changes into wage changes is 100 per cent. These examples, and others that occur repeatedly in economics, serve to illustrate the problem—which is essentially whether an estimated parameter value is consistent with a given hypothetical value or whether the two are so different that doubt is cast on our basic hypothesis

In order to assess the compatibility of our two values (the estimate and the hypothetical) we must concern ourselves again with the process by which the estimate is generated. Our description of the process, given in Chapters two and three, is that the dependent variable is generated by a (fixed) independent variable and by a random term. The consequence of this was that the dependent variable varied from sample to sample with the error term. This in turn meant that the value of the least squares estimator (or of any other reasonable estimator) varied also from sample to sample. Under certain assumptions about the error term the average value of the estimator in repeated samples was the true value. In a practical situation however, we have only the one sample and hence we must recognize that, although the method of generating the estimator may on average yield the true value, the sample value will not necessarily equal the true value. Hence even if our hypothetical value is indeed correct, the estimated value of the parameter will not necessarily equal the hypothetical value. There is therefore no possibility of a simple 'test' of the hypothesis according to whether or not the two values are equal—they can be unequal because the hypothesis is incorrect, and they can be unequal despite the correctness of the hypothesis because of the sampling variation.[1] Clearly a less decisive procedure of testing for compatibility will be required.

---

[1] Indeed they could be equal when the hypothesis was incorrect if the sampling variation happened to pull the estimated value to the value of the incorrect maintained hypothesis.

The key to designing a testing procedure lies in being able to analyse the *potential* variability of the estimated value—to be able to say whether a large divergence between it and the hypothetical value is better ascribed to sampling variability alone or whether it is better ascribed to the hypothetical value being incorrect. In order to analyse the variability of the estimate we can recall that for the basic one-variable model with an intercept, by (3.24):

$$\hat{\beta} = \beta + \Sigma x_t u_t / \Sigma x_t^2. \tag{5.1}$$

If the X values are fixed then the variability of the estimate $\hat{\beta}$ depends on the variability of the error term U (although this variability is weighted by the X values as the standard variance formula shows). We need then more information on the potential variability of the errors in order to assess the variability of the estimates. It is not enough merely to denote the possible range of variation of the errors—this would delimit the range of the $\hat{\beta}$ but tell us no more. In essence we need to know the distribution of $\hat{\beta}$ in repeated samples— what percentage will fall between any given values—so that we can assess what percentage of repeated samples could be expected to fall near to the hypothetical value. As is now apparent, in order to obtain the distribution of $\hat{\beta}$ we need to know the distribution of the error terms. We follow the usual definitions and take as a frequency function of a random variable (Z) the function ($f(Z)$) such that the *proportion* of values of Z obtained in repeated samples following between limits a and b is

$$\int_a^b f(Z) \, dZ \tag{5.2}$$

$$f(Z) \geq 0$$

Here we take as before the notion of a repeated sample to be the set of an indefinitely large number of trials.

There are infinitely many frequency functions and a large number have been analysed in detail. Nevertheless in regression type situations it is natural and convenient to concentrate on a principal case—the *normal* distribution. The main argument for concentrating on this particular distribution comes from our stylization of the error-generation process. When we discussed the nature of econometric modelling we argued that there was an important and measurable variable X or variables $(X_1, X_2 \ldots)$ and a series of unmeasurable influences. These latter factors could not be specified by the model builder but it would be recognized that many other, perhaps small, factors were involved in the determination of the dependent variable. If the influence of these factors were also additive and they were independent then the error term would in fact be the sum of a large number of independent unobserved variables. This formulation suggests that the distribution of the error term will be governed by a *central limit theorem*. There are several versions of this theorem but all would allow us to assume, under certain assumptions on the behaviour of the unobserved factors, that the error term will tend to follow a normal distribution irrespective of the distribution of the individual components making up the error term.

The normal distribution ranges from minus to plus infinity and can vary in specific shape according to two parameters—μ and σ. Its form is given by:

$$f(Z) = (2\pi\sigma)^{-1/2} \exp\left\{-\tfrac{1}{2}\left(\frac{Z-\mu}{\sigma}\right)^2\right\} \tag{5.3}$$

Once μ and σ have been specified then the whole distribution is known. It can be shown that the mean of the distribution is equal to μ and the variance to σ^2. The distribution has a symmetric bell shape around the mean value as in figure 5.1.

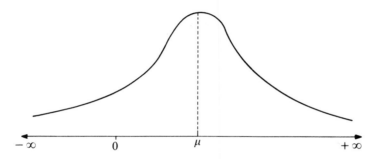

FIG. 5.1 The Normal Distribution

Formally we add to our list of assumptions:

Assumption (vi): the U_t are normally distributed for all t.

This can be combined with assumptions (i), (ii), and (iii) to yield the alternative:

Assumption (vii): the U_t are independently, identically normally distributed with zero mean, for all t. This is summarized as: the U_t are i.i.d. $N(0, \sigma^2)$ variables.

The mathematical properties of the normal distribution are such that the weighted sum of normal variables is also normal, as is a fixed multiple of a normal variate, i.e. if Z_1 is $N(\mu_1, \sigma_1)$ and Z_2 is $N(\mu_2, \sigma_2)$ and they are independent, then

(i) $\qquad c_1 Z_1$ is $N(c_1\mu_1, c_1^2\sigma_1^2)$ \hfill (5.4)

(ii) $\qquad Z_1 + Z_2$ is $N(\mu_1+\mu_2, \sigma_1^2+\sigma_2^2)$ \hfill (5.5)

(iii) $\qquad Z_1 - \mu_1$ is $N(0, \sigma_1^2)$ \hfill (5.6)

118 Hypothesis Testing

(iv) $$\frac{Z_1 - \mu_1}{\sigma_1} \text{ is } N(0,1) \qquad (5.7)$$

where c_1 and c_2 are constants. These results can be used to derive the results that the estimator $\hat{\beta}$ also follows a normal distribution

$$\hat{\beta} \sim N \frac{(\beta, \sigma^2)}{\Sigma x_t^2} \qquad (5.8)$$

or $$\hat{\beta} \sim N(\beta, \text{Var } \hat{\beta}). \qquad (5.9)$$

This derivation shows that if we are prepared to add the reasonable assumption of normality to our other assumptions we can derive the distribution of the estimate in repeated samples. This in turn would give us the opportunity to assess how typical a given estimated value might be. There is one obvious price that has been paid by enlisting the central limit theorem to obtain a general result—the possible range of the estimate is infinite so that all values are possible, even though large values may be rare. It is evidently not possible to use a given value of $\hat{\beta}$ to disprove the hypothesis that the true value is some other number—whatever the true value, the estimated value of $\hat{\beta}$ is logically consistent with it under the assumptions made.

This difficulty leads us to the ideas of statistical inference—instead of a logical incompatibility we can look for an implausibility.

5.2 Hypothesis testing

The strategy of the hypothesis test can be described by the following:

(i) formalize the model (e.g. $Y_t = \alpha + \beta X_t + U_t$);
(ii) specify the *null hypothesis* (H_0) on one (or more) of the parameters (e.g. $\beta = 1$);
(iii) estimate the model without regard to the hypothesis (in unrestricted form) to obtain $\hat{\alpha}$ and $\hat{\beta}$;
(iv) for hypothetical values of α and β, using the assumption of normally distributed errors, derive the sample distribution of the estimators;
(v) choose a decision rule to decide what value in (iv) would count as a 'very unlikely' value of the estimator given that the hypothetical value of the parameter were true;
(vi) check whether the actual estimate falls in this class of 'very unlikely' values; if it does not, then accept that the data is consistent with the null hypothesis (i.e. that the null hypothesis is true); if the estimate falls in the 'very unlikely' class then accept that the data and the null hypothesis are incompatible (reject the null hypothesis) and use the estimated parameter as being the best guide to the true value.

Hypothesis Testing 119

There are two practical difficulties with this scheme. The first and most important is that of formalizing the concept of a 'very unlikely' estimated value, while the second is that of deriving a usable formula for the distribution of estimates under the null hypothesis.

The choice of a 'very unlikely' or *significant* value of the estimate requires us to decide a *size* of the hypothesis test, which is usually denoted α (which corresponds to a level of significance of $1-\alpha$). For example, most econometric applications choose a 5 per cent test (a 95 per cent significance level). This means that an event is regarded as 'very unlikely' (significant) only if it falls in a set of values that would occur 5 per cent of the time in repeated sampling when the null is true. Figure 5.2 shows some such 5 per cent *critical regions* for a normal distribution with zero mean and unit variance. This diagram makes it clear that a normal (or any other) distribution will have infinitely many sub-regions which cover just 5 per cent of the total—so that the likelihood that a point (estimated value) chosen at random would fall in one of them is just 5 per cent. This ambiguity would remain whatever the size of the test chosen and so we need a second criterion to decide between these rival critical regions. Since our objective is clearly not just to support the null hypothesis when it is true (which we do by accepting 95 per cent of values that would be generated by it) but also to reject it if it is not true, we need to think about the distribution of estimated values that would occur if the null hypothesis were not true. Suppose that the true value of β is the value $\bar{\beta}$ whereas we hypothesize a value β^* (where $\bar{\beta} > \beta^*$), then figure 5.3 shows the hypothetical distribution and the actual distribution of estimates. The regions to the left of β_1 and the right of β_2 both contain 5 per cent of the hypothetical distributions (i.e. if H_0 is true a single estimated value would fall in these regions in 5 per cent of repeated trials). However, if H_0 is not true but H_1 is true ($\beta = \bar{\beta}$) then the estimated value $\hat{\beta}$ would fall in the region to the left of β_1 a very small fraction of the time but would fall in the area to the right of β_2 a very large fraction of the time. We can see that the *right-hand tail* of the null hypothesis gives the greatest chance to supporting the alternative hypothesis because the actual value will tend to fall there more often (given that H_1 is

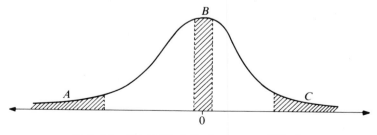

FIG. 5.2   5% Critical Regions on an $N(0, 1)$

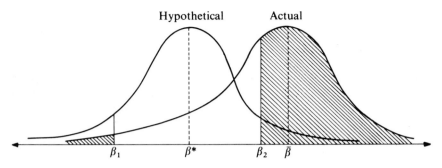

Fig. 5.3 Hypothetical and Actual Distributions for an Estimated Parameter

true). It can be seen that for any unimodal distribution the general result is true—when $\bar{\beta} > \beta^*$ the right-hand tail of the null hypothesis cuts off the maximum amount of the alternative hypothesis, all other 5 per cent regions for H_0 cut off smaller regions of H_1. Hence the right-hand tail would seem to be the best test (*best critical region*). However, although this is the case whenever $\bar{\beta} > \beta^*$ (however great or small the difference), this is not the case if $\bar{\beta} < \beta^*$. By analogous arguments the left-hand tail of the null hypothesis distribution would then be the best critical region. These arguments show that if, in the single parameter case, we can specify whether the actual value must be greater than or equal to the hypothetical value, or less than or equal to the hypothetical value, there is a best critical region for the test (of whatever size) in the distribution under the null hypothesis. For example we may take, in the consumption function model, the null hypothesis that the marginal propensity to consume is zero. As an alternative we are prepared to accept that it is positive if the evidence from the data is sufficiently convincing, but we would not be prepared to accept that the true value was negative, whatever emerged from a given set of data. A large negative value would be attributed to an extreme sampling deviation generated from the null hypothesis case.

Thus we can see that a test methodology can be derived in the *one-sided* case (where the alternative value is known to be on one side or the other of the hypothetical if the two are not equal). Such a test is said to be the *uniformly most powerful* in that it maximizes the power of the test (the probability of accepting the alternative hypothesis when it is in fact true) whatever the true value (given that it is on one side of the hypothetical value).

When we are prepared to accept an alternative estimated value either greater or lesser than the hypothetical value, if the evidence is sufficiently strong, then we require a *two-sided test* and matters become more difficult. There is clearly no single region which is most powerful whatever the true value happens to be (the switch from right- to left-hand tail as $\bar{\beta}$ shifts from greater than to less than β^* confirms this). In order to find a general principle a further restriction on the properties of the test procedure is usually imposed. One is the criterion of the *unbiasedness* of test. A hypothesis test is said to be unbiased if the probability of accepting the alternative hypothesis is always greater than or equal to the size of the test, whatever the true value of the parameter. If this condition does not hold then the test could have the property that we are more likely to accept the alternative hypothesis when it is false then when it is true. This is illustrated in figure 5.4, where a one-tail test is used for a two-sided hypothesis. The true value of the parameter is $\bar{\beta}$ while the hypothetical value is β^* ($\bar{\beta} < \beta^*$). The true and hypothetical sampling distributions for estimates of β are shown. If a right-tail test is used, such that the value β_1 cuts off 5 per cent of the hypothetical distribution to its right then we see that, if the hypothetical value β^* is true, the chance of rejecting it is 5 per cent, while if $\bar{\beta}$ is in fact true the chance of rejecting β^* is much less (the shaded area)—the power of the test is here less than the size of the test since we are less likely to reject the null hypothesis when false than when true (and more likely to reject the alternative hypothesis when true than when false). The test is said to be biased. If we restrict ourselves to unbiased tests then in the one-sided case the one-tailed test is unbiased and the test is uniformly most powerful unbiased (UMPU). In the two-sided case it can be shown that certain *two-tailed tests* are UMPU. Such a test places half of the critical region in each tail so that we have the situation shown in figure 5.5. The area to the left of β_1 contains 2.5 per cent of the total distribution and so does the area to the right of β_2. Hence if a sample value of $\hat{\beta}$ falls between β_1 and β_2 we conclude that the data supports the null hypothesis that β^* is true, otherwise we reject the hypothesis.

This discussion has established a methodology for testing and has described some principles upon which a good test can be designed.

Principles of hypothesis testing

The general principle that we have been following is that we wish our test procedure to maximize the power of the test for a given size, that is, for a

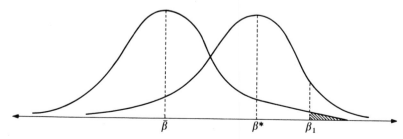

FIG. 5.4 A One-tailed Test for a Two-sided Hypothesis

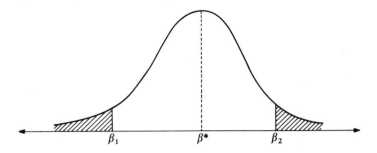

FIG. 5.5 A Two-tailed Test for a Two-sided Hypothesis

given chance of rejecting the null hypothesis when it is in fact true (*type I error*) we wish to have the maximum chance of accepting the alternative hypothesis when it is true (i.e. to maximize the power). This latter is of course equivalent to minimizing the chance of rejecting the alternative hypothesis when it is true (*type II error*). In the case of a one-sided hypothesis test we argued that a one-tailed test will achieve this (be uniformly most powerful), but in the case of a two-sided test we could not achieve this condition. However if we limit ourselves to tests that are unbiased—i.e. tests that are not more likely to accept the alternative hypothesis when false than when true—then a two-tailed test will often prove to be uniformly most powerful unbiased (i.e. have a smaller type II error and higher power at any value of the alternative hypothesis). The final problem that has to be solved is the distribution of the sample parameter and we turn to this in the next section.

5.3 Test statistics

We discussed in the beginning of this chapter the reason for assuming that the error terms would be normally distributed and showed that the parameter estimates would therefore also be normally distributed:

$$\hat{\beta} \sim N(\beta, \text{Var } \hat{\beta}) \qquad (5.9)$$

This result is not in itself usable to test hypotheses or the compatibility of an estimate $\hat{\beta}$ with an assumed true value, because the distribution depends on the variance of the OLS estimate, which in turn depends on the true error variance σ^2. Since this is unknown and has to be estimated the distribution (5.9) cannot be used as it stands. There is a second reason for wanting to move away from (5.9): although in a multiple regression each parameter estimate is normally distributed we may wish to consider hypotheses concerning several parameters, e.g. that all the β_i are zero (none of the X variables has a significant impact on the dependent variable). To do this we clearly need a statistic which simultaneously involves all the hypothetical values at the same time.

These two aspects can be handled by Fisher's F test. This test is related to the normal distribution by a series of manipulations. We state, without proof, the following classical results:

(i) if $\varepsilon_1, \varepsilon_2, \ldots \varepsilon_T$ are a series of independent identical normal variables $N(0, \sigma^2)$ then the distribution of $\Sigma \varepsilon_t^2/\sigma^2$ is a $\chi^2(T)$ variable where the mathematical form of the distribution is known and the sole parameter is T;

(ii) if Z_1 is distributed as $\chi^2(T_1)$ and Z_2 is distributed as $\chi^2(T_2)$ and Z_1 and Z_2 are independent then $(Z_1/T_1)/(Z_2/T_2)$ is distributed as $F(T_1, T_2)$; where the mathematical form of F is known and the sole parameters are T_1 and T_2.

The usefulness of the F distribution is that if the variables Z_1 and Z_2 were based on χ^2 variables obtained from normal variables *with the same variance*, then we would not need to know that variance in order to calculate the F statistic (since they cancel out). This suggests a way round the difficulty we observed with the distribution of $\hat{\beta}$. The actual mechanics of the F test, in linear regression and with linear hypotheses, are very straightforward. First we estimate the unrestricted model (i.e. the model in its most general form without trying to impose any hypothesis). From the least squares fit we can obtain a residual sum of squares

$$\sum_1^T \hat{U}_t^2$$

(the unrestricted residual sum of squares). Second we impose the linear restriction(s) on the model and re-estimate subject to these restrictions by ordinary least squares and obtain the restricted residuals and their sum of squares $\Sigma \hat{\hat{U}}_t^2$. From these form the statistic:

$$\frac{(\text{RSSQ}_R - \text{RSSQ}_U)/q}{\text{RSSQ}_U/T - K} \quad (5.10)$$

where RSSQ_R is the restricted sum of squares, RSSQ_U the unrestricted sum of square, q the number of restrictions imposed, T the total number of observations, K the number of parameters estimated in the unrestricted model. Equivalently we use

$$\frac{(\Sigma \hat{\hat{U}}_t^2 - \Sigma \hat{U}_t^2)/q}{\Sigma \hat{U}_t^2/T - K}. \quad (5.11)$$

Under the null hypothesis these statistics follow an F distribution with q and $T-K$ as parameters. Before we discuss how to estimate the model subject to linear restrictions we need to say a little more to justify the transition from the general notion of an F statistic as the ratio of two independent χ^2 variables, to the form given here. The residuals ($\hat{U}_t$ or $\hat{\hat{U}}_t$) from least squares are not independent of each other (see exercise 3.7) but it can be shown that their squared sum divided by the error variance is a χ^2 variable but with a parameter equal to the degrees of freedom for the regression rather than the number of observations, i.e.

$$\frac{\sum_1^T \hat{U}_t^2}{\sigma^2} \text{ is } \chi^2(T-K) \qquad (5.12)$$

$$\frac{\Sigma \hat{\hat{U}}_t^2}{\sigma^2} \text{ is } \chi^2(T-K+q). \qquad (5.13)$$

However, these two distributions themselves are not independent of each other—this can be seen most plausibly by noting that the unrestricted sum of squares can never be greater than the restricted sum of squares. Both are obtained by choosing parameters to minimize the residuals and the former can always choose the values of the restricted equation if this is optimal (while the restricted case cannot guarantee to choose the values of the unrestricted equation for those parameters which are restricted). Hence one sum of squares tells us something about the value of the other. However, the differences between the two χ^2 variables is also a χ^2 variable with parameter equal to the difference of parameters, and it is independent of both χ^2 variables—once we have climbed a certain way up a hill we know that the height of the hill is greater than our present height but the distance between the two is unrelated to either our present height or the total height.

$$\frac{\Sigma \hat{\hat{U}}_t^2 - \Sigma \hat{U}_t^2}{\sigma^2} \sim \chi^2(q) \qquad (5.14)$$

and hence result (5.10).

The F test as outlined is UMPU for two-sided tests on a single linear restriction and indeed is uniformly most powerful for a one-sided test on a single restriction. For cases where there is more than one linear restriction involved, weaker optimality properties hold. In the case above, the right-hand tail of the F distribution is the critical region. This can be seen by imagining that the null hypothesis is true—in such a case the estimated parameters with the restriction (under the null hypothesis) would tend to be near the estimated parameters without the restriction (under the alternative hypothesis). Hence the residuals would tend to be similar and the sums of squares (adjusted for the degrees of freedom) would be similar. Low values of the F statistic would tend to occur. As the restriction becomes more incorrect the difference between the unrestricted parameter estimates (the 'true' values) and the restricted values would tend to increase—the sums of squares would become progressively divergent (with the restricted sum of squares ever larger than the unrestricted) and the value of the F statistic when the alternative hypothesis is true would tend to be large and positive more often than when the null hypothesis is true. The squaring function together with the fact that one case is 'nested' (a subset) of the other produces this result.

We can finally turn to the actual application of this test. The various cases can be illustrated by the two-variable models we have already explored in Chapter 4, although they generalize to the many-variable cases.

For a two-sided test in a single parameter in a two-variable model we have:

$$Y_t = \alpha + \beta_1 X_{1t} + \beta_2 X_{2t} + U_t \qquad (5.15)$$

where U_t are i.i.d. $N(0, \sigma^2)$. The hypothesis we wish to test is:

$$H_0: \beta_1 = c \qquad (5.16)$$
$$H_1: \beta_1 \neq c$$

where c is a known constant.

Our strategy is first to estimate the model in unrestricted form (i.e. under H_1), which we do by carrying out OLS on (5.15) to obtain $\hat{\alpha}$, $\hat{\beta}_1$, $\hat{\beta}_2$ by the formulae (4.6), (4.9), and (4.10). The residual sum of squares $\Sigma \hat{U}_t^2$ has $T-3$ degrees of freedom. The second step is to rewrite the equation subject to the restriction and then estimate the restricted form. We have

$$Y_t = \alpha + c X_{1t} + \beta_2 X_{2t} + U_t. \qquad (5.17)$$

The technique, as explained in Chapter 4, is to collect all terms with known parameters on the left-hand side of the equation leaving just unknown parameters on the right (the number of explanatory variables will thus be reduced by the number of restrictions). Here we obtain

$$Y_t - c X_{1t} = \alpha + \beta_2 X_{2t} + U_t \qquad (5.18)$$

or

$$Y_t^* = \alpha + \beta_2 X_{2t} + U_t \qquad (5.19)$$

where

$$Y_t^* = Y_t - c X_{1t}.$$

We know that if the restriction is correct (i.e. β_1 does really equal c) then the properties of the error term are undisturbed and the equation can be estimated by OLS. We accordingly regress Y_t^* on X_{2t} to obtain

$$\hat{\hat{\beta}}_2 = \Sigma y_t^* x_{2t} / \Sigma x_{2t}^2 \qquad (5.20)$$

$$\hat{\hat{\alpha}} = \bar{Y}^* - \hat{\hat{\beta}}_2 \bar{X}_2 \qquad (5.21)$$

and

$$\hat{\hat{U}}_t = Y_t^* - \hat{\hat{\alpha}} - \hat{\hat{\beta}} X_{2t} \qquad (5.22)$$

$$= Y_t - c X_{1t} - \hat{\hat{\alpha}} - \hat{\hat{\beta}} X_{2t}. \qquad (5.23)$$

The residual sum of squares $\Sigma \hat{\hat{U}}_t^2$ has $T-2$ degrees of freedom so that the F statistic for the hypothesis $\beta_1 = c$ is:

$$\frac{(\Sigma \hat{\hat{U}}_t^2 - \Sigma \hat{U}_t^2)/1}{\Sigma \hat{U}_t^2/(T-3)}. \qquad (5.24)$$

126 Hypothesis Testing

The statistic (5.24) will, if the null hypothesis is correct, follow an $F(1, T-3)$ distribution in repeated samples. We can find the value in the right-hand tail of such a distribution that produces a test of the size we require (from standard tables) and if the sample 'F' score is greater than the critical value then we decide that the evidence is against the null hypothesis and that the alternative hypothesis must be accepted. If the score is less than the critical value we accept the null hypothesis. A value of β_1 which was much greater or much less than c would produce a large and significant F score so that we would be able to react to both cases. This is illustrated with our model of the consumption function.

Example 5.1: Hypothesis on a multiple regression

(a) Two-sided hypothesis test on a single parameter

We take the multiple regression of consumption on income and unemployment as reported in Chapter 4.

Our null hypothesis is that unemployment has no effect on consumption (conditional on the assumption that income does affect consumption). We write this

$$H_0: \beta_2 = 0$$
$$H_1: \beta_2 \neq 0. \tag{5.25}$$

The unrestricted equation, with $\hat{\beta}_2$ free to take any value that will minimize the RSSQ, is of course the multiple regression. The unrestricted sum of squares is obtained:

$$\Sigma \hat{U}_t^2 = 4.45 \times 10^7.$$

The restricted equation becomes:

$$Y_t = \alpha + \beta_1 X_{1t} + U_t \tag{5.26}$$

i.e. the single variable equation of Chapter 2. The restricted sum of squares was:

$$\Sigma \hat{\hat{U}}_t^2 = 4.77 \times 10^7.$$

The F statistic is thus

$$F(1, 19) = \frac{(0.32 \times 10^7)/1}{4.45 \times 10^7/19}$$

$$= 1.367.$$

If we choose a test of 5 per cent (significance level of 95 per cent) the critical value that cuts off 5 per cent of the right-hand tail of the hypothetical distribution with parameters 1 and 19 is 4.380. Clearly the sample value is not 'unlikely' on the null hypothesis and therefore we accept that the data supports the null hypothesis that unemployment does not affect consumption, given that income is already used to explain consumption.

(b) A one-sided test on a single parameter

Our null hypothesis this time is that in the two-variable model the marginal propensity to consume is unity with the alternative hypothesis that it is less than unity (given that the model includes unemployment as an explanatory variable).

$$H_0: \beta_1 = 1$$
$$H_1: \beta_1 < 1. \tag{5.27}$$

Again the unrestricted sum of squares is taken from OLS on the two-variable model (4.45×10^7). The restricted sum of squares is taken from the multiple regression subject to the restriction. This was given as example 4.2. (26.30×10^7). Hence the F statistic is

$$F(1, 19) = \frac{(21.85 \times 10^7)/1}{4.45 \times 10^7/19}$$
$$= 93.3.$$

The difference from the two-sided test is in the treatment of the right-hand tail. A 5 per cent right-hand tail includes under the null hypothesis values of $\hat{\beta}_1$ that are greatly in excess of unity and values which are much below it. If we are only prepared to treat as extreme values such below unity then under the null hypothesis only 2.5 per cent of the time would we actually reject the NH (the values much greater than unity we would count as consistent with the NH). Hence to find a critical value that rejects the NH 5 per cent of the time for one-sided values of β_1 we need to find the 10 per cent critical region (but remember only to count as a rejection those values of F which fall in it *and* which have $\hat{\beta}_1 < 1$).

The 10 per cent value of the F distribution with 1 and 19 degrees of freedom is 2.99. So that we see that if the null hypothesis is true ($\beta_1 = 1$) a most unlikely event has occurred. Hence we decide that, using a 95 per cent significance level, the data rejects the hypothesis that the m.p.c. is equal to unity (given that consumption is also related to unemployment).

(c) Two-sided tests on two restrictions

In our basic consumption function model we wish to test the hypotheses that unemployment has no impact on consumption and that the intercept is zero (given that income is included in the equation).

$$H_0 : \alpha = 0$$
$$\quad : \beta_2 = 0 \qquad (5.28)$$
$$H_1 : \alpha \neq 0$$

and/or
$$\quad : \beta_2 \neq 0 \qquad (5.29)$$

Once again the unrestricted sum of squares is 4.45×10^7.

The restricted model is now the regression of consumption on income which, using the standard formulae, gives

$$C = 0.894 \ Y$$
$$(0.0047)$$

$$\text{SEE} = 2879, \qquad \text{RSSQ} = 17.4 \times 10^7.$$

The F statistic is thus

$$F(2, 19) = \frac{(12.85 \times 10^7)/2}{(4.45 \times 10^7)/19}$$
$$= 27.43.$$

The critical value using a 5 per cent test of $F(2, 19)$ is 3.52 and so we must reject the null hypothesis that consumption is proportional to income and unrelated to unemployment. The F test of course does not distinguish between which of these two are rejected by the data—it just shows that as between the restricted and unrestricted model there is a sufficiently big difference in the two parameters in question to alter the residual sum of squares by a significant amount. This could be caused by a large change in one or other parameter or by both being shifted a lesser amount.

Other tests, for example those involving the relationship of one parameter to others, can be applied in the same way. There are several important features of this class of tests of linear restrictions on the parameters.

1. The inference on the hypothesis is conditional on the unconstrained part of the model being correct. In the previous example it was shown that the contribution of unemployment to the explanation of consumption was insignificant *given that income was also being used as an explanatory variable*. Had just unemployment and an intercept been in the unrestricted model then a test on the effect of unemployment on consumption would have indicated a highly significant effect. This changing inference, depending on what the rest of the model is assumed to be, is dominated by the correlation between the variable in question (U) and the rest of the explanatory variables (Y). If U and Y are uncorrelated then whether or not Y is included the coefficient of U

will be the same and its impact on the sum of squares will be equal; however, if the variables are strongly correlated then the coefficients of a variable in the restricted and unrestricted models will be very different and the test statistic will be affected by this.

2. The use of an F test depends on the unrestricted model being correct so that the error terms are indeed independently identically normally distributed with zero mean. Were the mean not to be zero, or were there to be serial dependence or non-constant variance, then the distribution of the statistic, even if the null hypothesis were correct, would not be that of an F distribution. Hence when we choose a value that cuts off 5 per cent of an F distribution this might cut off more or less of the true sampling distribution for our statistic. The size of the test could be very different from what we state. This could have important results for an inferential procedure. A value of the statistic might be greater than the critical 5 per cent value for an F distribution, so that we would claim the data did not support the null hypothesis, while in fact the value of the statistic might be substantially less than the critical 5 per cent value of the true sampling distribution, so that at a 95 per cent level of significance we should have concluded that the data supported the null hypothesis. It is clear that we have to attempt to specify the correct model since leaving out a variable effectively puts it into the error terms and thus produces conditions under which the null hypothesis would not generate a statistic following an F distribution.

3. In general one-sided tests for more than one restriction are not easy to apply because of the correlation between variables. Hence all multiple restriction tests are in practice two sided.

4. As we have pointed out the particular statistic is chosen from rival tests of the same size because of power considerations. For example the statistic:

$$F(q, T-K+q) = \frac{(\text{RSSQ}_R - \text{RSSQ}_U)/q}{\text{RSSQ}_R/(T-K+q)}$$

also follows an F distribution if the null hypothesis is true. However, if the NH is false the numerator tends to increase but so does the denominator and so the statistic is less sensitive to departures from the NH (has lower power), and is therefore not utilized.

5. The test statistics have particularly interesting forms for two classes of null hypotheses.

    (a) tests on a single parameter which can be reformulated as a 't' statistic which can be shown to have a known distribution and to give exactly the same critical values as the corresponding F test would do.

    (b) tests that some, or all, parameters are zero can be shown to have a test statistic that can be reformulated in tems of multiple correlation coefficients.

We will sketch these approaches in the next section not because they are logically distinct from the approach already used, but because the actual forms used occur so commonly in applied work.

6. If we carry out a series of tests on a given equation we may obtain what appear to be contradictory results, but which in fact are not so. For example in a two-variable regression it is possible for the test that variable one is significant (non-zero parameter) to be rejected given the inclusion of variable two; for the test that variable two is significant (non-zero parameter) to be rejected given the inclusion of variable one; but for the test that both variable one and two are zero (neither is non-zero) also to be rejected given that nothing else is included in the regression. Essentially the two variables are strongly correlated and the 'common' part is correlated with the dependent variable. At the same time the components of the independent variable which are not in common with the other variable are only weakly correlated with the dependent variable. Hence either variable contains all the significant material to relate to the dependent variable, and the addition of the other variable adds nothing. In short there is not enough information in the data to allow us to identify separate significant effects.

7. As we have seen the inferences we make are conditional on the model used. This idea has an important and difficult extension. The model we choose as the unrestricted form may itself be the result of the outcome of significance tests applied to even more general models, and if this is so then the properties of our current test depend on the nature of that earlier test. The simplest version of the argument is known as 'data-mining'. Suppose that our procedure is to regress variable Y on a plausible independent variable (X_1). If the correlation is significant then we accept this as 'the model'. However, if it is not we find a second variable (X_2) and try again. Eventually we are bound, purely by random sampling, to find a variable which has a correlation with Y that appears 'significant'. Nevertheless, given the procedure as a whole, we must be very much less confident than usual that this 'unlikely value' is due to the null hypothesis of no association between Y and the X variable being wrong, and much more convinced that the chance of finding something, if we search hard, is greater than 5 per cent (the nominal size of the test). It is possible to refine these ideas and to adjust the test size to allow for the complete procedure. However this is difficult and rarely done. We must be aware though that the whole strategy of obtaining the result has to be taken into account in assessing its importance.

5.4. Relations between test statistics

There are some relations between test statistics and alternative forms of calculating test statistics that are commonly employed. We begin with the most important of these: Student's 't' test on a single linear restriction.

Hypothesis Testing

Let us consider the model

$$Y_t = \alpha + \beta_1 X_{1t} + \beta_2 X_{2t} + U_t \qquad (5.30)$$

There are three common forms of a hypothesis involving a single linear restriction:

(i) $H_0: \beta_1 = 0$ (5.31)

(ii) $H_0: \beta_1 = C$ (where C is known) (5.32)

(iii) $H_0: \beta_1 + \beta_2 = C$ (where C is known) (5.33)

All three tests can be one or two sided in the alternative hypothesis and the third type of test is clearly generalizable to weighted linear sums of any subset of parameters.

The general form of the 't' test is to form the statistic

(i) $$\frac{\hat{\beta}_1}{\sqrt{\widehat{\text{Var}\,\hat{\beta}_1}}} \qquad (5.34)$$

(ii) $$\frac{\hat{\beta}_1 - c}{\sqrt{\widehat{\text{Var}\,\hat{\beta}_1}}} \qquad (5.35)$$

(iii) $$\frac{\hat{\beta}_1 + \hat{\beta}_2 - c}{\sqrt{\widehat{\text{Var}(\hat{\beta}_1 + \hat{\beta}_2)}}} \qquad (5.36)$$

where in each case the parameters are estimated from the (unrestricted) multiple regression by OLS and the variances are estimated in the same way. To obtain the variance of a sum we use

$$\widehat{\text{Var}(\hat{\beta}_1 + \hat{\beta}_2)} = \widehat{\text{Var}\,\hat{\beta}_1} + \widehat{\text{Var}\,\hat{\beta}_2} + 2\widehat{\text{Covar}(\hat{\beta}_1 \hat{\beta}_2)} \qquad (5.37)$$

all of which are generated from OLS by the formulae of Chapter 4 (see problem 5.2). The general pattern is plain. For a linear restriction involving J parameters

$$H_0: \sum_1^J f_k \beta_k = c \qquad (5.38)$$

where the weights f_k and c are known, the statistic to be constructed is:

$$t = \frac{(\Sigma f_k \hat{\beta}_k) - c}{\sqrt{\widehat{\text{Var}(\Sigma f_k \hat{\beta}_k)}}}. \qquad (5.39)$$

If the null hypothesis is in fact correct then in repeated sampling this statistic follows a known distribution. The distribution is known as a 't' distribution and has a single parameter—the degrees of freedom $(T-K-1)$ for the unrestricted multiple regression with an intercept.

It can be seen that, in particular, hypotheses such as equation (5.34) have a simple statistic, being the ratio of the estimated parameter to the standard error of the parameter estimate. This ratio is often reported in the applied work rather than the standard error itself (although given the parameter estimate we can obviously use the statistic to calculate the standard error of the parameter estimate).

Given that hypotheses of the type (5.39) are also suitable for testing by the F statistic we need to know which, if either, is superior. In fact it can be shown that the two tests are exactly equivalent—for any regression where for a given test size the 't' test would just reject the null hypothesis, an F test of the same size (corrected for being one sided if necessary) would also just reject the null hypothesis. The choice between tests is then based purely on convenience. The relevant considerations are:

1. The t test requires only one regression (the unrestricted) to be calculated while the F test requires two regressions (restricted and unrestricted).

2. For tests involving a single parameter most regression packages available even on micro-computers will give all the information needed to build up the statistic (5.39). For tests involving several parameters the covariances between estimated parameters are not always given in the computer output. For F tests the residual sums of squares are always given (sometimes as a multiple correlation together with the total sum of squares).

3. The F test must be used in tests of more than one restriction so that preference for a unified approach would lead to its use even in single-restriction cases.

The equivalence between the tests can be demonstrated for the following case:

$$Y_t = \alpha + \beta X_t + U_t \tag{5.40}$$
$$H_0: \beta = 0.$$

The F statistic uses the restricted sum of squares:

$$\Sigma \hat{U}_t^2 = \Sigma(Y_t - \bar{Y})^2 \tag{5.41}$$

(remembering that $\bar{Y}$ is the OLS estimator in this case), and the unrestricted sum of squares:

$$\Sigma \hat{U}_t^2 = \Sigma(Y_t - \hat{\alpha} - \hat{\beta} X_t)^2 \tag{5.42}$$

which reduces to:

$$\Sigma \hat{U}_t^2 = \Sigma(y_t - \hat{\beta} x_t)^2. \tag{5.43}$$

Hence the statistic is

$$F(1, T-2) = \frac{[\Sigma y_t^2 - \Sigma(y_t - \hat{\beta} x_t)^2]/1}{\Sigma(y_t - \hat{\beta} x_t)^2/T - 2} \qquad (5.44)$$

$$\cdot = \frac{\hat{\beta}^2 \Sigma x_t^2}{\hat{\sigma}^2} \qquad (5.45)$$

(where $\hat{\sigma}^2$ is the unbiased estimator of the error variance in the unrestricted model). The 't' statistic is in this case the ratio of the parameter estimate to its estimated standard error:

$$t(T-2) = \hat{\beta} \Big/ \left(\frac{\hat{\sigma}^2}{\Sigma x_t^2}\right)^{1/2} \qquad (5.46)$$

Hence the value of the squared t statistic is the same as that of the F statistic based on a single restriction and with the same degrees of freedom in the unrestricted equation. Such an identity can be shown to hold also in the general case. At the same time the distribution of the square of any $t(K)$ statistic is identical to that of an $F(1, K)$ statistic. We can illustrate the identity with our two-variable examples given above. The test that the coefficient of unemployment was zero (given the inclusion of income as an explanatory variable) had a critical value of F, for a two-sided test of size 5 per cent, at 4.38 (with 19 degrees of freedom). The critical value of a t test (with 2.5 per cent in each tail) with 19 d.f. is 2.093 which has a squared value of 4.38. The critical values of the two tests are the same and the actual numerical values of the test statistics when calculated from the data will also be the same (on squaring the t score) so that the same decision would be reached by either test.

A group of important tests are those with restrictions that some or all of the coefficients of a regression are zero. Let us suppose that the variables have been numbered so that the test is that the first J (out of the set of K variables) are zero. We also include the intercept in the set of unrestricted variables. The unrestricted sum of squares is as usual $\Sigma \hat{U}_t^2$ and this is related to the multiple correlation for the restricted model by the formula:

$$\frac{\Sigma \hat{U}_t^2}{\Sigma y_t^2} = 1 - R^2. \qquad (5.47)$$

The restricted model leaves out the first J variables (all of which are hypothesized to have zero coefficients) and so becomes a 'standard' regression of the remaining $(K-J)$ variables plus intercept. The residual sum of squares is $\Sigma \tilde{U}_t^2$ and this is related to the multiple correlation for the restricted model (with J variables deleted):

$$\frac{\Sigma \tilde{U}_t^2}{\Sigma y_t^2} = 1 - R_J^2 \qquad (5.48)$$

(the TSSQ is the same for both regressions).
The F statistic for the J 'zero' restrictions is

$$F(J, T-K-1) = \frac{(\sum \hat{U}_t^2 - \sum \hat{U}_t^2)/J}{\sum \hat{U}_t^2/(T-K-1)} \tag{5.49}$$

which can be shown to be:

$$F(J, T-K-1) = \frac{(R^2 - R_J^2)/J}{(1-R^2)/(T-K-1)}. \tag{5.50}$$

Hence the F statistic can in this case be calculated from the increment in the multiple correlation coefficient. This formula specializes further in two common cases. First, when we have the null hypothesis that all the coefficients are zero (but there is an intercept) the formula (5.50) reduces to

$$F(K, T-K-1) = \frac{R^2/K}{(1-R^2)/(T-K-1)} \tag{5.51}$$

(since the conventional correlation for a model fitted just to a constant is zero, and the ESSQ is zero). This statistic can be calculated from the single unrestricted regression and it is frequently used as a minimal check on the adequacy of the model since it tests whether some or all coefficients are non-zero. Were it to be non-significant then the model as a whole produces no evidence of any systematic influence on the variable under consideration (see problem 5.1).

The second case of interest is where we wish to test that just one coefficient is zero (say the Kth). Using (5.50) we obtain

$$F(1, T-K-1) = \frac{R^2 - R_K^2}{(1-R^2)/(T-K-1)} \tag{5.52}$$

where R_K^2 is the multiple correlation with variable K deleted. However from (4.57) we see that this is expressible in terms of the partial correlation between Y and X (given all the variables $1 \ldots K-1$).

$$F(1, T-K-1) = \frac{r_K^2}{(1-r_K^2)/(T-K-1)} \tag{5.53}$$

Using the fact that for this hypothesis the F statistic is the square of the appropriate t statistic we have

$$t_K^2 = \frac{r_K^2}{(1-r_K^2)/(T-K-1)} \tag{5.54}$$

or

$$r_K^2 = \frac{t_K^2}{t_K^2 + (T-K-1)} \tag{5.55}$$

This formula allows us immediately to calculate partial correlations from computer output which gives only the 't' statistics for each of the variables.

Example 5.2: 't' scores and partial conclusions

We can see from our previous results for the multiple regressions that

$$t_1 = 29.31 \qquad r_1^2 = 0.978$$
$$t_2 = 1.17 \qquad r_2^2 = 0.067$$

which correspond to the values derived by an alternative route in example 4.1.

5.5 Confidence intervals

The problem that we have analysed in the earlier sections of this chapter is that of how to decide whether the data is compatible with given hypothetical values for the parameters. Often it is useful to invert this process and ask for what range of hypothetical values *would* the data be compatible. That is, we may wish to establish a region for the hyptothetical values which would not be rejected by the set of data in question.

In order to solve this problem all we need to do is to reverse the procedure of the significance tests. In our formula for the significance tests we need to find the limiting values of the hypothetical parameters that would produce test statistics that are just on the margin of being insignificant (data compatible with the hypothetical values). The range of values between these limits would be a set of values of the hypothetical parameters all of which were 'consistent' with the data at the level of significance being used for the test— such a set of values is called a 'confidence interval' for the parameter. Clearly if the model is correct then 95 per cent of such intervals in repeated sampling (given that the size of the test is 5 per cent) would cover the single true value. Only those intervals constructed on the extreme values would not in fact cover the true value. We can illustrate these ideas best with a 't' test on a single parameter in our two-variable model. The true value of the parameter is β while we estimate it with the standard formula $\hat{\beta}$. The two-sided test statistic against a hypothetical value for β is

$$t(T-K-1) = \frac{\hat{\beta} - \beta^\circ}{(\text{Var } \hat{\beta})^{1/2}}. \tag{5.56}$$

This statistic will be on the border between accepting and rejecting the null hypothesis when

$$|t| = \text{Critical value in one tail} \tag{5.57}$$

e.g. for a test with 19 degrees of freedom the critical value is 2.093. Were the t

136 Hypothesis Testing

statistic to be larger than this positive value or smaller than this negative value we would conclude that the data and the null hypothesis were incompatible. We can now solve for the range of hypothetical values.

$$|\hat{t}| = t(\text{critical})$$
$$|\hat{\beta} - \beta^{\circ}| = t(\text{critical}) \times (\widehat{\text{Var } \hat{\beta}})^{1/2} \qquad (5.58)$$

So the limiting values for the null hypothesis to be accepted are given by:

$$\beta^{\circ} = \hat{\beta} \pm t(\text{critical}) \times (\widehat{\text{Var } \hat{\beta}})^{1/2} \qquad (5.59)$$

The range of values for acceptable β° lying between the larger and smaller of these limits is the confidence interval. Each sample would generate a different interval but only for those sample values of $\hat{\beta}$ which lie in the tails of the true distribution would it generate an interval which would fail to cover the true mean. Hence if the size of the test is 5 per cent then 95 per cent of confidence intervals calculated by (5.59) would include the true value. Figure 5.6 illustrates these ideas. The true value of the parameter is β. Sample values of $\hat{\beta}$ follow a normal distribution centred at β. A sample value of $\hat{\beta}_1$ leads to interval 1, while the value $\hat{\beta}_2$ yields interval 2 (the intervals will be of different lengths because of the estimated variance). The former includes the true value while the second does not. The transformation to the t statistic form (which allows for the estimated variance) gives a statistic whose distribution is known and hence for which we can identify the bounds of the confidence interval.

Example 5.3: Confidence interval for a regression coefficient

The confidence interval for the coefficient of unemployment (given that income is also used as a determinant) in the consumption function using a 95 per cent level is (see example 4.1) $0.930 \pm 2.093 \times 0.798$ or $-0.740 < \beta_2^{\circ} < 2.600$. This interval covers zero and hence the data is compatible with a true value of the parameter being zero as the significance test ($H^{\circ}: \beta_2 = 0$) has indicated (see problem 5.3)

In all cases confidence intervals will need to be two sided since we are asking for all hypothetical values (larger and smaller than the estimated value) that would be compatible with the data.

The use of the t statistic illuminates the circumstances under which tests on a parameter or the associated confidence intervals will be of most help. The statistic was given by (5.56) and clearly this will be large:

(i) when the estimated value is far from the hypothetical value;

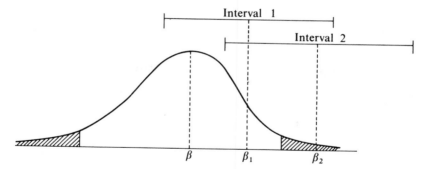

FIG. 5.6 Confidence Intervals for a Single Parameter

(ii) when the variance as estimated is small.

A large t score for a given hypothetical value means that the hypothetical value can be rejected at a higher level of significance (a smaller size test). Similarly the confidence interval based on a high t score will tend to be short—thus delimiting a narrower range of values compatible with the data. The role of the estimated variance, which itself depends on the variability of the data, is the key factor. With highly variable X values the variance of the parameter estimate will be small and for a test of given size the power will be high—there will be a better chance of accepting the alternative hypothesis when it is correct. Reversing the process the high variance of the X (low parameter variance) produces a short confidence interval which is much more helpful in locating the region of the true parameter value than a wider interval.

For a generalized test on a single linear restriction a confidence interval for the linear combination of the parameters can be constructed from the appropriate t statistic (5.39). The limiting values of the hypothetical restriction which would be compatible with the data are obtained from solving

$$t \text{ critical} = \frac{\Sigma f_k \hat{\beta}_k - C}{\{\widehat{\text{Var}(\Sigma f_k \hat{\beta}_k)}\}^{1/2}} \tag{5.60}$$

This generates an interval $M_1 < C < M_2$ in which the hypothetical linear sum $\Sigma f_k \beta_k^o$ could fall and still be consistent with the data. This is an (open) region in the space of all the β_k and it is unlikely in practice that all combinations would be economically acceptable. For inferences on several parameters (not linked by any restrictions) the general approach is the same. We would find those values of the appropriate F statistic which would just be compatible with the data and then relate these to a set of hypothetical parameter values that could generate such a value of the F statistic. In the single cases of two parameters (considered separately) it can be shown that the boundary of the region of hypothetical values consistent with the actual values is an ellipse centred on the estimated values. The higher the significance level the further away the ellipse is from the estimated values (the more points are consistent with them).

For more than two parameters the regions of compatible hypothetical values can be theoretically described by multidimensional analogues to ellipses but in practice this is too cumbersome to be useful.

5.6 Tests for structural change

An important use of F tests is for checking whether or not the values of the parameters of an equation have changed at a given point in time. For example we might expect that the relationship between the use of energy and the level of GDP would be different pre- and post-1974 as the shifts in behaviour induced by the oil price rise became a permanent feature of the demand for energy. To test this we would want to test whether or not the two elasticities for the two periods were equal. Many other examples of this type can be thought of. A related case is one where the relationship shifts for a sub-period (perhaps only for one observation) and then shifts back again—a supply function might be disturbed by strikes in the supply industry.

The general technique for modelling the unrestricted equation, i.e. one which allows for the possibility of shifts in parameters, is to use a so called 'dummy' variable. Consider an equation relating Y to X without an intercept which may have shifted after period τ. There are in fact two equations:

$$Y_t = \beta X_t + U_t \qquad t = 1 \ldots \tau$$

$$Y_t = \beta^1 X_t + U_t \qquad t = \tau + 1 \ldots T. \tag{5.61}$$

These two equations can be 'stacked' in a single long equation of T observations

$$\begin{pmatrix} Y_t \\ Y_t \end{pmatrix} = \beta \begin{pmatrix} X_t \\ 0 \end{pmatrix} + \beta^1 \begin{pmatrix} 0 \\ X_t \end{pmatrix} + \begin{pmatrix} U_t \\ U_t \end{pmatrix} \qquad \begin{matrix} t = 1 \ldots \tau \\ t = \tau = 1 \ldots T \end{matrix} \tag{5.62}$$

and we can write this in the form:

$$Y_t = \beta(XD_1) + \beta^1(XD_2)_t + U_t \qquad t = 1 \ldots T \tag{5.63}$$

where
$$D_{1t} = 1 \qquad t = 1 \ldots \tau \tag{5.64}$$
$$\phantom{D_{1t}} = 0 \qquad = \tau + 1 \ldots T$$
$$D_{2t} = 0t \qquad = 1 \ldots \tau \tag{5.65}$$
$$\phantom{D_{2t}} = 1 \qquad = \tau + 1 \ldots T.$$

The variables D_{it} are dummy variables which when multiplied by X_t generate a variable which takes the values of X_t for sub-period i and a value of zero for the other sub-period.

The two separate regressions with different parameters on the same variable are now shown as a single regression with two different parameters. If there is no structural change then $\beta = \beta^1$. This hypothesis of no structural change can now be tested by a standard F test procedure for testing a linear restriction. The unrestricted equation is the two-variable equation (5.63) while the restricted form is

$$Y_t = \beta(XD_1 + XD_2)_t + U_t \quad t = 1 \ldots T \qquad (5.66)$$
$$= \beta X_t + U_t$$

—a single-variable model where the variable becomes the basic X variable. From the restricted OLS regression we can also derive a sum of squares and as usual, providing that the U_t are i.i.d. normal *for the whole period*, the statistic:

$$F = \frac{(\text{RSSQ}_R - \text{RSSQ}_U)/1}{\text{RSSQ}_U/(T-2)} \qquad (5.67)$$

follows Fisher's F distribution under the NH: $\beta = \beta^1$ (no structural change). An alternative way of writing (5.62) is

$$\begin{pmatrix} Y_t \\ Y_t \end{pmatrix} = \beta \begin{pmatrix} X_t \\ X_t \end{pmatrix} + \delta \begin{pmatrix} 0 \\ X_t \end{pmatrix} + \begin{pmatrix} U_t \\ U_t \end{pmatrix} \quad \begin{matrix} 1 \ldots \tau \\ \tau+1 \ldots T \end{matrix} \qquad (5.68)$$

where $\delta = (\beta^1 - \beta)$. The test for no structural change is now H_0: $\delta = 0$ and this, for the single parameter case, can be carried out by the t statistic form of the test. The same test is sometimes given in a different form. For the unrestricted form we can run two separate equations for the sub-periods giving RSSQ_{1U} (with $\tau - 1$ d.f.) and RSSQ_{2U} (with $T - \tau - 1$ d.f.). The test statistic is

$$F = \frac{(\text{RSSQ}_R - \text{RSSQ}_{1U} - \text{RSSQ}_{2U})/1}{(\text{RSSQ}_{1U} + \text{RSSQ}_{2U})/(T-2)}. \qquad (5.69)$$

The estimate of the error variance based on the two separate (and independent) sums of squares has the same d.f. as the sum of d.f. for the two sub-regressions. The difference between the restricted and the unrestricted sums of squares (from separate regressions) is independent of this unrestricted sum of squares. In fact, providing there are adequate d.f. to carry out the two regressions on the sub-periods it is easy to show (see question 5.5) that the two statistics (5.67) and (5.69) are identical so that it does not matter which form of the test is used.

These forms of the test can then be used for any situation where the number of observations in each sub-period are greater than the number of parameters to be estimated for the sub-periods.[1] In cases where the shift is very short lived, e.g. a dock strike affecting an imports equation, there may be more parameters than observations in the period of the shift. Clearly OLS would break down in such a case since there will be perfect collinearity between any set of variables where there are more variables than non-zero observations. The strategy is simply to estimate the equation in restricted form for the whole period and obtain the RSSQ_R with $T-K$ degrees of freedom (for K variables). The equation is then re-estimated for the sub-period (T_1 say) where there are more observations than parameters to give RSSQ_{U1} with $T_1 - K$ degrees of freedom. The parameters for the second sub-period cannot be estimated so we form the statistic:

$$F = \frac{(\text{RSSQ}_R - \text{RSSQ}_{U1})/(T-T_1)}{\text{RSSQ}_{U1}/(T_1-K)}. \quad (5.70)$$

We can make this exactly analogous to the form of test (5.69) by arguing that the RSSQ_{U2} would have to be zero (a perfect fit) and the degree of freedom zero, when there are more parameters than observations. Hence we can test whether a shift has occurred when several variables are involved even though there may be fewer observations than variables in the shift period. We cannot, however, estimate what the individual coefficients would have been *if* the shift had taken place (see problem 5.10).

A particularly useful application of these techniques occurs when the shifts are periodic and of equal magnitude. Suppose that we have semi-annual data and that the intercept term is always higher (by a constant amount) in the first half of the year relative to the second half of the year. This can be modelled by a seasonal dummy variable as in equation (5.71).

$$Y_t = \alpha(1) + \delta(D_t) + \beta X_t \qquad t = 1 \ldots T \quad (5.71)$$

where
$$D_t = 1 \qquad \text{for} \quad t = 1, 3, 5 \ldots \quad (5.72)$$
$$= 0 \qquad t = 2, 4, 6 \ldots$$

This equation can be estimated by a multiple regression with an intercept and two explanatory variables (D_t and X_t) and a test for the significance of the seasonal effect is

$$H_0 : \delta = 0 \quad (5.73)$$

which can be carried out by an F (or t test) as usual. It is very important to notice that since the shift in the intercept is *relative to the second half of the year* we do not need a second dummy variable to pick up the seasonal value of the second half of the year (it is of course equal to α). With quarterly data we might expect that, for example, the intercept would be different in every

---

[1] When there is equality between the two there is still a unique solution for OLS and the RSSQ and the d.f. are both zero.

quarter. This could be modelled by a constant term normalized to (say) the first quarter and three dummy variables, each taking a value of unity in its own quarter and values of zero in the other three quarters. If the effect of the explanatory variable was also thought to be different in different quarters we could add extra dummy slope variables of the form $(D_{it}X_t)$ where $D_i = 1$ in the ith quarter, $= 0$ otherwise. A test for seasonal variation in a quarterly context would then use the equation:

$$Y_t = \alpha + \delta_1(D_1) + \delta_2(D_2) + \delta_3(D_3) + \beta X + \gamma_1(D_1 X)$$
$$+ \gamma_2(D_2 X) + \gamma_3(D_3 X_3). \tag{5.74}$$

This would also be estimated in restricted form for the hypothesis of no seasonal variation: $\delta_1 = \delta_2 = \delta_3 = \gamma_1 = \gamma_2 = \gamma_3 = 0$ (see problems 5.8 and 5.9).

5.7 Tests for predicted values

In Chapter 3 we showed that making a prediction from an equation was in fact entirely analogous to estimating a parameter from that equation, and it follows that it is possible also to carry out significance tests on a forecast value in the same way. The first way would be to compare the forecast value with the actual outcome (as a hypothetical value of that outcome) once it is known. This is rarely possible, since we do not know the actual outcome unless we have divided the data into sub-samples as in tests for structural stability—see problem (5.10).

The second way is to construct an analogue to the confidence interval for a parameter, i.e. to construct a range in which there is a high degree of probability that the true value lies. As before the key assumption is that regarding the statistical distribution of the errors. We assume that the error terms, in both estimation and forecasting periods, are independent identical normal variables (with zero mean). It follows that the predictor also has a known distribution, such as for the single-variable model without an intercept:

$$\hat{\beta} X_{T+1} \sim N\left(\beta X_{T+1}, \frac{\sigma^2 X_{T+1}^2}{\Sigma X_t^2}\right) \tag{5.75}$$

or

$$\frac{\hat{\beta} X_{T+1} - \beta X_{T+1}}{(\sigma^2 X_{T+1}^2)/\Sigma X_t^2} \sim N(0, 1) \tag{5.76}$$

and analogously:

$$\frac{\hat{\beta} X_{T+1} - Y_{T+1}}{(\sigma^2 + \sigma^2 X_{T+1}^2/\Sigma X_t^2)^{1/2}}. \tag{5.77}$$

The formula (5.77) is the more useful in practice since it relates to the actual outcome and not just the systematic component of that outcome. We know that, under our hypothesis about the errors the forecast statistic (5.75) in repeated samples follows a normal distribution and that, for example, 95 per cent of the realizations of such a statistic should fall in the interval -1.96 to $+1.96$, i.e.

$$-1.96 < \frac{\hat{\beta}X_{T+1} - Y_{T+1}}{(\sigma^2 + \sigma^2 X_{T+1}^2/\Sigma X_t^2)^{1/2}} < 1.96 \tag{5.78}$$

will occur in 95 per cent of repeated samples (of both the past data Y_t and the forecast Y_{T+1}). This can be manipulated to give:

$$\hat{\beta}X_{T+1} - 1.96 \text{ Var } Y_{T+1}^* < Y_{T+1} < \hat{\beta}X_{T+1} + 1.96 \text{ Var } Y_{T+1}^* \tag{5.79}$$

where Var Y_{T+1}^* is the denominator in (5.78). Under the NH this inequality will be satisfied in 95 per cent of random drawings. This interval (known as a 'tolerance interval') must be carefully interpreted—95 per cent of intervals constructed on such a basis will include the actual outcome (which is itself varying from sample to sample)—a given interval would cover only 95 per cent of the varying outcomes (were we able to separate the randomness in the estimation and in the forecasting periods). Again, since we do not know σ^2 in practice we need to use a t (or F) distribution based on the estimated error variance:

$$\frac{\hat{\beta}X_{T+1} - Y_{T+1}}{\widehat{(\text{Var } Y^*)}^{1/2}} \sim t(T-1) \tag{5.80}$$

or

$$\frac{(\hat{\beta}X_{T+1} - Y_{T+1})^2}{\widehat{\text{Var } Y^*}} = F(1, T-1). \tag{5.81}$$

These formulae can be used to give a range for the forecast with the desired probability of including the actual outcome—obviously the higher we set this probability the wider will be the interval.

The generalization to the multiple regression situation is relatively simple—it follows the test on a linear combination of parameters (5.60). In the two-variable model the best predictor is

$$Y_{T+m}^* = \hat{\beta}_1 X_{1,T+m} + \hat{\beta}_2 X_{2,T+m} \tag{5.82}$$

and the variance is

$$\text{Var } Y_{T+m}^* = \text{Var } \hat{\beta}_1 X_{1,T+m}^2 + \text{Var } \hat{\beta}_2 X_{2,T+m}^2$$
$$+ 2 \text{ Cov } \hat{\beta}_1 \hat{\beta}_2 X_{1,T+m} \cdot X_{2,T+m} + \sigma^2 \tag{5.83}$$

where the variances and covariances come from the standard OLS formulae.

The confidence interval is then based on

$$\frac{(Y_{T+m}^* - Y_{T+m})}{(\text{Var } Y_{T+m}^*)^{1/2}} \sim t(T-2). \tag{5.84}$$

In the case where the actual Y_{T+m} are known (perhaps data have become available since estimation was carried out) we can carry out a significance test on whether the forecast is compatible with the actual value. The actual statistic (5.80) or (5.81) can be calculated and then compared with the critical values from the hypothetical distribution. An extreme value of the statistic would lead us to reject the null hypothesis. This hypothesis is usually formulated to be that the parameters (β_i) are equal in the forecast and estimation period, but could also refer to the equality of error variances (for example). Clearly this test is exactly the same as a test of structural stability and can be generalized in an obvious way to the case of several forecasts—see (5.71) and problem (5.9).

Result

For an equation $Y_t = \beta X_t + U_t$ where data are available in both series for $t = 1 \ldots T$, and on the independent variable for other periods ($t = T+1 \ldots T+M$), provided that the equation obeys the conditions sufficient for the Gauss–Markov theorem to hold for estimation, the best linear unbiased predictor is

$$\hat{Y}_{T+m} = \hat{\beta} X_{T+m}$$

with variance:

$$\text{Var } \hat{Y}_{T+m} = \text{Var } \hat{\beta} \cdot X^2_{T+m}.$$

This estimated has a 'variance' relative to the actual outcome (Y_{T+m}) of

$$\text{Var } Y^*_{T+m} = \sigma^2 + \text{Var } \hat{\beta} X^2_{T+m}.$$

Assuming additionally that the error terms are normally distributed: we have:

$$\frac{(\hat{\beta} X_{T+m} - Y_{T+m})}{(\text{Var } Y^*_{T+m})^{1/2}} \sim t(T-1)$$

where $\text{Var } Y^*_{T+m}$ is the estimated value of the variance using s^2 for a estimator of σ^2.

A 'tolerance interval' for the random outcome of the equation in period $T+m$ is given by:

$$\hat{\beta} X_{T+m} + t^*_{\alpha/2} (\text{Var } Y^*_{T+m})^{1/2}$$

where t^* is the (two-sided) critical value of 'students' t distribution with the confidence level set to $1 - \alpha$ (α being the size of the critical region generated by the critical values)

Problems 5

5.1 In a regression model with several variables show that the goodness of fit as measured by $\bar{R}^2$ (4.49) will fall when a subset of variables are removed and the equation re-estimated by OLS if the value of the F statistic, for testing the null hypothesis that the coefficients of all these variables are zero, is greater than 1. In the case of the deletion of a single variable show that $\bar{R}^2$ will fall on deletion if the t statistic for that parameter is greater in absolute terms than unity (see also question 4.15).

If our testing procedure were to include any extra variable (variables) if this caused $\bar{R}^2$ to increase, discuss the implications for the size and power of the test.

5.2 Using the model

$$C_t = \alpha + \beta_1 Y_t + \beta_2 U_t$$

where C is consumption, Y is disposable income, and U is unemployment, test the hypothesis

$$H_0: \beta_1 + \beta_2 = 1$$
$$H_1: \beta_1 + \beta_2 \neq 1.$$

Do this by a t test (data are given in example 4.1).

5.3 For the model of the previous question construct a 90 per cent confidence interval for the parameter β_1 (data are given in example 4.1).

5.4 In a two-variable model, what is the relationship between the power and the size of a test on a hypothesis concerning a single parameter and the degree of correlation between the two variables?

5.5 There is an equation

$$Y_t = \beta X_t + U_t \qquad t = 1 \ldots T$$

where the U_t are i.i.d. normal. We wish to test the hypothesis that the value of the parameter β shifts after the τth observation. The restricted sum of squares is based on a regression of Y on X for all T observations. The unrestricted sum of squares is obtained by two different procedures: (i) estimating separate regressions for the sub-periods and adding the residual sums of squares and degrees of freedom for the two separate periods; (ii) rewriting the equation in the form:

$$Y_t = \beta(X_t D_{1t}) + \beta^1(X_t D_{2t}) + U_t$$

where
$$D_{1t} = 1, D_{2t} = 0 \qquad t = 1 \ldots \tau$$
$$D_{1t} = 0, D_{2t} = 1 \qquad t = \tau + 1 \ldots T$$

carrying out the multiple regression of Y on the two constructed variables for the full sample period, with the residual sum of squares being derived from this estimated equation. Show that the F statistics for these two forms of the test are equal.

Show that the same result would be obtained if the unrestricted equation was written in the form:

$$Y_t = \beta X_t + \delta(X_t D_{2t}) + U_t$$

where $\quad \delta = \beta^1 - \beta.$

5.6 The variable Y is related to variable X by a function which may have a 'kink' or 'knot' at the value $X = X^\circ$. Show how to test whether or not this kink exists (the function is linear in all the segments and is continuous with the linear segments meeting where $X = X^\circ$).

5.7 The rate of change of wages (Y) is known to depend linearly on the level of unemployment (X). However in periods $\tau + 1$ to $\tau + \sigma$ an incomes policy was applied. It is hypothesized that the effect of the incomes policy was the same in all these periods, depressing the rate of inflation by a constant amount. Describe a test for the null hypothesis that this effect was zero.

Suppose it is accepted that incomes policy was significant and we add the hypothesis that in the first period after the removal of the incomes policy there was a 'catching up' effect which increased the rate of inflation above what would normally be expected at that level of unemployment. Describe how to test whether there was complete catching up (i.e. as great as the total effect on inflation during the period of the operation of the incomes policy).

If incomes policies had been used in several different sub-periods, how would you test that they had been of equal severity?

5.8 A model with quarterly data relates variable Y to variable X together with an intercept. It is also expected that the intercept will vary systematically from quarter to quarter. The equation is estimated in two different forms:

(i) $\quad Y_t = \alpha(1) + \delta_1(D_{1t}) + \delta_2(D_{2t}) + \delta_3(D_{3t}) + \beta X_t + U_t$

(ii) $\quad Y_t = \varepsilon_1(D_{1t}) + \varepsilon_2(D_{3t}) + \varepsilon_3(D_{3t}) + \varepsilon_4(D_{4t}) + \beta X_t + U_t$

where $D_{it} = 1$ in quarter $i = 0$ in other quarters. What is the relation between the α, δ_i and ε_i? Discuss what would be the relation between the estimated $\hat{\alpha}$, $\hat{\delta}_i$ and $\hat{\varepsilon}_i$: would the two estimates of β be the same?

What would happen if we added the term $\delta_4 D_{4t}$ to the first equation?

5.9 In a cross-section of data we have information on expenditure (Y) and income (X) by household. We can also categorize the household by the number of people $p = 1, 2, 3 \ldots P$ in the household (a one-way classification). Show how to test the null hypothesis that the marginal propensity to spend is equal for all household sizes (allowing for the fact that there is an intercept in the relation).

Further data become available which give us a cross-classification on whether the household has children or not for each household ($i = 0$, no

146 Hypothesis Testing

children, 1 = children). How would you test that the marginal propensity to spend is constant irrespective of household structure?

5.10 A model $Y_t = \beta X_t + U_t$ is estimated on data from $1 \ldots T$. There is in addition one extra observation on both variables for period $T+1$. Construct the test statistics, on the assumption that the U_t are i.i.d. $N(0, \sigma^2)$:

  (i) for a test that the predictor $\hat{Y}_{T+1}$ is compatible with the actual outcome;
  (ii) that there is no structural change in the equation form in the $T+1$th period.

Show that those two tests are exactly the same.

5.11 For our model of the consumption function (see example 4.1) new data on income and unemployment became available for 1983 (see also example 4.3):

$$\text{Income 1983} = £170{,}000 \text{ million}$$
$$\text{Unemployment 1983} = 2{,}900 \text{ (thousands)}$$

Construct a 95 per cent confidence interval for a forecast of consumption in 1983.

Actual consumption later was known to be £155,000 million. Does this support the hypothesis that there was no structural change in the relation after 1982?

Answers 5

5.1 The F test statistic for the deletion of a set of q variables, i.e. coefficients restricted to zero, is

$$F = \frac{(\text{RSSQ}_R - \text{RSSQ}_U)/q}{\text{RSSQ}_U/(T-K)}$$

where K is the number of parameters estimated in the full model (5.10). From the special case (5.50) developed in the text (dividing throughout by TSSQ),

$$F = \frac{\{(1 - R_R^2) - (1 - R_U^2)\}/q}{(1 - R_U^2)/(T-K)}$$

where R_R^2 is the squared multiple correlation coefficient in the restricted model (variables excluded), and R_U^2 is the same in full model. From the relation between R^2 and $\bar{R}^2$ for a model (4.53):

$$(1 - \bar{R}^2) = \frac{T-1}{T-K}(1 - R^2)$$

assuming there is an intercept which is included in the K parameters. Substituting in for the two different correlation coefficients, dividing by

$(T-1)$ and rearranging we have have:

$$\frac{qF+T+K}{T+K+q} = \frac{1-\bar{R}_R^2}{1-\bar{R}_U^2}$$

If $F=1$, $\bar{R}_U^2 = \bar{R}_R^2$ while if $F \gg 1$, $\bar{R}_U^2 > \bar{R}_R^2$. Since $F(1, N) = t^2(N)$, it also follows that if F is greater than unity for a single restriction so that $\bar{R}^2$ falls on removing a single variable then if t^2 is greater than unity the same happens. Thus if the absolute value of t is greater than unity then $\bar{R}^2$ will fall if the variable is omitted.

If we used this $|t|>1$ criterion for including or not a particular variable ($F>1$ for a set of variables), this can be seen from standard statistical tables to be equivalent to using a test of a very large size. Tables of the 't' distribution show that for large number of degrees of freedom the relation between the significance level and the statistic is:

Significance level.   0.70   0.80   0.90   0.95   0.975
Statistic: 0.52   0.84   1.28   1.64   1.96

Hence, by interpolation, a value of the statistic of unity corresponds to a significance level of around 84 per cent. Given that a two-sided test is being used the size of the test procedure is 32 per cent, i.e. we are prepared to reject the null hypothesis when true (extra variables have zero effect) roughly one-third of the time. With such a large size of test the power will be correspondingly high. Even small departures from zero effect will be detected very often.

5.2 The statistic is

$$\frac{\hat{\beta}_1 + \hat{\beta}_2 - 1}{(\widehat{\text{Var}\,\hat{\beta}_1} + \widehat{\text{Var}\,\hat{\beta}_2} + 2\widehat{\text{Cov}\,\hat{\beta}_1 \hat{\beta}_2})^{1/2}} \sim t(19).$$

From the values given in example 4.1 we have $t = 0.880$. The critical value using a two-sided test with a significance level of 95 per cent is 2.093.

Hence the data are consistent with the null hypothesis that the sum of the coefficients is unity.

5.3 The estimate value of $\hat{\beta}_1$ is 0.752 with a standard error of 0.026. Using equation (5.59) we have the interval generated by

$$\beta_1^o = \hat{\beta}_1 \pm (\widehat{\text{Var}\,\hat{\beta}_1})^{1/2} \times t_{0.1}(\text{critical})$$

For a two-sided interval of 10 per cent and 19 degrees of freedom the critical value of the 't' statistic is 1.73. Hence the confidence interval for our data is:

$$0.752 \pm 0.026 \times 1.73$$
$$= 0.707 - 0.797$$

This interval does not include either zero or unity and so we could conclude both that there is a (non-zero) effect of income on consumption but that the m.p.c. is less than unity.

5.4 In the model $Y_t = \alpha + \beta_1 X_{1t} + \beta_2 X_{2t} + U_t$ where the errors are 'well-behaved', consider the null hypothesis

$$H_0: \beta_1 = \beta_1^o$$
$$H_1: \beta_1 \neq \beta_1^o$$

which is chosen by the investigator. If H_0 is true then the test statistic

$$t = (\hat{\beta}_1 - \beta_1^o)/(\widehat{\text{Var } \hat{\beta}_1})^{1/2}$$

is distributed as Student's 't' with $T-3$ degrees of freedom (the equation is estimated by OLS).

Clearly, the larger the variance, the greater must the difference between $\hat{\beta}_1$ and β_1^o be for the NH to be rejected; the statistic has to be larger than the critical value, which is a fixed number for a given size of test. Thus the chance of accepting AH when true will be reduced as the variance increases. Thus, for a given size of test, power falls as the variance increases; also for a given power (chance of accepting AH) the size of test increases—a smaller value of the test statistic must be used.

Since in the two-variable model the variance is proportional to the correlation between the explanatory variables—

$$\text{Var } \hat{\beta}_1 = \frac{\sigma^2}{\Sigma x_1^2 (1 - r_{12}^2)}$$

—the trade-off between size and power worsens as the multicollinearity increases.

5.5 Denote the regression coefficients in the models estimated separately for the two sub-periods as $\hat{\beta}_1$ and $\hat{\beta}_2$. The sums of squares are then added to yield

$$\sum_{1}^{\tau} (Y_t - \hat{\beta}_1 X_t)^2 + \sum_{\tau+1}^{T} (Y_t - \hat{\beta}_2 X_t)^2$$

with

$$(\tau - 1) + \{T - (\tau + 1) - 1\} = T - 1 \text{ degrees of freedom.}$$

Denote the regression coefficients in the multiple regression $\hat{\hat{\beta}}_1$ and $\hat{\hat{\beta}}_2$ with variables X_1 and $X_2 (X_{1t} = X_t D_{1t}$, etc). The residual sum of squares is then

$$S = \sum_{1}^{T} (Y_t - \hat{\hat{\beta}}_1 X_{1t} - \hat{\hat{\beta}}_2 X_{2t})^2.$$

But for the first τ observations $X_{2t} = 0$ and for the subsequent observations $X_{1t} = 0$

$$\therefore \quad S = \sum_{1}^{\tau} (Y_t - \hat{\hat{\beta}}_1 X_{1t})^2 + \sum_{\tau+1}^{T} (Y_t - \hat{\hat{\beta}}_2 X_{2t})^2.$$

Since $X_{1t} = X_t$ for the first set and $X_{2t} = X_t$ for the second set, we need to prove finally that $\hat{\beta}_1 = \hat{\hat{\beta}}_1$ and $\hat{\beta}_2 = \hat{\hat{\beta}}_2$. Now,

$$\hat{\beta}_1 = \sum_1^\tau Y_t X_t \Big/ \sum_1^\tau X_t^2$$

$$\hat{\hat{\beta}}_1 = \frac{\Sigma YX_1 \Sigma X_2^2 - \Sigma YX_2 \Sigma X_1 X_2}{\Sigma X_1^2 \Sigma X_2^2 - (\Sigma X_1 X_2)^2}$$

But from definitions summed from 1 to T,

$$\Sigma X_1 X_2 = 0$$

and

$$\sum_1^T X_1^2 = \sum_1^\tau X_t^2$$

(last observations in X_{1t} being all zero).

∴ $\hat{\beta}_1 = \hat{\hat{\beta}}_1$ ($\hat{\beta}_2 = \hat{\hat{\beta}}_2$ follows similarly).

The two approaches yield identical sums of squares for the unrestricted model and identical degrees of freedom. Since

$$Y_t = \beta(X_t D_{1t}) + \beta^1 X_t D_{2t} + U_t$$
$$= \beta X_t D_{1t} + \beta X_t D_{2t} + \beta^1 X_t D_{2t} - \beta X_t D_{2t} + U_t$$
$$= \beta X_t + (\beta^1 - \beta) X_t D_{2t} + U_t$$

which is identical to the new form, the residual sum of squares will be unchanged, as will the degrees of freedom and the test statistic.

5.6 The variable Y can be represented by the equation

$$Y_t = \beta X_t + \delta W_t + U_t$$

when

$$W_t = X_t - X_0 \text{ for } X_t \geq X_0$$
$$= 0 \qquad \text{otherwise}$$

or

$$Y_t = \beta X_t + U_t \qquad X < X_0$$
$$= \beta X_t + \gamma(X_t - X_0) \quad X \geq X_0$$

The multiple regression of Y on X and W yields the unrestricted sum of squares, while the regression of Y on X (over the whole range of values) yields the restricted sum of squares ($\delta = 0$). The F statistic is based on 1 and $T-2$ degrees of freedom and corresponds as usual to a 't' test on the coefficient of W in the multiple regression.

5.7 The model can be formulated

$$Y_t = \alpha + \beta X_t + \delta D_t + U_t$$

where
$$D_t = 1 \text{ for } t = \tau+1 \text{ to } \tau+\sigma$$
$$= 0 \text{ otherwise.}$$

Assume the errors are well behaved and normally distributed. The test procedure would be to estimate the unrestricted equation (multiple regression of Y on the variables X_t and D_t) and carry out a 't' test on the coefficient δ with $H_0: \delta = 0$, $H_1: \delta \neq 0$. The test should be a one-sided test, since even if δ were a very large positive value we would be unwilling to infer that the imposition of an incomes policy would increase the rate of inflation.

If there is the possibility of catching up in period $\tau+\sigma+1$ we need another dummy variable: $E_t = 1$ for $t = \tau+\sigma+1$, $E_t = 0$ otherwise. The model then becomes:

$$Y_t = \alpha + \beta X_t + \delta D_t + \varepsilon E_t + U_t$$

and the hypothesis of complete catching up is $H_0: \varepsilon = -\sigma\delta$ (σ being the number of periods the policy was in operation). This can be tested by the F test strategy. The model is estimated in three-variable unrestricted form as above and then the restriction is imposed:

$$Y_t = \alpha + \beta X_t + \delta(D_t - \sigma E_t) + U_t$$

and the resulting two-variable regression estimated. The test statistic will have an $F(1, T-4)$ distribution.

If more than one incomes policy had been imposed (but there was no catching up) we would have the model

$$Y_t = \alpha + \beta X_t + \delta_1 D_{1t} + \delta_2 D_{2t} \ldots \delta_N D_{Nt} + U_t$$

where $D_{nt} = 1$ for $\tau_n \ldots \tau_n + \sigma_n$ ($n = 1 \ldots N$) $= 0$ otherwise (there are N separate episodes).

The test for equal severity is based on the hypothesis

$$H^0: \delta_1 = \delta_2 = \ldots = \delta_N.$$

The unrestricted model is as given above with $N+2$ parameters to be estimated while the restricted model is

$$Y_t = \alpha + \beta X_t + \delta_1(D_{1t} + D_{2t} + \ldots + D_{Nt}) + U_t$$

(the composite variable has zero for all non-incomes policy periods and unity for all 'policy-on' periods). The test statistic based on the two residual sums of squares follows an $F(N-1, T-N-2)$ distribution if the null hypothesis is correct.

5.8 In the first form the model has been 'normalized' on the fourth quarter—when all the D_i (1 ... 3) are zero in the fourth quarter then $Y_t = \alpha + \beta X_t + U_t$. Hence δ_1 is the effect of the first quarter relative to the fourth quarter etc.

In the second equation each quarter is assigned its own intercept. Hence

$$\alpha = \varepsilon_4$$
$$\alpha + \delta_1 = \varepsilon_1$$
$$\alpha + \delta_2 = \varepsilon_2$$
$$\alpha + \delta_3 = \varepsilon_3.$$

Since one equation is just a linear transformation of the other the estimated parameters are linked by similar equations and the residual sum of squares, and $\hat{\beta}$ will be identical; i.e. $\hat{\alpha} = \hat{\varepsilon}_4$ etc.

The tests on individual coefficients must be carefully interpreted. In model (i), $H_0: \delta_1 = 0$ is the hypothesis that the intercept of the first quarter is equal to that of the fourth quarter (i.e. does not deviate from it—$\hat{\alpha}$—by a significant amount). In model (ii), $H_0: \varepsilon_1 = 0$ is the hypothesis that the intercept of the first quarter is zero.

If the term $\delta_4 D_{4t}$ were added to the first equation there would be perfect multicollinearity: $1 = D_{1t} + D_{2t} + D_{3t} + D_{4t}$ all t (i.e. we have already normalized three-quarters on the fourth quarter so that to add the fourth quarter is redundant) and no solution to OLS exists.

5.9 Suppose there are P distinct household groups (up to a maximum of P people = a household). If the m.p.c. is different for each group we might be tempted to write:

$$Y_l = \alpha + \beta X_l + \sum_1^P \beta_p D_{lp} X_l + U_l.$$

β_p is the additional m.p.c relative to a general value β for group p. D_{lp} are P dummy variables defined over all households. They take value 0 if the household is not of size p and l if the household is of size p. However this formulation is perfectly multicollinear—

$$X_l = \sum_{p=1}^P D_{lp} X_l$$

—because any household l must be in one of the P groups and has a unit value for that 'dummy' variable and zero for all others. Clearly we must either drop the general variable so that each β_p is the m.p.c. for a different group, or we must drop any one group dummy and interpret β as the m.p.c. of that group and the β_p as the deviation in m.p.c. from that of the reference group. Both models give the same results once this relationship is taken into account.

The test for equal m.p.c. compares the RSSQ for this unrestricted model with that of the restricted model $Y_1 = \alpha + \beta X_1 + U_1$. The test statistic of the conventional F form will follow an $F(P-l, T-P-l)$ distribution under the null hypothesis.

For the 'cross-classification' we might be tempted to add two dummy variables for the possession or not of children (since no general variable has been included to normalize against), e.g.

$$Y_l = \alpha + \sum_1^P \beta_p D_{lp} X_l + \sum_1^2 \gamma_i F_{ei} X_l + U_l.$$

However this form is also perfectly multicollinear: Since

$$\sum_1^P D_{lp} = \sum_1^2 F_{li}$$

for all l, any household will score one value of unity on the household-size variables and one value of unity on the two possession-or-not-of-a-child variable. We can equally well omit a household size or a child possession dummy. Suppose we omit E_2 (children in family) then the value of E_1 measures the increment in the m.p.c. associated with non-possession of children relative to possession. This increment is equal at all household sizes.

5.10 The test for the equality of the prediction ($\hat{Y}_{T+1}$) and the actual outcome is $H_0^1: \hat{\beta} X_{T+1} = Y_{T+1}$ ($\hat{\beta}$ based on OLS for $t = 1 \ldots T$). The test for no structural shift is $H_0^2: \hat{\beta} = \hat{\hat{\beta}}$ ($\hat{\hat{\beta}}$ is based on OLS for $t = 1 \ldots T+1$). The test statistics are

$$H_0^1: \frac{(Y_{T+1} - \hat{\beta} X_{T+1})^2}{\sigma^2(1 + X_{T+1}^2/\Sigma X_t^2)} \sim F(1, T-1)$$

(the square of the usual 't' test—5.77)

$$H_0^2: \frac{\sum_1^{T+1}(Y_t - \hat{\hat{\beta}} X_t)^2 - \sum_1^T (Y_t - \hat{\beta} X_t)^2}{\Sigma(Y_t - \hat{\beta} X_t)^2/(T-1)} \sim F(1, T-1).$$

Consider $\hat{\beta} = \hat{\hat{\beta}}$. This is

$$\hat{\beta} = \frac{\sum_1^{T+1} X_t Y_t}{\Sigma X_t^2} = \frac{\Sigma X_t Y_t + X_{T+1} Y_{T+1}}{\Sigma X_t^2 + X_{T+1}^2}$$

$$\therefore \hat{\beta} = \frac{\hat{\beta} + \dfrac{X_{T+1} Y_{T+1}}{\Sigma X_t^2}}{1 + \dfrac{X_{T+1}^2}{\Sigma X_t^2}}.$$

Equality requires $Y_{T+1} = \hat{\beta} X_{T+1}$. Hence the two tests do test the same hypothesis.

The test statistics are also equal. Replace $\hat{\hat{\beta}}$ and $\hat{\beta}$ in the H_0^2 formula and expand squares. Replace terms of form

$$\sum_1^{T+1} W_t = \Sigma W_t + W_{T+1}$$

to give:

$$\frac{Y_{T+1}^2 - \frac{(\Sigma Y_t X_t + Y_{T+1} X_{T+1})^2}{\Sigma X_t^2 + X_{T+1}^2} + \frac{(\Sigma Y_t X_t)^2}{\Sigma X_t^2}}{\hat{\sigma}^2}$$

(all summation from 1 to T). Divide the second and third terms by ΣX_t^2

$$= \frac{Y_{T+1}^2 - \frac{(\Sigma Y_t X_t + Y_{T+1} X_{T+1})^2 / \Sigma X_t^2}{1+K} + (\Sigma Y_t X_t)^2 / \Sigma X_t^2}{\hat{\sigma}^2}$$

where $K = X_{T+1}^2 / \Sigma X_t^2$. Expanding the middle term and reducing to a common denominator

$$= \left\{ Y_{T+1}^2(1+K) - \frac{(\Sigma Y_t X_t)^2}{\Sigma X_t^2} - \frac{2Y_{T+1} X_{T+1} \Sigma Y_t X_t}{\Sigma X_t^2} - \frac{Y_{T+1}^2 X_{T+1}^2}{\Sigma X_t^2} \right.$$

$$\left. + (1+K)(\Sigma Y_t X_t)^2 / \Sigma X_t^2 \right\} \bigg/ (1+K)\hat{\sigma}^2.$$

Reintroduce $\hat{\beta} = \Sigma Y_t X_t / \Sigma X_t^2$ and using definition of K

$$= \frac{Y_{T+1}^2 - 2Y_{T+1} X_{T+1} \hat{\beta} + X_{T+1}^2 \hat{\beta}^2}{(1+K)\hat{\sigma}^2}$$

$$= \frac{(Y_{T+1} - \hat{\beta} X_{T+1})^2}{\hat{\sigma}^2 \left(1 + \frac{X_{T+1}^2}{\Sigma X_t^2}\right)}$$

which is the test statistic for H_0^1. Hence the test statistic and the hypothesis are identical to test the equality of a forecast and the actual outcome, and to test for stability of the regression parameter.

5.11 From the data of example 4.3 we have the estimated forecast of consumption: $\hat{Y}_F = £148{,}417$ million; with standard error of forecast (including that of the error term): $SE^* Y_F = £1{,}963$ million. The confidence interval is (5.75)

$$\hat{Y}_F = Y_F \pm SE^* \hat{Y}_F \cdot t \text{ (critical)}.$$

With 19 degrees of freedom and a 95 per cent significance level the critical value of t (two sided) is 2.093 so that the interval is $£144{,}432 \leq Y_F^0 \leq £152{,}402$ million. Since a confidence interval tests against a particular NH, in this case $Y_F^0 = £155{,}000$ million, we see that the forecast interval is not consistent with the actual outcome. This is equivalent (problem 5.10) to rejecting the null hypothesis that there was structural stability in the equation.

6
Maximum Likelihood Estimation and Inference

6.1 Introduction

In all the models we have so far considered we have relied on the assumption that the X data are 'fixed in repeated samples'. For example in the basic model

$$Y_t = \beta X_t + U_t \tag{6.1}$$

we interpreted the equation as implying that in repeated trials a value of X could conceptually be held constant (as in the agricultural experiment described in Chapter 3). At the same time, for a given level of X_t, the error term U_{tt} would vary from sample to sample and hence, as a result, Y_{tt} would vary. In this way it becomes clear that certain economic variables cannot be assumed to be fixed from sample to sample (any endogenous variable). However, this assumption is particularly convenient in deriving optimality properties of OLS estimates. Consider the case of (6.1) in which the OLS estimator is:

$$\hat{\beta} = \frac{\Sigma Y_t X_t}{\Sigma X_t^2}. \tag{6.2}$$

First, if the X_t are fixed in repeated samples, the estimator is linear (in Y) in the sense that it can be written:

$$\hat{\beta} = \Sigma w_t Y_t \tag{6.3}$$

where the w_t are weights *fixed from sample to sample*. Second, when we come to assess whether the estimator is unbiased we have:

$$E(\hat{\beta}) = E\left(\frac{\Sigma Y_t X_t}{\Sigma X_t^2}\right) \tag{6.4}$$

and by substituting in (6.1)

$$E(\hat{\beta}) = \beta + E\left(\frac{\Sigma X_t U_t}{\Sigma X_t^2}\right). \tag{6.5}$$

In the case where the X_t are fixed, so that since the expectation is taken, for each term, over a fixed weight times the random variable U_t, we can rewrite (6.5)

$$E(\hat{\beta}) = \beta + \frac{\Sigma X_t E(U_t)}{\Sigma X_t^2} \tag{6.6}$$

and this, using the assumption that $E(U_t)=0$, yields the result that OLS is unbiased. From this line we can further derive the Gauss–Markov results that (given certain other assumptions on the behaviour of the U_t) OLS is BLUE. The role of the fixity of the X_t is central in allowing this development.

As we have already seen by reference to the variables Y_t, and as several important special cases dealt with in later chapters make explicit, the variable X_t can in fact rarely be thought to have this fixity outside of a controlled experiment, except for variables such as a trend ($X_t = t$) or a seasonal dummy ($X_t = 1$ in winter quarter, $=0$ otherwise) or the unit variable ($X_t = 1$ all t). Accordingly we need to be able to estimate models where the X_t can conceptually vary from sample to sample, and also to be able to derive the properties of the estimators we use.

The first variation on our basic assumption is to recognize that X_t is also random but to assume that it varies independently of U_t, i.e.

$$E_\tau(X_{t\tau}U_{t\tau}) = E_\tau(X_{t\tau})E_\tau(U_{t\tau}) \qquad (6.7)$$

This assumption of an independent stochastic explanatory variable will be shown in section 6.2 to have nearly the same properties for OLS as in the basic case and it follows that it does not give rise to a need for a different technique of estimation or of inference.

However, situations often arise in economics where even (6.7) does not hold, and so we require an alternative to OLS. The technique of maximum likelihood estimation is introduced in section 6.4 after some new concepts of estimator properties for large samples have been given in 6.3. In particular, the concept of a probability *limit*, which is used to replace the expectation operator as a method of evaluating an estimator, is explained. Associated with the techniques of maximum likelihood estimation are certain hypothesis testing procedures which utilize new test statistics and procedures—this is covered in section 6.5. A general note of warning must be issued at this stage: the mathematical foundations of maximum likelihood estimation and inference are considerably more difficult than those used elsewhere in this book. A simple account of these techniques is presented but it should not be regarded as a rigorous statement of proofs or procedures. For such a treatment more advanced texts must be consulted.

6.2 Estimation and inference with independent stochastic explanatory variables

We start with the simplest possible model, since this illustrates all the arguments we wish to make, and use (6.1): $Y_t = \beta X_t + U_t$. We assume:

(i) $\qquad\qquad E(U_t) = 0$

(ii) $\qquad\qquad E(X_t U_t) = E(X_t)E(U_t) \quad$ for all t

(iii) $\quad E(U_t^2) = \sigma^2 \quad$ all t

(iv) $\quad E(U_t U_s) = 0 \quad s \neq t$

(v) $\quad E(\Sigma X_t^2) > 0$

and thus replace the 'fixed in repeated samples' assumption by one of independence between X and U for each value of t. The OLS estimator

$$\hat{\beta} = \frac{\Sigma Y_t X_t}{\Sigma X_t^2}$$

is, in the context of all outcomes, no longer a linear (in Y) estimator since the weights vary for sample to sample. However, for the set of weights $(X_t/\Sigma X_t^2)$ that we observe for a particular outcome, it is a linear estimator (i.e. *conditional* on the values of X_t). The estimator is also unbiased whatever the sample size since (6.5) can be rewritten:

$$E(\hat{\beta}) = \beta + \frac{E(\Sigma X_t U_t)}{E\Sigma(X_t^2)} \tag{6.8}$$

$$= \beta + \Sigma E\left(\frac{X_t}{\Sigma X_t^2}\right) E(U_t) \tag{6.9}$$

$$= \beta$$

(since the second term is zero using (i) and (v)). In a similar way

$$\operatorname{Var}(\hat{\beta}) = \frac{\sigma^2}{E(\Sigma X_t^2)} \tag{6.10}$$

The true variance of the OLS estimator depends now on the expectation of the sum of squared X values rather than on their realization in a particular sample. Thus the problem of needing to estimate the variance of the estimator is extended to requiring to estimate the error variance as well as the expected value of the squared X terms. It seems natural to estimate this variance using the residuals and X values obtained in the single sample i.e.

$$\widehat{\operatorname{Var} \hat{\beta}} = \frac{\sigma^2}{\Sigma X_t^2} \tag{6.11}$$

and indeed it can be shown that if the error assumptions are true then

$$E(\widehat{\operatorname{Var} \hat{\beta}}) = \operatorname{Var} \hat{\beta} \tag{6.12}$$

so that the usual estimator of the variance of the estimator continues to be unbiased.

In a special sense the Gauss–Markov optimality properties continue to hold. For any given set of the X values which do not violate (v), the usual proof holds (i.e. $\hat{\beta}$ is BLUE conditional on the set of Xs). Since this holds for all sets of Xs we are justified in regarding OLS as having its optimality unimpaired.

Similar remarks relate to hypothesis testing. For *any given set of* Xs the standard test statistics used in Chapter 5 will continue to follow F or t distributions when the null hypothesis is true and so the NH will be rejected (say) 5 per cent of the time even when true. Since this percentage is the same whatever the values of X, we can then argue that, whatever the pattern of Xs, a test which is based on the using of the sample values of X as if they were fixed will reject the NH 5 per cent of the times when it is actually true. Confidence intervals constructed with the available X values (whatever they are) will also cover the true value 95 per cent (say) of all samples (even though conceptually the X values and hence the size of the interval would vary from sample to sample).

In summary, we can continue as before to use OLS and standard significance tests based on it, when we can assume that the X_t and U_t are independent (considered over all their possible outcomes). No extra problems are raised by the generalization to many variables except that we must modify (v) so that there is no exact linear relation between the *expected values* of the different explanatory variables, i.e. in the two-variable case we cannot have:

$$E_\tau(X_{1t,\tau}) = K\, E_\tau(X_{2t,\tau}) \text{ all } t. \tag{6.13}$$

If this were true then the *expected value* of the estimator would be indeterminate. This generalization of the *rank* condition for the existence of OLS estimations needs to be carefully interpreted—it is possible, for a given sample of data, that the variables are not perfectly correlated and that a corresponding value for the estimator exists. However, in repeated samples there must be data combinations which are perfectly multicollinear (in order for (6.13) to be satisfied in aggregate) and for these samples no estimator exists and the variance would be infinite. Hence its expectation does not exist. A similar argument means that OLS could not be considered the 'best' estimator even in the new conditional sense. The analogous case, where there is no variation in the X data (for a given t), means that in every sample the data is perfectly collinear (which in a way is a more restrictive case) and that the estimated value *never* exists.

6.3 Asymptotic properties of estimators

We have already seen that only in the special case of an independent stochastic variable can a simple extension of optimality justify the use of OLS. There will be circumstances in which we cannot assume independence, so that we cannot take expectations term by term as before. There are also situations, which we have not discussed in detail, in which a particular estimator is biased for a given sample size, e.g. the estimator of the error variance from (6.1):

$$\hat{\sigma}^2 = \frac{1}{T}\Sigma \hat{U}^2 \qquad (6.14)$$

which has

$$E(\hat{\sigma}^2) = \frac{T-1}{T}\sigma^2. \qquad (6.15)$$

The technique of maximum likelihood estimation is often preferred in such situations. It is a technique whose optimality properties are defined for the limiting sample size (as T becomes indefinitely large). This large sample focus also allows us to use a very important statistical result on products or ratios of random variables even when they are not independent and this is clearly useful in the case under discussion.

We begin, not with the estimation technique, but with some concepts which are useful in this content. These are all asymptotic or large-sample properties. An estimator, $\tilde{\theta}$, of a true parameter θ, is said to be *asymptotically unbiased* if

$$\lim_{T\to\infty} E(\tilde{\theta}) = \theta \qquad (6.16)$$

We do not give a precise definition of a limit here but the general sense is clear. If we considered a series of situations with larger and larger samples and the expected value of the estimator, as the sample grew in size, eventually tended towards the true value, then the estimator would asymptotically be unbiased. The example of (6.14) and (6.15) is clearly asymptotically unbiased—since $(T-1)/T$ tends to unity as T becomes large. Such an estimator can conceptually still have a large variance (over the repeated samples) even as the sample size becomes large and so it would not be true that a particular realization would necessarily be near the true value. To have this additional property we would require in addition that the variance over repeated samples (each of size T) tended to zero as the typical sample size T increased. If the bias disappears and the variance also tends to zero the estimator is *consistent*. This property of requiring that virtually all realizations of an estimator will be very close to the true value (and on average equal to it) as the sample tends to infinity is often written

$$\text{probability} \lim_{T\to\infty} \tilde{\theta} = \theta \qquad (6.17)$$

or

$$\text{plim } \tilde{\theta} = \theta \qquad (6.18)$$

Although a probability limit can be evaluated using the theoretical distribution of the estimator and then taking the limiting distribution as the sample size increases, it is often sufficient to evaluate the limiting values of the estimator and its variance.

There are two general results concerning probability limits which are very helpful.

1. *Slutsky's Theorem.* If $\tilde{\theta}$ is an estimator of θ and $g(\tilde{\theta})$ is a continuous function of θ, then if plim $(\tilde{\theta}) = \theta$ it follows that plim $g(\tilde{\theta}) = g(\theta)$.

This can be interpreted as saying that if $\tilde{\theta}$ is a consistent estimator of θ then $g(\tilde{\theta})$ is consistent for $g(\theta)$. The transformation is often used in logs or the exponential function. Such a property clearly does not generally hold for the expectation operator dealing with a finite sample.

2. *Cramer's Theorem.* If plim $(\tilde{\theta}) = \theta$ and plim $(\tilde{\mu}) = \mu$ then in general plim $(\tilde{\theta}\tilde{\mu}) = \theta\mu$.

This result does not restrict $\tilde{\theta}$ and $\tilde{\mu}$ to being independent variables and so is also of great usefulness.

These two theorems, although only applicable in the limit as T approaches infinity, do allow results to be used which are not true in finite samples, and are therefore attractive tools to use when standard finite sample techniques (e.g. OLS) are inapplicable.

Armed with these definitions and concepts we can next consider a completely different approach to estimation which has the property of consistency.

6.4 Maximum likelihood estimation

Consider again the model (6.1) together with the assumptions used in section 6.2. If we add to this the assumption that the error terms follow a known probability distribution $f(U)$ then we can give an expression for the probability of observing our data set. For the set of U_t (which in practice we do not observe) the probability of that particular sequence is given by:

$$L(U_1 \ldots U_T) = \prod_1^T f(U_t) \tag{6.19}$$

i.e. the product (expressed by the symbol Π) of the T probabilities. This formula is of course critically dependent on the assumptions that the errors are independent (no serial correlation) and all have the same probability distribution. If the probability function is continuous then this formula would need to be adjusted in order to make it a strict probability, but for estimation purposes this is not necessary. It is known as the *likelihood value*, being proportional to the likelihood of observing the particular set of U variables given f. In its present form the likelihood value is not very useful since it is expressed entirely in terms of the unobservable errors. However we can substitute in (6.1) to obtain:

$$L = \prod_1^T f(Y_t - \beta X_t) \tag{6.20}$$

Thus the probability of observing the Y_t, given the values of X_t, the true parameter β, and the distribution f (and its associated parameters) is (6.20). From this concept which is a true, but unknowable value, we can invert the problem in order to estimate β and the parameters of f. We examine the likelihood value for all possible values of β (denoted by β^*) and choose that value of β^* which maximizes (6.20), i.e. we are choosing as an estimate that value of β^* which would have given the highest probability of observing the actual set of outcomes $Y_1 \ldots Y_T$. This is the technique of maximum likelihood estimation (ML) and we shall illustrate its use below. First we need to be clear what the properties of such a technique are. The choice of value for β^* which gives the highest probability to observing the actual set of data does not imply that we are assuming that nature is in fact so kind to us. We are not assuming that with the true β the errors generated are the most probable ones—clearly in practice there can be unusual events and unlikely (low probability) values of the data. However, it can be shown that as the sample size tends to infinity *maximum likelihood estimators are consistent.* The possibility of nature presenting us with an untypical sample of data, so that when we treat it as very typical by the ML procedure and hence choose very misleading parameter estimators, becomes so unlikely in the large sample that the average estimator converges to the true estimator, and has a variance which tends to zero. This general property of ML estimators, which holds under rather unrestrictive assumptions, means that when using this technique we do not have to prove the optimality of the estimator in each separate case.

To use the technique of ML estimation there are two important problems to be faced. The first is methodological while the second is technical. In order to maximize the likelihood function we need to know the probability distribution of the error terms, i.e. the form of f. The conventional assumption in standard econometric models is that errors follow a normal distribution. This is assumed for the reasons given in Chapter 5: the central limit theorem, under certain assumptions about the origin of the error terms as the sum of a large number of independent random variables, allows us to make this assumption. Hence we have:

$$f(U_t) = (2\Pi\sigma)^{-1/2} \exp(-U_t^2/2\sigma^2) \qquad (6.21)$$

where σ is the variance of the normal distribution. Substituting into (6.20) we obtain

$$L = \prod_1^T (2\Pi\sigma)^{-1/2} \exp\{-(Y_t - \beta X_t)^2/2\sigma^2\}. \qquad (6.22)$$

This value, which is exact for the unknown but true values of σ and β, is used to generate a *function* of the estimators σ^* and β^*, which relates the chance of observing Y_t for the observed x_t as a function of σ^* and β^*.

$$L(\sigma^*, \beta^*) = \Pi(2\Pi\sigma^*)^{-1/2} \exp\{-(Y_t - \beta^* X_t)^2/2\sigma^{*2}\}. \qquad (6.23)$$

Maximum Likelihood Estimation and Inference   161

This function has now to be maximized for β^* and σ^*. Since the maximum of a function of β^* will occur at the same value as the maximum of the log of that function (because the log is a strictly increasing transformation) we equivalently take the log of (6.23) and maximize the log-likelihood function. This converts the product into a sum, as well as removing the exponential functions:

$$l = \ln L(\sigma^*, \beta^*) = \ln\{(2\Pi\sigma^{*2})^{-T/2}\} - \frac{1}{2\sigma^{*2}}\Sigma(Y_t - \beta^* X_t)^2. \quad (6.24)$$

We can now differentiate with respect to the variables and set the results equal to zero

$$\frac{\partial l}{\partial \beta^*} = -\frac{1}{2\sigma^{*2}}\Sigma - 2(Y_t - \beta^* X_t)X_t = 0. \quad (6.25)$$

For the second argument we find the maximum for σ^{*2} since this will correspond to the maximum for σ^*.

$$\frac{\partial l}{\partial \sigma^{*2}} = -\frac{T}{2}\left(\frac{1}{\sigma^{*2}}\right) - \frac{1}{2(\sigma^{*2})^2}\Sigma(Y_t - \beta^* X_t)^2 = 0 \quad (6.26)$$

The first equation can be seen, in this case, to be identical to the normal equation of OLS (once the term $1/\sigma^{*2}$ has been cancelled). Hence the ML estimator is

$$\tilde{\beta} = \frac{\Sigma Y_t X_t}{\Sigma X_t^2}. \quad (6.27)$$

This shows that in our particular case, with the errors also normally distributed, OLS is also an ML estimator and as such is consistent.

The second equation yields

$$\sigma^{*2} = \frac{\Sigma(Y_t - \beta^* X_t)^2}{T} \quad (6.28)$$

so that substituting in for the ML value for β^* we have

$$\tilde{\sigma}^2 = \frac{\Sigma(Y_t - \tilde{\beta} X_t)^2}{T} \quad (6.29)$$

$$= \Sigma \tilde{U}_t^2 / T \quad (6.30)$$

where $\tilde{U}_t$ is the residual from ML estimation (which in this case coincides with $\hat{U}_t$ obtained from OLS estimation). However the ML estimator of the residual variance is not the same as used in OLS to obtain an unbiased estimator of the variable, which is:

$$\hat{\sigma}^2 = \Sigma \hat{U}_t^2 / (T-1).$$

The difference between the two, which also serves to illustrate an example where an ML estimator is biased in small samples, disappears in the large sample. The ML estimator being consistent must be asymptotically unbiased and hence tends to the same value as the OLS estimator in this particular example. This example shows how the choice of the normal distribution together with the standard linear regression model leads to an estimator which is identical to OLS. Of course, with other models the solution is not as easy and often the ML and OLS estimations will be completely different.

The second problem in fact arises in cases where the derivatives of the (log) likelihood functions are not the same as the 'normal' equations and indeed do not have analytic solutions; i.e. we cannot write down an explicit expression for the solution. In such cases it is necessary to find the optimum by some approximation or search technique. Since ML is particularly useful when OLS is not valid it tends to be the case that these techniques are frequently required. In a first course on econometrics we can do no more than draw the reader's attention to this possibility and point out that modern computer programmes often contain options to find the optimum by a variety of approximation techniques.

A second very important feature of ML estimators is that in the large sample they tend to be normally distributed (the estimator is in effect a weighted sum of random events and hence the central limit theorem is applicable). The variance of the distribution can be estimated by the variance of the ML estimator and so a test statistic based on the normal distribution can be used. However for more general testing a series of tests analogous to F statistics are used and we present these in the next section.

The final problem is to obtain the variance of an ML estimator. For finite samples this is often difficult, but there are large sample results available. However these are beyond the scope of this book. Again computer programmes will often supply the values required.

6.5 Large sample test statistics

The procedures for testing hypothesis when working with ML estimators will tend to appeal to properties that hold for large samples (since it is only for the large sample that the optimality of ML estimation is ensured). This approach has the advantage of being applicable in a much wider range of situations than exist where classical F tests are valid, but against this we must remember that the properties of the tests are often only approximate for small samples (and that the degree of approximation is often unknown). There are three separate types of test, each of which makes different demands of the user. The choice of test will sometimes depend on the advantages to be gained from not meeting the demands made by other tests in the particular situation, and sometimes just by the convenience of what is available in a computer programme.

All three tests will be illustrated for simplicity with a basic model for which a classic F test would also be appropriate. However this will serve to illustrate the lines of the approach. The model we use is

$$Y_t = \beta_1 X_{1t} + \beta_2 X_{2t} + U_t \qquad (6.31)$$

for which the U_t are assumed to be independently, identically distributed normal variables with zero mean and variance σ^2. We first need to derive the ML estimators since all three tests use them. The log likelihood function of the estimators is obtained analogously to (6.24):

$$l(\sigma^*, \beta_1^*, \beta_2^*) = \ln\{(2\Pi\sigma^{*2})^{-T/2}\} - \frac{1}{2\sigma^{*2}} \Sigma(Y_t - \beta_1^* X_{1t} - \beta_2^* X_{2t})^2. \qquad (6.32)$$

Differentiating and equating the derivates to zero we again obtain the 'normal' equations for β_1^* and β_2^* to yield the ML estimators (which of course again coincide with the OLS values):

$$\tilde{\beta}_1 = \frac{\Sigma Y X_1 \Sigma X_2^2 - \Sigma Y X_2 \Sigma X_1 X_2}{\Sigma X_1^2 \Sigma X_2^2 - (\Sigma X_1 X_2)^2} \qquad (6.33)$$

while $\tilde{\beta}_2$ is obtained by using symmetry and interchanging X_1 and X_2 in (6.33). We also have:

$$\tilde{\sigma}^2 = \frac{1}{T} \Sigma(Y_t - \tilde{\beta}_1 X_{1t} - \tilde{\beta}_2 X_{2t})^2. \qquad (6.34)$$

In these cases where the ML estimators coincide with OLS or RLS we also know the finite (and large) sample variances of the ML estimator. The true variance of ML must be the same as that for OLS since each representation over which the variance is defined would give identical estimators. For example,

$$\text{Var } \tilde{\beta}_1 = \frac{\sigma^2 \Sigma X_2^2}{\Sigma X_1^2 \Sigma X_2^2 - (\Sigma X_1 X_2)^2} \qquad (6.35)$$

(conditional on a certain set of Xs). The variance needs itself to be estimated and as in section 6.2 we use the actual values of the sums of squares and products of the Xs to estimate their expectations. The ML estimator of σ is used (rather than the OLS value $\hat{\sigma}$) to give

$$\text{Var } \tilde{\beta}_1 = \frac{\tilde{\sigma}^2 \Sigma X_2^2}{\Sigma X_1^2 \Sigma X_2^2 - (\Sigma X_1 X_2)^2}. \qquad (6.36)$$

The other variances and covariances in both unrestricted and restricted forms follow analogously.

In order to carry out a test we need to specify the nature of the hypotheses. By writing the model as (6.31) and estimating it in that form we imply that there is no known restriction on the β_i or σ. The alternative is that there are known restrictions on the parameters. In the general case such restrictions could be linear or non-linear and could involve any of the parameters (including σ). We take the simple linear restriction of equality of the β_i to yield:

$$H_0: \beta_1 = \beta_2 \tag{6.37}$$
$$H_1: \beta_1 \neq \beta_2.$$

The three testing procedures approach this from different angles. The first approach exploits the idea that if the restriction(s) is (are) correct then the value of the likelihood function at the optimum, estimated subject to the restrictions, would be very close to that value obtained under free estimation, since in large samples the ML estimates tend toward the true values and hence would tend to obey the restriction. The second test looks at the estimated parameter values obtained from the unconstrained model and looks to see by how much they fail to obey the restriction. Again in the large sample the values of estimators should nearly obey the restriction if it is true. The third test looks at the derivatives of the unrestricted likelihood function. These will be zero if we evaluate them at the optimal unrestricted values (since that is how the unrestricted estimators are obtained). Thus if we evaluate these derivatives using the restricted estimators (from the likelihood function maximized subject to the constraint) then the derivatives should be approximately zero if the restrictions were correct (since the two sets of estimators would be very similar). Each test procedure of course needs to provide a method of assessing whether these test statistics are significant or not. We now look at the procedures in more detail.

(a) Likelihood Ratio (LR) tests

This test compares the values of the log likelihood function at its maximum when there is no constraint placed on the estimated parameters, with the value at the maximum obtained subject to the restriction being imposed on the estimated values. We denote the restricted values $\tilde{\tilde{\beta}}_1, \tilde{\tilde{\beta}}_2, \tilde{\tilde{\sigma}}$ and they are the solution to

$$\text{Max} \quad l(\beta_1^*, \beta_2^*, \sigma^*) \tag{6.38}$$

$$\text{Subject to} \quad \beta_1^* = \beta_2^*. \tag{6.39}$$

This optimization can be carried out either by substituting (6.39) into (6.38) and optimizing over the two remaining unknowns (which is particularly simple when the restriction is linear) or by using a Lagrangian technique. We can see that in this case the restricted ML estimators again coincide with the restricted least squares estimators because the maximand becomes:

$$l^R(\beta_1^*, \sigma^*) = \ln\{(2\Pi\sigma^{*2})^{-T/2}\} - \frac{1}{2\sigma^{*2}} \Sigma \{Y_t - \beta_1^*(X_{1t} + X_{2t})\}^2 \tag{6.40}$$

The first-order conditions yield the same normal equation as would occur for RLS and hence at the restricted optimum:

$$\tilde{\tilde{\beta}}_1 = \frac{\Sigma YX^*}{\Sigma X^{*2}} \qquad (6.41)$$

where
$$X_t^* = X_{1t} + X_{2t} \qquad (6.42)$$

$$\tilde{\tilde{\sigma}} = \frac{1}{T}\Sigma(Y_t - \tilde{\tilde{\beta}}_1 X_t^*)^2 \qquad (6.43)$$

and using the restriction

$$\tilde{\tilde{\beta}}_2 = \tilde{\tilde{\beta}}_1. \qquad (6.44)$$

Substituting the unrestricted optimum estimators into the likelihood function we denote the unrestricted value $l(\tilde{\beta}_1, \tilde{\beta}_2, \tilde{\sigma})$ and the value at the set of restricted estimators $l(\tilde{\tilde{\beta}}_1, \tilde{\tilde{\beta}}_2, \tilde{\tilde{\sigma}})$. It can be shown that the following statistic,

$$LR = -2[l(\tilde{\tilde{\beta}}_1, \tilde{\tilde{\beta}}_2, \tilde{\tilde{\sigma}}) - l(\tilde{\beta}_1, \tilde{\beta}_2, \tilde{\sigma})] \qquad (6.45)$$

if the null hypothesis is correct, is distributed in large samples as a $\chi^2(q)$ variable where q is the number of restrictions. In this case we substitute in for our estimated values:

$$LR = -2\left\{-\frac{T}{2}\log(2\Pi\tilde{\tilde{\sigma}}^2) - \frac{1}{2\tilde{\tilde{\sigma}}^2}\Sigma(Y_t - \tilde{\tilde{\beta}}_1 X_{1t} - \tilde{\tilde{\beta}}_2 X_{2t})^2 \right.$$
$$\left. + \frac{T}{2}\log(2\Pi\tilde{\sigma}^2) + \frac{1}{2\tilde{\sigma}^2}\Sigma(Y_t - \tilde{\beta}_1 X_{1t} - \tilde{\beta}_2 X_{2t})^2\right\} \qquad (6.46)$$

However:
$$\Sigma(Y_t - \tilde{\tilde{\beta}}_1 X_{1t} - \tilde{\tilde{\beta}}_2 X_{2t})^2 = \Sigma \tilde{\tilde{U}}_t^2 \qquad (6.47)$$

etc., so that using $\tilde{\tilde{\sigma}}^2 = \Sigma \tilde{\tilde{U}}_t^2 / T$ and collecting terms we have

$$LR = T\log(\tilde{\tilde{\sigma}}^2/\tilde{\sigma}^2) \sim \chi^2(1) \qquad (6.48)$$

This in fact is the general form of the LR statistic whenever we are estimating a single equation (except that the degrees of freedom will depend on the number of restrictions imposed). However in more complex cases the LR test continues to approximate a χ^2 statistic and is therefore generally applicable. The major disadvantage with this test is that both the restricted and unrestricted likelihood functions have to be optimized and that this can sometimes be difficult for one or other. In the single equation case the fact that we often use a normal distribution for the errors allows us to derive the particularly simple statistic for the test:

$$LR = T\log(\tilde{\tilde{\sigma}}^2/\tilde{\sigma}^2) \sim \chi^2(q) \qquad (6.49)$$

(b) Wald (W) tests

The second approach to testing is to concentrate not on the likelihood function in restricted and unrestricted cases but on the restriction itself. Suppose the restriction is again as in (6.35) with all parameters written on one side of the equation:

$$\beta_1 - \beta_2 = 0. \tag{6.50}$$

We can estimate the unrestricted equation by ML techniques and obtain the 'estimated' restriction

$$\tilde{\beta}_1 - \tilde{\beta}_2 = \hat{\delta} \tag{6.51}$$

i.e. $\hat{\delta}$ is the difference between the estimators of the two parameters. Since ML is consistent it should be true that, as the sample size increases, $\hat{\delta}$ should tend to zero if the null hypothesis is correct. The idea of the Wald test is to derive a statistic for the distribution of $\hat{\delta}$ so that inferences on its value (whether it is a 'significantly large' value or not) can be carried out. The test statistic for a single restriction is again very simple in form.

$$W = \frac{\hat{\delta}^2}{\text{Var } \hat{\delta}} \sim \chi^2(1) \tag{6.52}$$

where $\text{Var } \hat{\delta}$ is the estimated variance of the sample restriction. For linear restrictions the variance of a sum (or difference) is particularly simple, and utilizes the fact that the variance of ML estimators is the same as for OLS, so that in our case we have:

$$\widehat{\text{Var } \hat{\delta}} = \widehat{\text{Var } \tilde{\beta}_1} + \widehat{\text{Var } \tilde{\beta}_2} - 2\widehat{\text{Cov } \tilde{\beta}_1 \tilde{\beta}_2}. \tag{6.53}$$

The ML estimator of the error variance is however used:

$$\tilde{\sigma}^2 = \frac{1}{T} \Sigma \tilde{U}_t^2. \tag{6.54}$$

The important feature of the Wald test is that it merely requires us to estimate the model and hence the restriction from the unrestricted equation—the 'true' restriction is known via the null hypothesis. In more complex cases with more than one restriction, or a model which is not in the linear regression format, or with non-linear constraints the principle can be generalized to obtain a statistic which is distributed, when the NH is true, as $\chi^2(q)$ in large samples for q restrictions.

(c) Lagrange Multiplier (LM) tests

The third test concentrates on the restricted model. It can be given a number of interpretations, all of which are rather complicated, but a particularly useful form occurs when the normal distribution assumption is used and the model can be put into the standard linear regression form. We first carry out restricted ML estimation to obtain the restricted estimators and the restricted residuals $\tilde{U}_t$. These residuals are then regressed (OLS) on all the explanatory variables in the model (including the intercept) to obtain the coefficient of multiple correlation R^2. The Lagrange Multiplier statistic is

$$\text{LM} = T.R.^2 \sim \chi^2(q) \qquad (6.55)$$

where q is the number of restrictions. We can see instinctively why this is a plausible statistic. If the restriction is correct then imposing it on the model will not disturb the values of the estimated residuals in the large sample from the true error terms and they will then tend to be uncorrelated with the explanatory variables, thus yielding a low R^2 and an insignificant test statistic. However an incorrect restriction forces part of the actual variables into the estimated residuals, and these will then tend to be correlated with the explanatory variables, leading to large and significant values of LM. This test is particularly useful when the restricted model is simple to estimate.

It is important to remember that all three test statistics only follow χ^2 distributions in the large sample. They are often used in small samples, but the effect of this is that the choice of a critical region has to be interpreted with care. The 5 per cent critical region of a χ^2 distribution may cover more or less than 5 per cent of the LR, W, or LM test statistics for small samples. In general there is no way to calculate the exact size of the test that we apply and so we can only claim that our inference is 'asymptotically correct', i.e. we are applying a procedure that would in the large sample reject the NH, when true, 5 per cent of the time.

Example 6.1. ML estimation and tests

We consider again our model of consumption related to income and unemployment, $C = \alpha + \beta Y + \gamma U + v$, with the assumption that the error terms are independently identically normally distributed. We wish first to obtain maximum likelihood estimators for α, β and γ and their variances, and then we wish to test the hypothesis that

$$H_0 : \beta = \gamma.$$

Since the model is of the special type where the ML estimators are equal to OLS estimators we have for the unrestricted case: $\tilde{\beta} = 0.752$, $\tilde{\gamma} = 0.930$. The ML estimator for the standard error of estimates differs from OLS because there is no correction for degrees of freedom: $\tilde{\sigma} = 1397$. Accordingly the standard errors of the estimated coefficients change being based also on $\tilde{\sigma}$ (but otherwise are identical to those for OLS).

$$\text{Var } \tilde{\beta} = 0.00057$$
$$\text{Var } \tilde{\gamma} = -0.550$$
$$\text{Cov } (\tilde{\beta}, \tilde{\gamma}) = -0.0136.$$

The restricted ML estimator corresponds to RLS, because on substituting in the restriction to the likelihood function we will find the derivative at zero to be the same as the 'normal' equation for RLS. The values are:

$$\tilde{\tilde{\alpha}} = 17473$$
$$\tilde{\tilde{\beta}} = 0.757$$
$$\tilde{\tilde{\gamma}} = 0.757$$
$$\tilde{\tilde{\sigma}} = 1423$$
$$\text{Var } \tilde{\tilde{\beta}} = 0.00022.$$

We now can use the three large sample procedures:

(a) LR test

From the standard errors of estimates we have:

$$\text{LR} = 22 \log \left[\left(\frac{1423}{1397} \right)^2 \right]$$
$$= 0.811$$

The critical value of a χ^2 statistic with 1 degree of freedom at a 95 per cent level of significance is 3.841 so that the LR statistic provides no evidence against the NH of equal parameter values.

(b) Wald test

The statistic $\hat{\delta}$ is the difference of the two (unrestricted) estimators $\hat{\delta} \sim \tilde{\beta} - \tilde{\gamma}$ and the estimated variance of $\hat{\delta}$ is thus:

$$\widehat{\text{Var } \hat{\delta}} = \widehat{\text{Var } \tilde{\beta}} + \widehat{\text{Var } \tilde{\gamma}} - 2 \widehat{\text{Cov } \tilde{\beta}, \tilde{\gamma}}.$$

Hence

$$W = \frac{\hat{\delta}^2}{\widehat{\text{Var } \hat{\delta}}} = 0.055.$$

This statistic is also compared to a $\chi^2(1)$ distribution, which has critical value 3.841, so that the W test also supports the null hypothesis of parameter equality.

(c) LM test

The restricted model is estimated by ML (as in the LR test the RLS equation is utilized in this example) and the residuals are then regressed on *all* the explanatory variables (the unit variable, income, and unemployment). The resulting multiple correlation coefficient is $R^2 = 0.00025$, so that the test statistic is

$$LM = TR^2 = 0.055.$$

As before this is compared to a $\chi^2(1)$ variable and the null hypothesis is again accepted. For comparison with these large sample tests we also illustrate testing this hypothesis with a standard F test. This is an exact test because the model can be estimated in both restricted and unrestricted forms by least squares. The model is linear with normally distributed errors.

The unrestricted and restricted sums of squares are:

$$RSSQ_U = 44{,}477{,}100$$
$$RSSQ_R = 44{,}580{,}980.$$

Hence the test statistic based on a single restriction is $F(1, 19) = 0.044$. The critical value of the F distribution with 1 and 19 degrees of freedom based on a 95 per cent significance test is 4.38 so that the NH is again accepted when the exact test is utilized.

Problems 6

6.1. Consider the model $Y_t = A + \beta X_t^\gamma + U_t$, where the U_t are i.i.d. normal. Show why OLS cannot be used to estimate the equation and derive an implicit expression for the ML estimators.

6.2. Let the variable Y_i take the value unity if a household owns a car and the value zero if it does not. The probability of a randomly selected household owning a car is related to the income (X) level of the household by the function,

$$f(X) = 1/\{1 + \exp(-\beta X)\}$$

where β is a parameter. What is the general shape of f? Show that the expected value of Y for a randomly selected household of income X_i is:

$$E(Y_i) = f(X_i).$$

Consider the regression model:

$$Y_i = \frac{1}{1 + \exp(-\beta X_i)} + U_i$$

What are the properties of the error term U_i? Would OLS be a suitable estimator? Show how ML could be used to estimate β.

6.3 Consider the model $Y_t = \beta_1 X_{1t} + \beta_2 X_{2t} + U_t$, where the U_t are i.i.d. normal. We wish to test the hypothesis that X_2 has no effect on Y, i.e.

$$H_0: \beta_2 = 0$$
$$H_1: \beta_2 \neq 0.$$

Find the LR statistic and relate this statistic to that used for a conventional F test for a zero restriction.

6.4 For the model and hypothesis of problem 6.3 construct and compare the Wald and LM test statistics with the standard F test statistic. (Hint: show that the residuals from the second stage regression in the LM test procedure equal those from the unrestricted ML estimation.)

6.5 For the model relating consumption to income and unemployment (data in Chapters 2 and 4) we wish to test the hypothesis that the coefficient for unemployment is zero. Do this using ML estimation and LM, W, and LR test procedures.

Answers 6

6.1 The equation cannot be transformed to keep it additive in all the parameters and U_t, and so we cannot find an expression from the residuals which is amenable to simple differention to yield an explicit optimum. We could of course attempt to solve the non-linear least squares function

$$S(\hat{A}, \hat{B}, \hat{\gamma}) = \Sigma (Y_t - \hat{A} - \hat{B} X_t^{\hat{\gamma}})^2$$

but the solutions will be hard to find and the BLUE optimality property cannot be assumed to hold. In such a case ML estimation, with its guarantee of consistency, is attractive. Because the errors are normal the log likelihood function, when differentiated, yields the non-linear least squares normal equations:

$$\frac{\partial}{\partial \tilde{A}} \Sigma (Y_t - \tilde{A} - \tilde{B} X_t^{\tilde{\gamma}})^2 = 0$$

$$\frac{\partial}{\partial \tilde{B}} \Sigma (Y_t - \tilde{A} - \tilde{B} X_t^{\tilde{\gamma}})^2 = 0$$

$$\frac{\partial}{\partial \tilde{\gamma}} \Sigma (Y_t - \tilde{A} - \tilde{B} X_t^{\tilde{\gamma}})^2 = 0.$$

These equations can be solved by approximation techniques to yield the ML estimators.

6.2 For positive values of β we can see that at large negative X the value of $f(X)$ is near zero while at large positive X the value of $f(X)$ tends to unity. To obtain the expected value of Y we write

$$E(Y_i) = 1 \times \text{prob}(Y=1) + 0 \times \text{prob}(Y=0)$$
$$= f(X_i).$$

Considering the regression model

$$Y_i = \frac{1}{1+\exp(-\beta X_i)} + U_i$$

we see that since Y_i can take only the values 1 or 0 the error term will take the value complementary to $f(X)$. When Y_i is 1 then U_i must be $1 - f(X_i)$, while when Y_i is 0 then U_i must be $-f(X_i)$. Moreover the expected value of U_i is:

$$E(U_i) = \text{prob}(Y=1) \times \{1-f(X)\} + \text{prob}(Y=0) \times \{-f(X)\} = 0$$

The variance of U_i is

$$E(U_i^2) = \text{prob}(Y=1) \times \{1-f(X)\}^2 + \text{prob}(Y=0) \times \{-f(X)\}^2$$
$$= f(X)\{1-f(X)\}^2 + \{1-f(X)\}\{-f(X)\}^2$$
$$= f(X)\{1-f(X)\}$$

Clearly the variance is heteroskedastic being different at the different levels of household income. The model is not suitable for estimation by OLS but ML can be used. If we order the observations so that the first S all own cars and the rest of the sample of I do not then the log likelihood function is

$$l(\tilde{\beta}) = \sum_1^S \log\left\{\frac{1}{1+\exp(-\tilde{\beta} X_i)}\right\} + \sum_{S+1}^I \log\left\{1 - \frac{1}{1+\exp(-\tilde{\beta} X_i)}\right\}$$

This equation does not have a derivative which is explicitly soluble for $\tilde{\beta}$, but it can be approximated by standard techniques.

The general model in which a household responds to a stimulus following a cumulative probability function (one ranging between zero and one) and where we observe just whether or not the reaction has taken place (Y is zero or one) is known as a probit-type model. This term probit is generally reserved for the case where the distribution f is a cumulative normal, while the case we have examined, of the logistic function, is known as a logit model.

6.3 The LR statistic for a linear model with a linear restriction and normal errors is of the form

$$\text{LR} = T \log\left(\frac{\tilde{\tilde{\sigma}}^2}{\tilde{\sigma}^2}\right)$$

where $\tilde{\tilde{\sigma}}$ is the estimated variance from the restricted ML equation (which coincides with RLS) and $\tilde{\sigma}$ is the estimated variance from the unrestricted ML equation (which coincides with OLS)

The F statistic for exclusion is

$$F = \frac{(\text{RSSQ}_R - \text{RSSQ}_U)/1}{\text{RSSQ}_U/(T-2)}.$$

Now
$$\text{RSSQ}_R = \Sigma \hat{U}_t^2 = T\tilde{\tilde{\sigma}}^2$$
$$\text{RSSQ}_U = \Sigma \hat{U}_t^2 = T\tilde{\sigma}^2$$

(remembering that the ML estimators of the error variances divide the residual sums of squares by the total number of observations).

$$\therefore \quad F = \frac{\tilde{\tilde{\sigma}}^2 - \tilde{\sigma}^2}{\tilde{\sigma}^2/(T-2)}$$

$$= (T-2)\left[\frac{\tilde{\tilde{\sigma}}^2}{\tilde{\sigma}^2} - 1\right]$$

$$\therefore \quad \frac{F}{T-2} + 1 = \left(\frac{\tilde{\tilde{\sigma}}^2}{\tilde{\sigma}^2}\right).$$

$$\therefore \quad T\log\left(\frac{F}{T-2} + 1\right) = \text{LR}.$$

Hence given the value of the LR statistic we can immediately calculate the associated F statistic and vice-versa. Furthermore since the transformation is monotonic we could use the critical value of the F statistic for the given degrees of freedom and size of test (call it F^*) to derive the exact critical value for the LR statistic using the above equation rather than relying on the $\chi^2(1)$ distribution for LR. In such a case however, it is simplest to use the F test directly.

6.4 (a) The Wald test requires us to estimate the unrestricted model by ML (which coincides here with OLS) and then substitute the estimated values into the restriction.

$$\therefore \quad \tilde{\delta} = \tilde{\beta}_2$$

where
$$\tilde{\beta}_2 = \frac{\Sigma YX_2 \Sigma X_1^2 - \Sigma YX_1 \Sigma X_1 X_2}{\Sigma X_1^2 \Sigma X_2^2 - (\Sigma X_1 X_2)^2}$$

$$\widehat{\text{Var }\tilde{\delta}} = \widetilde{\text{Var }\tilde{\beta}_2}$$

$$= \frac{\tilde{\sigma}^2 \Sigma X_1^2}{\Sigma X_1^2 \Sigma X_2^2 - (\Sigma X_1 X_2)^2}.$$

Now
$$W = \tilde{\delta}^2 / \widehat{\text{Var }\tilde{\delta}}$$
$$= \tilde{\beta}_2^2 / \widehat{\text{Var }\tilde{\beta}_2}.$$

However this is almost exactly the square of the t statistic that is used to test the exclusion of variable X_2 from the model (see Chapter 5). The only difference is that in the formula:

$$t = \hat{\beta}_2 / \widehat{\text{Var }\hat{\beta}_2},$$
$$\hat{\beta}_2 = \tilde{\beta}_2$$

but
$$\widehat{\text{Var }\hat{\beta}_2} \neq \widehat{\text{Var }\tilde{\beta}_2}$$

because
$$\hat{\sigma}^2 \neq \tilde{\sigma}^2$$

(the same RSSQ is divided by $T-2$ in the case of OLS and T for ML). The test statistics will be almost equal for large samples. Furthermore since any t statistic squared, based on q degrees of freedom, is $F(1, q)$ statistic the Wald statistic is again almost equal to the F statistic for the exclusion of a variable.

It is important to notice that although this Wald statistic is distributed in large samples as $\chi^2(1)$, we can again find the exact significance points in the small sample by working back from the critical values of the appropriate t or F distributions.

(b) The LM test requires us to estimate the restricted model by ML (which equals OLS here) and obtain the residuals. These are also equal to the RLS residuals from the regression of Y on X_1 alone.

$$\tilde{U}_t = Y_t - \tilde{\beta}_1 X_{1t}$$

(where $\tilde{\beta}_1$ is the restricted estimator). These residuals are regressed on X_{1t} and X_{2t} to yield residuals $\hat{e}_t$ and coefficients $\hat{\gamma}_1$ and $\hat{\gamma}_2$. Now,

$$\hat{\gamma}_1 = \frac{\Sigma(Y - \tilde{\beta}_1 X_1) X_1 \Sigma X_2^2 - \Sigma(Y - \tilde{\beta}_1 X_1) X_2 \Sigma X_1 X_2}{D}$$

$$\therefore \quad \hat{\gamma}_1 = \hat{\beta}_1 - \tilde{\beta}_1$$

where $\hat{\beta}_1$ is the unrestricted ML (= OLS) estimator

$$\hat{\beta}_1 = \frac{\Sigma Y X_1 \Sigma X_2^2 - \Sigma Y X_2 \Sigma X_1 X_2}{D}.$$

Similarly $\quad \hat{\gamma}_2 = \hat{\beta}_2$.

$\therefore$ The second stage residual which is

$$\hat{e}_t = (Y - \tilde{\beta}_1 X_1) - \hat{\gamma}_1 X_1 - \hat{\gamma}_2 X_2 = Y - \hat{\beta}_1 X_1 - \hat{\beta}_2 X_2$$

which is the residual from the unrestricted equation $\hat{U}_t$. Now defining the LM statistic as the correlation between dependent variable ($\tilde{U}_t$) and the explanatory variables we have by the usual decomposition of sums of squares

$$TSSQ = ESSQ + RSSQ$$

$$R^2 = \frac{ESSQ}{TSSQ} = 1 - \frac{RSSQ}{TSSQ}$$

(provided all sums are here measured around the origin).

$$\therefore \quad R^2 = 1 - \frac{\Sigma \hat{e}_t^2}{\Sigma \tilde{U}_t^2}$$

$$= \frac{\Sigma \hat{U}_t^2 - \Sigma \tilde{U}_t^2}{\Sigma \hat{U}_t^2}$$

(using the result that $\hat{e}_t = \tilde{U}_t$). Therefore, the LR statistic is

$$LR = \frac{T(\Sigma \hat{U}_t^2 - \Sigma \tilde{U}_t^2)}{\Sigma \hat{U}_t^2}.$$

This can be compared to the standard F statistic, which also uses residuals from the restricted and unrestricted equations.

$$F = (T-2)\left(\frac{\Sigma \hat{U}_t^2 - \Sigma \tilde{U}_t^2}{\Sigma \tilde{U}_t^2}\right).$$

Again we can see that knowing the critical value of F we can in this case derive the exact critical value at the LR statistic even for the small sample (the difference in denominators merely means that we should have to make a transformation to go from one to the other).

6.5 We first require the restricted and unrestricted ML estimators. In this case, the linear model with linear restrictions and a normally distributed error term and the ML and OLS estimators of the parameters are the same. The estimated variances differ solely in the estimator of the error variance. Using the results of examples 3.1, we can arrive at Table 6.1.

TABLE 6.1 Restricted and Unrestricted ML Estimates

	Unrestricted ML	Restricted ML
Income	0.752	0.775
(Standard error)	(0.024)	(0.016)
Unemployment	0.930	0
(Standard error)	(0.741)	
σ	1423	1472

(a) *LR test.* The LR statistic is based on one restriction so we have

$$LR = T \log\left(\frac{\tilde{\tilde{\sigma}}^2}{\tilde{\sigma}^2}\right) \sim \chi^2(1)$$

$$= 1.49.$$

The critical value of $\chi^2(1)$ with a 95 per cent significance level is 3.841 so that the null hypothesis that unemployment does not have an influence on consumption is accepted.

(b) *Wald Test.* The estimated restriction is $\hat{\delta} = 0.930$, $\widehat{\text{Var } \hat{\delta}} = 0.549$ so that

$$W = \frac{\hat{\delta}^2}{\widehat{\text{Var } \hat{\delta}}} = 1.58.$$

The Wald test also continues to accept the null hypothesis.

(c) *LR test.* Here we can either calculate the residuals from the restricted regression (they are given in table 2.2) and regress then on an intercept, income, and unemployment to obtain the value of R^2, or we can work backwards from the relationship shown in the answer to problem 6.4. We illustrate the latter approach.

We have

$$LR = T\left(\frac{RSSQ_R - RSSQ_U}{RSSQ_R}\right)$$

$$= T(\tilde{\tilde{\sigma}}^2 - \tilde{\sigma}^2)/\tilde{\tilde{\sigma}}^2$$

$$= 1.44.$$

Again the null hypothesis that unemployment has no influence on consumption is accepted (since we compare the LR value with the critical value from a $\chi^2(1)$ distribution).

7
Generalized Least Squares: Serial Correlation and Heteroskedasticity

In all the models we have so far analysed, both when estimating and when testing hypotheses, we have rested the results on a series of assumptions about the error terms and about the independent variables. It is clearly important to be able to relax these assumptions if possible, since there is no a priori reason why a particular model should conform to this basic specification.

This chapter concentrates on the second and third assumptions which concern the variances and covariances of the error terms. There are two separate issues raised by a failure of either of these assumptions: (i) if we continue to use it, what are the properties of OLS, given that we cannot guarantee its optimality by the Gauss–Markov theorem; (ii) if OLS is not optimal, then what is the optimal estimator? The first question can be initially explored without making specific alternative assumptions about the error terms but the second needs to postulate specific cases in order to be operational. Hence we begin with a general treatment and then move to specific and separate treatments of the two types of failure.

We begin with the simplest model, since the problems are identical for more complex cases. The model is the single-variable case without an intercept:

$$Y_t = \beta X_t + U_t \tag{7.1}$$

where

(i) $\qquad E(U_t) = 0$

(ii$a$) $\qquad E(U_t^2) = \sigma_t^2$

(iii$a$) $\qquad E(U_t U_s) = \sigma_{ts} (\neq 0 \; t \neq s)$

(iv) $\qquad X_t$ fixed in repeated samples

(v) $\qquad \Sigma X_t^2 > 0.$

As in Chapter 3 we can derive the ordinary least squares estimator:

$$\hat{\beta} = \Sigma X_t Y_t / \Sigma X_t^2. \tag{7.2}$$

We can investigate its properties under the new set of assumptions. It is clearly still linear (linearity does not depend on the assumptions) and it is still unbiased. The proof of unbiasedness uses assumptions (i), (iv), and (v), but not (ii$a$) so that variations (ii$a$) and (iii$a$) do not affect this property. However the variance of $\hat{\beta}$ is no longer the smallest among linear unbiased estimates. This can be shown by first deriving the variance and then retracing the steps of the Gauss–Markov proof. From (7.2) we have as usual because of the unbiasedness property:

$$\text{Var } \hat{\beta} = E\left\{\frac{(\Sigma X_t U_t)^2}{(\Sigma X_t^2)^2}\right\} \quad (7.3)$$

$$= \frac{\Sigma \sigma_t^2 X_t^2 + \overset{s \neq t}{\Sigma\Sigma} \sigma_{ts} X_t X_s}{(\Sigma X_t^2)^2} \quad (7.4)$$

This new formula illustrates a first important consequence of the change in assumptions—the variance of OLS is given by a different expression than in the basic case, so that attempts to estimate this variance by the standard formula would lead to an incorrect assessment of the true variability of OLS (which is given by 7.4). This divergence stems from two sources which can be seen by considering the standard estimator of the variance:

$$\widehat{\text{Var } \hat{\beta}} = \frac{\hat{\sigma}^2}{\Sigma X_t^2}. \quad (7.5)$$

This now differs from (7.4), not only in that the multiplier of the error variance is different from the new formula (and might be either larger or smaller) but also in that the expected value of $\hat{\sigma}^2$, if calculated from the OLS residuals in the usual fashion, is no longer equal to the true variance. The exact degree of bias can be evaluated for certain special cases (see problem 7.4). Thus the first consequence of the use of OLS is that, were we also to continue to use the conventional variance formula (because we had not noticed the change in the error properties), the estimated variance would be biased in general. In particular if the bias were downwards the results could be serious for inferential problems—the apparent small variance of parameters would be likely to make them look more significantly different from any values under the null hypothesis than they actually are. The size of the significance test would therefore be much larger than we state (i.e. we would be likely to reject the NH when true more often) and hence we would tend to accept models as estimated by the data too readily.

Were we to be conscious of the alteration in the error assumptions then we would know that (7.4) should be used and we could attempt to base an estimate of the variance upon it. However, we will next show that since OLS is not the smallest variance linear unbiased estimator, we would abandon it and look for an estimator whose variance is genuinely less than (7.4).

We proceed as if attempting to prove the Gauss–Markov theorem: consider the OLS estimator

$$\hat{\beta} = \Sigma w_t Y_t \text{ where } w_t = X_t / \Sigma X_t^2. \quad (7.6)$$

The variance of this estimator is

$$\operatorname{Var} \hat{\beta} = E\{(\Sigma w_t u_t)^2\} \tag{7.7}$$

$$\neq \sigma^2 \Sigma w_t^2$$

(unlike the case where the original error properties hold). Now consider any other linear unbiased estimator:

$$\beta^+ = \Sigma(w_t + c_t) Y_t \tag{7.8}$$

where for unbiasedness the following restriction holds:

$$\Sigma c_t X_t = 0 \tag{7.9}$$

$$\therefore \quad \Sigma c_t w_t = 0. \tag{7.10}$$

The variance of this general LUE is

$$\operatorname{Var} \beta^+ = E[\{\Sigma(w_t + c_t) U_t\}^2] \tag{7.11}$$

$$= E\{(\Sigma w_t U_t)^2\} + E\{(\Sigma c_t U_t)^2\} + E\{(\Sigma w_t U_t)(\Sigma c_t U_t)\} \tag{7.12}$$

The third term is however not necessarily zero as in the standard case. It has two components:

$$(\Sigma w_t c_t \sigma_t^2) + \left(\sum\sum_{s \neq t} w_t c_s \sigma_{ts}\right) \tag{7.13}$$

—the first is no longer zero from (7.10) because of the presence of σ_t^2 inside the summation and the second cannot be guaranteed to be zero because the term $\sigma_{ts} \neq 0$ and we have no restriction on the values $w_t c_s$ that makes zero. Indeed by a suitable choice of the c_s we can make part of the term larger and negative and so it could outweigh the positive effect of the second term in (7.12), making the whole expression smaller than the first term, which is the OLS variance. Hence we are not guaranteed that OLS is BLUE.

This result means that we would need to find a BLU estimator rather than working with $\hat{\beta}$ and Var $\hat{\beta}$. In order to be able to find the BLU we shall need to assume knowledge of the terms σ_t and σ_{ts}. Although this may seem a heavy demand, it is what we would need to know in order to evaluate the variance of OLS correctly. In fact the most fruitful way to approach the problem is by considering particularly simple error processes which might underlie the failure of the basic assumptions to hold.

7.1 Heteroskedasticity

We have seen that non-constancy of the error variance disturbs the optimality of OLS and so this defines one distinct class of problems. It is helpful first to consider what sorts of situations might lead to such a phenomenon.

Non-constancy of the error variance, commonly called *heteroskedasticity*, is most often associated with so-called 'cross-section' data. We can imagine that a survey has interviewed a large number of households during the same period and has recorded their total expenditure on all items and their expenditure on food. Empirical experience (Engel's Laws) and standard demand analysis both point to a relationship between those two magnitudes that could be expected to hold across households. However it has often been found that the spread of points about the line is greater (negative and positive) the greater the level of total expenditure. Figure 7.1 illustrates such a scatter. Around the fitted line the points spread out more as X increases. It seems as if the households have more flexibility to depart from a relationship as their total resource increases even though *on average* their behaviour is the same. Even more dramatic versions of this picture are obtained with industry cross-section data where the largest firms in an industry may be more than one hundred times as large as the smallest.

A second but much less commonly used example is in a 'time series' context where there is a sequence of observations over time on a single agent (e.g. a firm or household). If there is some form of learning process involved then the agent itself may effectively reduce the error variance over time.

A third case of heteroskedasticity is where there is a structural shift in the error variance (as well as possibly in the coefficient on X). Pre- and post-war or pre- and post-first oil shock the size of the error variance may be very different. This type of structural shift, as opposed to the parameter shifts analysed in Chapter 5, cannot be handled by just including extra (dummy) variables.

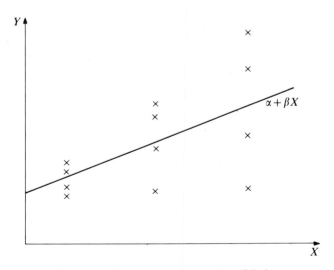

FIG. 7.1   Heteroskedastic Scatter of Points

To analyse the problem of heteroskedasticity in detail we continue to assume that the errors are pairwise independent so that we have:

(i) $\qquad E(U_t) = 0 \quad \text{all } t$

(iia) $\qquad E(U_t^2) = \sigma_t^2$

(iii) $\qquad E(U_t U_s) = 0 \quad s \neq t$

(iv) $\qquad X_t$ fixed in repeated samples

(v) $\qquad$ the explanatory variables are not linearly dependent.

We consider the basic model with an intercept:

$$Y_t = \alpha + \beta X_t + U_t. \tag{7.14}$$

As we have already shown the OLS estimator

$$\hat{\beta} = \frac{\Sigma x_t y_t}{\Sigma x_t^2} \tag{7.15}$$

has variance

$$\text{Var } \hat{\beta} = \Sigma \sigma_t^2 x_t^2 / (\Sigma x_t^2)^2 \tag{7.16}$$

which is no longer BLUE (noting that the presence of the intercept, as usual, puts (7.4) into deviation form).

Our problem is that if we demand the estimator to be linear (in Y) unbiased then the structure (7.14) does not have 'well-behaved' error terms. However, the generalization of the Gauss–Markov theorem by Aitken suggests what we should attempt to do. If we can find a linear (in Y) transform of the equation which produces error terms which are well-behaved, then OLS applied to the transformed variables will be best linear (in the transformed dependent variable) unbiased. However, the set of all linear combinations of linear transformations[1] of Y is the same as the set of all linear combinations of Y, and thus the new estimator is best in a class that includes OLS as a member. This strategy, which is remarkably simple in design, does however depend on knowledge of the error process. Without knowing the σ_t there is no way to transform the errors so that they become homoskedastic. As a first approach to the problem we assume that we know all the σ_t. By *inspection* we can see that if

$$E(U_t^2) = \sigma_t^2 \tag{7.17}$$

then the expected value of the transformed error

$$U_t / \sqrt{\sigma_t} = U_t^* \tag{7.18}$$

---

[1] Providing that the transformation is applied to all the Y_t and that the transformed variables are independent.

has the desired homoskedasticity property

$$E(U_t^{*2}) = 1 \quad \text{for all } t. \tag{7.19}$$

Dividing by the square root of the variance inflation factor produces constant variance. We must however check that this transformation does not disturb either of the other error properties. By inspection we can see that if σ_t is fixed in repeated samples then

$$E(U_t^*) = 0 \tag{7.20}$$

$$E(U_t^* U_s^*) = 0 \quad s \neq t. \tag{7.21}$$

Hence applying this transformation to (7.14) we have

$$\frac{Y_t}{\sqrt{\sigma_t}} = \alpha \cdot \frac{1}{\sqrt{\sigma_t}} + \beta \cdot \frac{X_t}{\sqrt{\sigma_t}} + \frac{U_t}{\sqrt{\sigma_t}} \tag{7.22}$$

or
$$Y_t^* = \alpha X_{0t}^* + \beta X_{1t}^* + U_t^*$$

where
$$X_{0t}^* = 1/\sqrt{\sigma_t}$$

$$X_{1t}^* = X_t/\sqrt{\sigma_t} \text{ etc.} \tag{7.23}$$

This is a well-behaved equation, so that OLS applied to (7.23) would be BLUE (in Y^*). The equation has changed form, from a model with an intercept and one variable, to a model with no intercept and two variables so that we have:

$$\hat{\hat{\beta}} = \frac{\Sigma Y^* X_1^* \Sigma X_0^{*2} - \Sigma Y^* X_0^* \Sigma X_0^* X_1^*}{\Sigma X_0^{*2} \Sigma X_1^{*2} - (\Sigma X_0^* X_1^*)^2} \tag{7.24}$$

$$\text{Var } \hat{\hat{\beta}} = \frac{\Sigma X_0^{*2}}{\Sigma X_0^{*2} \Sigma X_1^{*2} - (\Sigma X_0^* X_1^*)} \tag{7.25}$$

and similar estimators for α (remembering that the new error variance is unity). These *generalized least squares* estimators (GLS) are obviously BLUE in Y^*, by applying the Gauss–Markov theorem to (7.23) and using the properties of the transformed errors. It is obvious also that Y_t^* is linear in Y_t, so that since $\hat{\hat{\beta}}$ is the best of all estimators of the form

$$\beta^+ = \Sigma v_t Y_t^*$$

this is the same set as all estimators of the form

$$\beta^\circ = \Sigma w_t Y_t.$$

Any set of w_t that generate an estimator β° can be recreated by setting $v_t = w_t \sqrt{\sigma_t}$, and vice-versa, so that there is a 'one-to-one' correspondence between the two sets of estimators. GLS is thus also best linear (in Y) unbiased and as such is better than OLS which is a member of the same class.

182 Generalized Least Squares

These optimality properties are subject to the several qualifications, some of which are the same as for OLS in Chapter 3:
  (i) the estimator is of the class which is unbiased whatever the data or true parameter values;
  (ii) the error properties are sufficient for GLS to be BLUE;
  (iii) it is necessary to know the true values of the error variances in order to use GLS; in fact we need to know the error variances only up to a proportionality factor, i.e. we can normalize on any individual variance, such as that of the first observation.

Once the optimality of GLS has been established we can consider some practical details of its usage. First we need to estimate the variance of the estimator and we do this in the usual fashion:

$$\widehat{\text{Var } \hat{\hat{\beta}}} = \frac{\Sigma X_0^{*2}}{\Sigma X_0^{*2} \Sigma X_1^{*2} - (\Sigma X_1^* X_0^*)^2}. \tag{7.26}$$

If we had normalized on any individual variance (say the first) then this would appear in the numerator of (7.26) and we should estimate it from the residuals from the transformed equation:

$$\hat{\sigma}^2 = \frac{1}{T-2} \Sigma U_t^{*2} = \frac{1}{T-2} \Sigma (Y_t^* - \hat{\hat{\alpha}} X_{0t}^* - \hat{\hat{\beta}} X_{1t}^*)^2. \tag{7.27}$$

For the purposes of actually seeing how well the basic data is explained by the model we must work with the untransformed model (GLS is merely a technique of estimation). Hence the 'fitted values' are

$$\hat{\hat{Y}}_t = \hat{\hat{\alpha}} + \hat{\hat{\beta}} X_t \tag{7.28}$$

with
$$\hat{\hat{U}}_t = (Y_t - \hat{\hat{Y}}_t) \tag{7.29}$$

The derived $\tilde{R}^2$ is given by:

$$\tilde{R}^2 = (\Sigma y_t \hat{\hat{y}}_t)^2 / (\Sigma y_t^2 \Sigma \hat{\hat{y}}_t^2). \tag{7.30}$$

It should be noted that, if we attempted to base an RSSQ on (7.29) and from that construct a correlation from $(1 - \text{RSSQ/TSSQ})$, there is no longer any guarantee that the value will not be negative.

Certain special models for the heteroskedastic process have particularly simple forms. Consider the case:

$$\sigma_t = X_t^\gamma \tag{7.31}$$

where the degree of error spread is a direct function of the X values themselves (linked by a single parameter). The transformed equation becomes:

$$\hat{\hat{\beta}} = \frac{\Sigma(Y_t X_t / X_t^\gamma) \Sigma(1/X_t^\gamma) - \Sigma(Y_t / X_t^\gamma) \Sigma(X_t / X_t^\gamma)}{\Sigma(1/X_t^\gamma) \Sigma(X_t^2 / X_t^\gamma) - (\Sigma X_t / X_t^\gamma)^2} \tag{7.32}$$

which simplifies further for certain values of γ, e.g. for $\gamma = 2$ the estimator becomes:
$$\sigma_t = X_t^2 \tag{7.33}$$

$$\hat{\hat{\beta}} = \frac{\Sigma(Y/X)\Sigma(1/X^2) - \Sigma(Y/X^2)\Sigma(1/X)}{T\Sigma(1/X^2) - (\Sigma 1/X)^2}. \tag{7.34}$$

However by referring to the transformed equation

$$\frac{Y_t}{X_t} = \frac{\alpha}{X_t} + \beta + U_t^*. \tag{7.35}$$

we see that the GLS estimator corresponds to that which would be obtained in the single variable $(1/X)$ plus intercept model, so that

$$\hat{\hat{\alpha}} = \frac{\Sigma\left(\frac{y}{x}\right)\left(\frac{1}{x}\right)}{\Sigma\left(\frac{1}{x}\right)^2} \tag{7.36}$$

$$\hat{\hat{\beta}} = \overline{\left(\frac{Y}{X}\right)} - \hat{\hat{\alpha}}\overline{\left(\frac{1}{X}\right)} \tag{7.37}$$

(where deviations and means are for the transformed variables). This last example serves to show that if there is an intercept term in the basic equation, then in the GLS estimating form there should be no intercept, unless the heteroskedasticity is of the special form (7.33).

As well as relating the σ_t to an included exogenous variable it is quite common to relate it to another variable not appearing in the model but which is acceptable as a measure of the 'size' of the observation. Clearly relating σ_t to the variable Y would raise serious problems for the estimator's properties. The general form of GLS would no longer be linear in Y, the proof of unbiasedness could not stand since the σ_t would be random variables, and the optimality of the variance would not therefore be established by the Gauss–Markov theorem. If we had a size variable Z which was fixed in repeated samples and the heteroskedasticity was of the form:

$$\sigma_t^2 = AZ_t^\gamma \tag{7.38}$$

then if we knew A and γ we could define the deflation factor as $\sqrt{(AZ_t^\gamma)}$ and apply GLS as in (7.24). In fact since $\sqrt{A}$ is constant over all observations it would not be necessary to include it in the deflation; and hence if it were unknown (but γ were known) it could be estimated from the residuals from GLS (which would have true variance A).

Significance tests on the parameters can also be applied to the GLS equation. Assume that the model is of the form (say)

$$Y_t = \alpha + \beta X_t + U_t$$

with $E(U_t^2) = \sigma_t^2$ where σ_t is known. The null hypothesis could be of the form

$$H_0: \beta = 0$$
$$H_1: \beta \neq 0$$
(7.39)

(these being conditional on the model having an intercept and heteroskedastic errors). The errors U_t are also assumed to be normally distributed and to be independent with zero mean (but differing variances). Hence by standard properties of normal distributions the transformed errors of the Aitken GLS model are independently identically normally distributed and estimated sums of squares based on these errors follow standard distributions. Hence an F statistic based on the GLS equation in restricted and unrestricted form would have the distribution claimed for it, if the null hypothesis were true. In the particular case cited above the unrestricted estimators are obtained from regressing (Y_t/σ_t) on $(1/\sigma_t)$ and (X_t/σ_t) and the restricted from regressing (Y_t/σ_t) on $(1/\sigma_t)$. Generalizations to more complex null hypotheses follow analogously.

The final but very important estimation problem, which we will face in practice, is that we do not know the exact values of the σ_t (the different variances). These also will need to be estimated from the data. The extreme case would be one where we had no information as to the relationship between the σ_t but knew merely that they were different. In such a case there are effectively too many unknowns for the data—there being T of the σ_t plus α and β as against T observations. Clearly it is necessary to find a way of reducing this imbalance. As the discussion has suggested the way is to find the model which describes the σ_t. From considerations of economic theory we may be able, for instance, to argue that the error variance is a function of the size of a given measurable variable Z_t (which might be the variable X itself) and here only the parameters of the relationship are unknown. A typical form is

$$\sigma_t = f(Z_t).$$
(7.40)

The commonest function which provides flexibility is of the form

$$\sigma_t = Z_t^\gamma.$$
(7.41)

Thus, if we know a few values of σ_t, we could find γ by relating σ_t to Z_t in this exact equation. In practice we do not know σ_t and so we cannot hope to find γ directly and hence to be able to use GLS. There are several approaches to dealing with this problem, all similar in intention but different in execution. The first approach is to substitute (7.41) into the Aitken-transformed equation:

$$\frac{Y_t}{X_t^{\gamma/2}} = \alpha\left(\frac{1}{X_t^{\gamma/2}}\right) + \beta\left(\frac{X_t}{X_t^{\gamma/2}}\right) + \left(\frac{U_t}{X_t^{\gamma/2}}\right).$$
(7.42)

This can be used to form a sum of squared (transformed) residuals as a function of the three parameters α, β, and γ. Because the parameters no longer enter just additively as before there is no simple analytic solution to the first-order derivatives. However, there are well-established approximation routines. The use of such techniques is known as *non-linear least squares*. This is also a problem that lends itself to maximum likelihood estimation with estimators that will be consistent.

A second approach is that of a *search procedure*. We notice that if we knew γ then α and β could be estimated by OLS. Hence we delimit a range in which γ could conceivably fall (e.g. 0 to 4) and then try, over a large number of values of $\tilde{\gamma}$ within the range, to find the estimates $(\hat{\hat{\alpha}}, \hat{\hat{\beta}})$ which minimize the SEE (residual sum of squares) in the transformed equation (7.42). In practice the choice of the $\tilde{\gamma}$ can be made in such a way as to reduce the number of computations. We might first try all values of $\tilde{\gamma}$ from 0 to 4 in increments of 0.50. This might yield a graph of the estimate SEE as in figure 7.2 ($\tilde{\gamma}=0$ is OLS). We see that the function SEE ($\tilde{\gamma}$) appears to have a single local minimum (this is usually but not always the case) somewhere between 1.5 and 2.5. This interval could then be subdivided, say by increments of 0.1, and the procedure repeated. This can be iterated by finer subdivisions to obtain an estimated minimum of the SEE and hence its associated $\hat{\hat{\alpha}}$ and $\hat{\hat{\beta}}$, to the required number of significant figures.

A third procedure attempts to avoid the laborious searching over the whole range of values of γ by iterating in such a way that every step is an improvement on the previous step. We obtain a first estimate of $(\hat{\alpha}, \hat{\beta})$ by assuming $\gamma=0$. These are unbiased but inefficient estimators. Given these

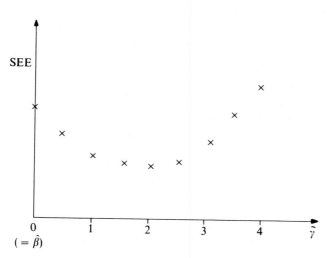

FIG. 7.2 Plot of estimated SEE Versus $\tilde{\gamma}$

values we can calculate the estimated residuals $\hat{U}_t$:

$$\hat{U}_t = Y_t - \hat{\alpha} - \hat{\beta} X_t.$$

Since the true U_t are related to σ_t we can attempt to estimate this relation using the imperfect $\hat{U}_t$. We now write the relation (7.41) in a stochastic form

$$U_t^2 = \sigma_t^2 e^{vt} \qquad (7.43)$$

where v_t is a random variable such that $E(e^{vt}) = 1$. Replacing U_t by $\hat{U}_t$, σ_t by Z_t^γ, and taking logs we have:

$$\log(\hat{U}_t^2) = \log \sigma^2 + \gamma \log Z_t + \delta_t$$

where δ includes both v_t and the measurement error between U_t and $\hat{U}_t$. We regress $\log(\hat{U}_t^2)$ on $\log Z_t$ to obtain an improved (over 0) estimate of γ (call this γ_2). Using this new estimate of γ we go back to the transformed structural equation and obtain the Aitken estimator. These new values $\hat{\alpha}_2$, $\hat{\beta}_2$ could then be used to obtain revised estimates of the structural equation errors U_t. The process can be iterated until two successive sets of estimates of α and β are effectively the same (within whatever tolerance limit is desired). This procedure tends to converge rapidly and may save in computing time. However, if there is more than one local minimum it may not always find its lower of the two while the search procedure is less likely to miss it.

We have sketched the various methods of estimation when the form, but not the values of parameters, of the heteroskedastic process is known. The question that remains is, of course, what are the statistical properties of these techniques. The strong properties of the Gauss–Markov theorem no longer hold and indeed in small samples such estimators can even be biased (whereas OLS is unbiased). However, as the sample size increases these estimators are *consistent*. This property, as we explained in Chapter 6, means that in large samples (the expected value of the estimator will tend to the correct value and the variance of the estimator will tend to zero) so that we can be increasingly confident that the estimator is near the true value. Hence there is a reasonable expectation that, unless the sample size is very small or the data very peculiar, the use of an approximation to GLS will be superior to OLS. A distinct advantage of using these techniques is that an approximation to the variance can be calculated (from the Aitken equation) whereas without an estimate of the σ_t the OLS *true* variance could not be correctly estimated (the conventionally calculated variance being incorrect as we saw earlier). This in turn allows us to carry out significance tests which will be approximately of the size claimed. In cross-section studies, where the problem of heteroskedasticity is often found, the sample size is usually so large that the gains from using an approximation to GLS are always worthwhile.

In most practical circumstances we do not know whether the errors are homoskedastic or heteroskedastic, and it is then important to be able to test for the presence of heteroskedasticity. Many tests have been suggested but we present two very useful approaches. The first test (F test) has as its virtue that it does not make any assumption about the nature of the heteroskedasticity if it is present, while the second test (*White test*) does make an assumption about the general nature of the heteroskedasticity if present.

(a) *F test for heteroskedasticity*

Let us consider the basic model

$$Y_t = \beta X_t + U_t \qquad (7.45)$$

where the usual assumptions hold and the errors are normally distributed. The null hypothesis is that of homoskedasticity

$$H_0 : E(U_t^2) = \text{constant all } t$$
$$H_1 : E(U_t^2) \neq \text{constant all } t. \qquad (7.46)$$

The alternative hypothesis is simply that the errors are not homoskedastic, but it says nothing about the nature of the heteroskedasticity if present. The test methodology is based on the approach of obtaining two independent estimates of the error variance which will have the same expectation if the errors are homoskedastic and using Fisher's F distribution to see whether the two estimates are significantly different. If they are unusually different then we conclude that the original errors are not homoskedastic and that this is why the two estimated variances are so dissimilar. The sole problem with this approach is to obtain the two independent estimates of the error variance. The technique used is simply to split the data into two groups and carry out a separate regression for each group and then from those regressions estimate the error variance in the usual fashion. Suppose we split the data at point τ giving

$$Y_t = \beta X_t + U_t \qquad 1 \ldots \tau \qquad (7.47a)$$
$$Y_t = \beta X_t + U_t \qquad \tau+1 \ldots T. \qquad (7.47b)$$

We then obtain the separate OLS estimators

$$\hat{\beta}_1 = \sum_1^\tau Y_t X_t \Big/ \sum_1^\tau X_t^2 \qquad (7.48a)$$

$$\hat{\beta}_2 = \sum_{\tau+1}^T Y_t X_t \Big/ \sum_{\tau+1}^T X_t^2 \qquad (7.48b)$$

and from these we obtain the residuals and the estimated variances.

$$\hat{\sigma}_1^2 = \sum_1^\tau (Y_t - \hat{\beta}_1 X_t)^2 / \tau - 1 \qquad (7.49a)$$

$$\hat{\sigma}_2^2 = \sum_{\tau+1}^T (Y_t - \hat{\beta} X_t)^2 / T - \tau - 1 \qquad (7.49b)$$

where the usual correction for degrees of freedom is made to each group in order to obtain unbiased estimations of the error variances. Under the null hypothesis the associated sums of squares are both based on the same true variance so that the F statistic

$$F = \frac{(\text{RSSQ}_1)/\text{d.f.}_1}{(\text{RSSQ}_2)/\text{d.f.}_2} \qquad (7.50)$$

where RSSQ_i is the residual sum of squares for the ith group, and d.f._i its degrees of freedom, is distributed according to Fisher's F, i.e. as $F(\text{d.f.}_1, \text{d.f.}_2)$. Thus we have:

$$\frac{\hat{\sigma}_1^2}{\hat{\sigma}_1^2} \sim F(\tau - 1, T - \tau - 1). \qquad (7.51)$$

There are several important practical points to be made with the use of the conventional F test.

1. There is no need for the ratio to indicate a failure of the hypothesis only in the right-hand tail of the distribution. The tests we have so far considered for linear restrictions will tend to yield increasingly large positive values of F the more the data diverges from the restriction. Very low values of F do not indicate that the restriction is more likely to be incorrect. Hence to achieve the maximum power for the test of a given size we used the right-hand tail to test the correctness of restrictions. In the heteroskedasticity case it is obvious that either sub-period could have the larger variance if there were heteroskedasticity and so both very large and very small values of F warn us of the likely departure from homoskedasticity. Accordingly we should use a two-tail test with (usually) 2.5 per cent of the test in each tail. Conventional F tables normally give only the upper tail critical value so we need to be able to find the value for the lower tail. To do this we use a classical statistical result that if a statistic θ follows an $F(n, m)$ distribution, then $1/\theta$ follows an $F(m, n)$ distribution. This allows us to obtain the right-hand tail value for $(1/\theta)$ which then yields the left-hand tail value for θ. We illustrate this below.

2. The choice of the point of sample separation τ is free (providing each subgroup has enough degrees of freedom to permit estimation). If we have some ideas about the nature of the heteroskedasticity that might be present (either from graphical analysis of the data or prior theory) then it will be sensible to choose a value of τ such that the estimated error variances will be as different as possible if heteroskedasticity were present, in order to maximize the power of the text. For example, if the variance is thought to be possibly a function of the size of the independent variable we could rearrange the data into two groups, one containing the larger values of X and the other the smaller. The numbers in each group are usually chosen to be the same but need not be. An even more extreme version of this procedure, designed to accentuate any estimated differences in variances, is to omit a number of observations on the middle-range values of X and compare regressions based

on (say) the largest one-third of the X values and the smallest one-third of the X values. The increase in power of this procedure (amplifying the likely chance of finding heteroskedasticity if present), more than compensates for the loss of overall degrees of freedom especially if the total sample size is large.

3. This test can be applied also when there is a possibility that the parameter β changes value between sub-samples. The separate estimates of β allow separately estimated error variances to be unbiased under the null hypothesis whether or not β is constant.

4. It might appear desirable, when the value of β is known not to shift, to wish to impose the restriction of constancy by using a single regression for the whole sample, but calculating two separate variance estimates, one based on the first τ residuals and the other based on the last $(T-\tau)$ residuals. However, the two resulting variance estimates will not then be independent since, even when there is no serial correlation or heteroskedasticity in the true residuals, the OLS residuals are serially correlated and heteroskedastic (see problem 3.7). Hence exact tests cannot be based on the residuals from a single regression.

(b) White test for heteroskedasticity

This test does make an explicit assumption about the nature of the heteroskedasticity and then checks to see whether the data actually show signs of such a relationship. If there is no evidence that they do so then we conclude that the errors are homoskedastic. The idea is best illustrated with a two-variable model

$$Y_t = \beta_1 X_{1t} + \beta_2 X_{2t} + U_t \qquad (7.52)$$

We consider that the variance of U_t, if not constant, could be a function of a general quadratic in the X variables, i.e.

$$E(U_t^2) = A_0 + A_{11} X_{1t}^2 + A_{12} X_{1t} X_{2t} + A_{22} X_{2t}^2. \qquad (7.53)$$

The test procedure is to carry out OLS on the basic equation (7.52) and obtain the residuals $\hat{U}_t$. We then construct the auxiliary regression of $\hat{U}_t$ on all the products and cross-products to yield $\hat{A}_0$, $\hat{A}_{11}$, $\hat{A}_{12}$ and $\hat{A}_{22}$. If there is no heteroskedasticity only $\hat{A}_0$ should be significant. It can then be shown that, given the value of the multiple correlation R^2 from the auxiliary equation, in large samples the statistic $TR^2 \sim \chi^2(3)$. There are in this case three degrees of freedom to the Chi square since there are three parameters (restrictions) imposed in the homoskedastic model. The technique has an obvious relation to LM tests in that we estimate the restricted form (no heteroskedasticity) and regress the restricted squared residuals on the full range of variables that would have been in the error term if the restriction were incorrect.

Example 7.1. Tests for heteroskedasticity

We begin with the single-variable regression of consumption on income and an intercept for the period 1961 to 1982. The first alternative hypothesis is that there is an increase in the error variance after 1972 corresponding to the greater volatility of the post-oil shock world. This is tested by the F test procedure based on the RSSQ from two separate regressions. The results are shown in table 7.1. From the SEE we can recover the RSSQ/d.f. simply by squaring. Hence the F statistic for variance equality is $F(10, 8) = 0.378$. Under the alternative hypothesis the left-hand tail of the distribution will be the rejection region (large values of the statistic would not be accepted as evidence that the variance in the second sub-period was the larger). The standard F tables however give only the right-hand tail figures for critical values so that we need to invert the above statistic to give $F(8, 10) = 2.645$, where the critical right-hand tail value, at a 95 per cent significance level test, is 3.07. Hence we accept the null hypothesis that the error variance does not increase post-1972.

The second test postulates that, if there is heteroskedasticity present, the variance of the errors is related to the square of income. We can carry out a White test based on the LM approach. The residuals are obtained from the basic OLS regression of consumption on income and an intercept (see chapter 2) and are squared. The squared residuals are then regressed on a constant and a squared term in income (here income was scaled down by a factor of 1000 so that its square should be manageable). The multiple R^2 in this regression is 0.036 so that the test statistic is $TR^2 = 0.792$. The large sample distribution of this statistic is $\chi^2(1)$ with the single non-constant factor relating to the error variance. The critical value of $\chi^2(1)$ with a 95 per cent significance level is 3.841. Hence we reject the AH that the error variance is relevant to the square of income and thus maintain the null hypothesis of homoskedasticity.

TABLE 7.1 Results for Two-sample Test for Variance Equality

Period	Intercept	Coefficient on Income	SEE
1961–1972	3375	0.887	976
	(3171)	(0.028)	
1973–1982	21092	0.737	1587
	(9852)	(0.066)	

7.2 Serial Correlation

The breakdown of the Gauss–Markov assumptions can also come from the lack of pairwise independence between the error terms. Because this is most often encountered with data ordered in a temporal sequence the problem is usually referred to as 'serial correlation'. The cause of such correlation is often taken to be some omitted factor which itself is serially correlated.

We can analyse the model for the error assumptions:

(i) $\qquad E(U_t) = 0 \qquad$ all t

(ii) $\qquad E(U_t^2) = \sigma^2 \qquad$ all t

(iiia) $\qquad E(U_t U_s) = \sigma_{ts} \neq 0 \qquad$ some $t = s$.

Again we take the simple model:

$$Y_t = \alpha + \beta X_t + U_t \qquad (7.55)$$

The OLS estimator $\hat{\beta} = \Sigma x_t y_t / \Sigma x_t^2$ has variance:

$$\text{Var } \hat{\beta} = \frac{\sigma^2}{\Sigma x_t^2} + \frac{\overset{t \neq s}{\Sigma \Sigma} \sigma_{ts} x_t x_s}{(\Sigma x_t^2)^2} \qquad (7.56)$$

which is again no longer BLUE. Attempts to estimate the variance by the conventional formula (based on the first term alone) will be misleading and can lead to underestimation and incorrect inferences.

We have seen how Aitken's technique can generate a BLUE estimator if we can find a linear transform that has well-behaved error terms. Again, in order to make the transformation, we must know the exact process generating the errors. The commonest and easiest process analysed is the first-order Markov process (or autoregressive process—$AR(1)$).

$$U_t = \rho U_{t-1} + \varepsilon_t \qquad (7.57)$$

where $E(\varepsilon_t) = 0$, $E(\varepsilon_t^2) = \sigma^2$ all t, $E(\varepsilon_t \varepsilon_s) = 0$ $s \neq t$, and $|\rho| < 1$ with ρ known. Such a process can be interpreted as having the total error in the current period (U_t) built up out of a fraction (ρ) of yesterday's (U_{t-1}) error plus a new component (ε_t). By repeated substitution we have:

$$U_t = \varepsilon_t + \rho \varepsilon_{t-1} + \rho^2 \varepsilon_{t-2} \ldots \qquad (7.58)$$

This confirms that $E(U_t) = 0$ if $E(\varepsilon_t) = 0$. It also allows the variance and autocovariances of the U_t to be expressed in terms of the variance of the $\varepsilon(\sigma^2)$ and the autocorrelation parameter (ρ). We have from (7.58) using the assumptions on the ε_t:

$$E(U_t^2) = \sigma^2 + \rho^2 \sigma^2 + \rho^4 \sigma^2 \ldots \qquad (7.59)$$

Now, provided that $|\rho| < 1$, we can express this as

$$E(U_t^2) = \frac{\sigma^2}{1-\rho^2} \qquad (7.60)$$

similarly

$$E(U_t U_{t-\tau}) = \frac{\rho^\tau \sigma^2}{1-\rho^2}. \qquad (7.61)$$

The correlation between pairs of the errors decreases steadily the further apart are the errors. These expressions allow us to evaluate the σ_{ts} and we could, if ρ were known, use them to calculate the true variance of OLS from (7.56). However since GLS will have a smaller variance than this, we can immediately turn to the problem of finding the linear transform of (7.55) that will make the errors well behaved. The key here is equation (7.57). Since the ε_t are well behaved we need to transform (7.55) so that the errors become the ε_t (equivalently $U_t - \rho U_{t-1}$). This is done by first taking the lagged relation

$$Y_{t-1} = \alpha + \beta X_{t-1} + U_{t-1}$$

multiplying it by ρ to give:

$$\rho Y_{t-1} = \alpha \rho + \beta \rho X_{t-1} + \rho U_{t-1}$$

and subtracting this from the basic equation to give:

$$(Y_t - \rho Y_{t-1}) = \alpha(1-\rho) + \beta(X_t - \rho X_{t-1}) + \varepsilon_t. \qquad (7.62)$$

This equation is just as correct as the equation (7.55). Such a transform can be applied from all t ranging from 2 to T to give a set of $(T-1)$ transformed variables:

$$Y_t^* = Y_t - \rho Y_{t-1}$$
$$X_t^* = X_t - \rho X_{t-1}. \qquad t = 2 \ldots T \qquad (7.63)$$

It might appear that the OLS regression of Y_t^* on X_t^* would yield BLUE estimates of β and $\alpha(1-\rho)$ since the error terms are well behaved. Indeed this is so far a particular class—of all estimators of the form

$$\beta^\circ = \sum_2^T v_t Y_t^* \qquad (7.64)$$

this is the smallest variance. It is also linear in Y_t because Y_t^* is itself a linear function of Y_t. However the set of estimators (7.64) is not the same as the set of all

$$\beta^* = \sum_1^T w_t Y_t \qquad (7.65)$$

i.e. every estimator of the form β° is a member of the set β^* but not vice-versa. This is easy to see—the coefficient on Y_1 in the former set is not free to vary but is always tied to that on Y_2, since the only way Y_1 enters the estimator is through the first observation: $Y_2^* = Y_2 - \rho Y_1$; the coefficient on Y_1 in the second set is not necessarily tied to those of other Y_t and hence the first set is a subset of the second set of estimators. Thus although OLS on the $T-1$ values Y_t^* is BLUE in the former set, there could be an estimator in the second set, different from basic OLS, which was better than both. To obtain the full efficiency of GLS it is clearly necessary to bring in an extra observation to the transformed data which allows independence for the first observation. We can see how to do this by considering the transformed observations for $T \geq 2$ and the untransformed for $T=1$:

$$Y_T - \rho Y_{T-1} = \alpha(1-\rho) + \beta(X_T - \rho X_{T-1}) + \varepsilon_T$$

$$\cdot \quad \cdot \quad \cdot \quad \cdot \quad \cdot$$

$$Y_2 - \rho Y_1 = \alpha(1-\rho) + \beta(X_2 - \rho X_1) + \varepsilon_2$$

$$Y_1 = \alpha(1) + \beta(X_1) + U_1. \tag{7.66}$$

Now the first untransformed observation has an error which is actually independent of all the other transformed errors—each U_t is made up of current and previous shocks (ε_t) but is independent of shocks that are later in time. However, although we have fully corrected for serial dependences we have induced heteroskedasticity. The variance of all the errors from $2 \ldots T$ is σ^2 while that of the first observation is $\sigma^2/(1-\rho^2)$. However, we know how to correct for heteroskedasticity—multiply both sides of the first observation equation by the factor $\sqrt{(1-\rho^2)}$. The resulting errors now are both serially independent and homoskedastic. The set of T transformed equations is

$$(Y_t - \rho Y_{t-1}) = \alpha(1-\rho) + \beta(X_t - \rho X_{t-1}) + \varepsilon_t \qquad t=2 \ldots T \tag{7.67}$$

$$(Y_t\sqrt{(1-\rho^2)}) = \alpha\sqrt{(1-\rho^2)} + \beta\sqrt{(1-\rho^2)}X_1 + U_1\sqrt{(1-\rho^2)}.$$

This is now a two-variable expression without a constant—the variables are Y_t^*, X_t^* and X_{0t}^* where

$$X_{0t}^* = 1 - \rho \qquad t=2 \ldots T \tag{7.68}$$

$$= \sqrt{(1-\rho^2)} \qquad t=1.$$

Applying OLS to this transformed set of data yields the BLUE estimator in the class

$$\beta^\circ = \sum_1^T v_t Y_t^*. \tag{7.69}$$

However since Y_t^* are linear in Y_t this class is identical with the class

$$\beta^* = \sum_1^T w_t Y_t \tag{7.65}$$

(i.e. for any set of v_t there is a set of w_t which will produce the same estimator and vice-versa) so that the Aitken GLS estimator is best linear (in Y) unbiased and hence superior to OLS. In practice, when there are a large number of observations, the gain from correcting the first observation is likely to be very small and the 'weighted difference' transform of the $(T-1)$ observations is utilized (often this is called the Cochrane–Orcutt transform).

These ideas can easily be extended to higher-order autoregressive schemes. For example, if we had quarterly data we might find that the errors were correlated only between like quarters

$$U_t = \rho U_{t-4} + \varepsilon_t \tag{7.70}$$

yielding the transform:

$$Y_t^* = Y_t - \rho Y_{t-4}. \tag{7.71}$$

There might be mixture of both first and second order

$$U_t = \rho_1 U_{t-1} + \rho_2 U_{t-2} + \varepsilon_t \tag{7.72}$$

yielding:

$$Y_t^* = Y_t - \rho_1 Y_{t-1} - \rho_2 Y_{t-2} \tag{7.73}$$

etc. Thus for autoregressive process the optimal estimator can be found if the autocorrelation parameters are known. It also follows that we can carry out significance tests on the parameters (α, β) if the ε_t are known to be normally distributed. We carry out GLS on both the restricted and unrestricted equations and use the F statistic in the usual way. Finally the generalization to the many-variable case is straightforward—each X variable is subject to the same weighted differencing.

As with heteroskedasticity we are faced with the fact that even if we know that the error process is first-order autoregressive it is most unlikely that we will know the value of ρ, so that we need to consider methods of jointly estimating α, β, and ρ.

The methods of approach are exactly the same as the heteroskedasticity—non-linear least squares (ML), a search procedure, and an iterative procedure. We discuss only the third method, that being the most commonly used approach to the problem. Given that we are sure that the error process is AR(1) then we have to estimate the parameter ρ as well as α and β. The first step is to carry out OLS on the untransformed data (equivalent to assuming $\rho = 0$). This allows us to calculate residuals $\hat{U}_t$:

$$\hat{U}_t = Y_t - \hat{\alpha} - \hat{\beta} X_t. \tag{7.74}$$

Given that the true errors are linked by the equation $U_t = \rho U_{t-1} + \varepsilon_t$, we can attempt to estimate ρ from the relation by use of OLS:

$$\hat{U}_t = \rho \hat{U}_{t-1} + \varepsilon_t^* \tag{7.75}$$

(where ε_t^* includes both ε_t and the measurement errors for the U_t). Hence we

have
$$\hat{\rho}_1 = \Sigma \hat{U}_t \hat{U}_{t-1} / \Sigma \hat{U}_{t-1}^2. \tag{7.76}$$

Given this estimate of ρ we can construct an approximation to the GLS transformed variables: $Y_t^* = Y_t - \hat{\rho}_1 Y_{t-1}$ etc., and then apply Aitken's method of using OLS on the transformed variables to obtain new estimates $\hat{\alpha}_2$, $\hat{\beta}_2$. These can in turn be substituted into the basic equation to yield revised estimates of the errors. The process can be iterated until it converges within the limits of tolerance laid down. The final estimators are again consistent but this technique is biased in general for small samples. The parameter variances estimated at the final round of the iteration should also be an accurate guide to the true variance of the estimator if the sample size is substantial, and so significance tests can be based on the results of approximate GLS.

Just as with heteroskedasticity we may be uncertain as to whether the errors are serially correlated and thus require a test to check whether or not we should depart from the use of OLS. The most important test is that developed by Durbin and Watson. This considers the null hypothesis of no serial correlation (against the AH of serial correlation) and uses the statistic:

$$d = \frac{\sum_{2}^{T} (\hat{U}_t - \hat{U}_{t-1})^2}{\sum_{1}^{T} \hat{U}_t^2}. \tag{7.77}$$

This is calculated from the OLS residuals. The important feature of the distribution of this statistic in repeated samples, when the NH is true, is that the distribution depends on the values taken by the X_t—each set of X_t gives a different sample distribution. It is possible to work out the exact distribution for any set of data but this is very complicated. Instead Durbin and Watson were able to prove that for one- or two-tailed tests the critical regions would always fall in a certain range whatever the data. By considering large samples where the $\hat{U}_t$ would tend to be very accurate we can see that if there were no serial correlation:

$$d \approx 2. \tag{7.78}$$

If however there were first-order serial correlation a value near zero would indicate positive serial correlation and a value near four would indicate negative serial correlation. A two-tailed test would be required if we were unsure as to the sign of the possible serial correlation. Given that a size for the test is fixed (say at 5 per cent) then Durbin and Watson were able to find upper and lower bounds for the critical values of d. We illustrate the procedure in figure 7.3. For a set of data with twenty-two observations, with a constant and one other variable, and with a 5 per cent significance test, the value of d_L is 1.24 and the value of d_u is 1.43. This means that *whatever the values taken* by the single X variable no critical value would have been less than d_L or greater than d_u (while less than 2) or greater than 4-d_L or less than

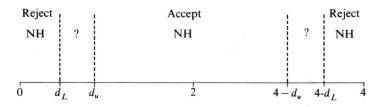

Fig. 7.3 Bounds for a Durbin–Watson Test

$4 - d_u$ (while greater than 2). Hence if the sample value of d is less than d_L or greater than $4 - d_L$ we can be sure that calculation using the exact distribution of d would reject the null hypothesis, while if the value lay between d_u and $4 - d_u$ we can be sure that an exact calculation would lead us to accept the null hypothesis. However, there are two 'inconclusive' regions—if d falls in either of these then we cannot say what would be the result of a test based on the exact distribution for the set of X in question. Fortunately these regions shrink with an increase in the number of observations. Generally a sample value falling in such a region is said to be 'inconclusive' and many practitioners then stick with OLS rather than using an approximate GLS. Durbin and Watson's test of course has the size stated whatever the alternative hypothesis since it is based on the distribution under the null hypothesis. However its power is likely to be high when there is first-order autocorrelation.

However in practical work the form of the serial correlation will often not be of a simple first-order autoregressive type. It may well be of simple higher order—either because of seasonality type considerations (e.g. fourth order with quarterly data) or of a general process including lags of several different lengths. Clearly it is useful to have tests which are able to have good performance in these more complex situations. The key to such testing will be in the specification of the serial correlation if present. Once we can specify the process then we can use the techniques associated with maximum likelihood estimation and inference. Let us consider the single model $Y_t = \beta X_t + U_t$ where there may be first order serial correlation $U_t = \rho_{t-1} + \varepsilon_t$. Using weighted first differences we have:

$$Y_t = \Pi_1 Y_{t-1} + \Pi_2 X_t + \Pi_3 X_{t-1} + \varepsilon_t \tag{7.79}$$

or
$$Y_t = \rho Y_{t-1} + \beta X_t - \beta \rho X_{t-1} + \varepsilon_t. \tag{7.80}$$

The coefficients on Y_{t-1}, X_t, and X_{t-1} are subject to the restriction:

$$\Pi_1 \Pi_2 = -\Pi_3. \tag{7.81}$$

Such a set-up is suitable for maximum likelihood estimation, subject to the non-linear constraint (7.81) which, if the errors are normally distributed, is also non-linear least squares. We can carry out a standard likelihood ratio test by estimating (7.79) with and without the constraint and forming T times the log-likelihood ratio. This will be distributed as $\chi^2(1)$ variable if the restriction is correct (i.e. there is first-order serial correlation.) This approach could be easily generalized when higher-order serial correlation is involved. Similarly a Wald test can be constructed but it requires the variance of the estimated restriction to be calculated, which is not straightforward here. The most useful test is often a form of the LM test—the residuals ($\hat{U}_t$) are taken from the restricted model (here OLS assuming no serial correlation). In the unrestricted case we know that the right-hand side explanatory variables will be X_t and U_{t-1} (ε_t is purely random) and so we regress $\hat{U}_t$ on X_t and $\hat{U}_{t-1}$ to yield a multiple correlation coefficient R^2. The statistic TR^2 is distributed as $\chi^2(1)$ if the restriction (first-order serial correlation) is correct. This test is extremely simple to generalize to many explanatory variables and higher-order serial correlation processes. In the second stage we include all the explanatory variables and as many separate lagged residuals as the order of the possible serial correlation process. The χ^2 statistic has the same number of degrees of freedom as the order of the serial correlation process.

Estimation and testing for serial correlation is given in example 7.2.

Example 7.2: Tests for serial correlation

We consider the single-variable regression of consumption on income and an intercept for the period 1961 to 1982. The basic OLS regression was given in Chapter 2 but we repeat it for convenience.

$$\text{Consumption} = 15750 + 0.775 \text{ Income}$$

$$R^2 = 0.991, \text{DWS} = 0.995.$$

The first test to check the null hypothesis of no serial correlation is the Durbin–Watson test. (Such a test is particularly sensitive to the presence of pure first-order serial correlation.) In our case there are twenty-two observations and one variable as well as the intercept, so that using a 5 per cent two-tail test ($2\frac{1}{2}$ per cent in each tail) we find $d_L = 1.24$ and $d_u = 1.308$. Hence if the actual value of d fell between 1.308 and $(4-1.308 =)2.692$ we would continue to accept the NH of no serial correlation. In fact the actual value (0.99) is less than d_L so that we reject the null hypothesis and note that the value in the left-hand tail suggests that there may be *positive* first-order serial correlation present.

Our next step is to assume that there is an error process of the form $U_t = \rho U_{t-1} + \varepsilon_t$ and to re-estimate the equation taking this into account by generalized least squares. Since ρ is not known it has to be estimated and we report the results of the iterative approach in which the first observation has been omitted from the transformed equation (Cochrane–Orcutt procedure):

$$\text{Consumption} = 20495 + 0.743 \text{ Income}$$
$$(5134) \quad (0.037)$$
$$R^2 = 0.993, \text{ DWS} = 1.62, \text{ SEE} = 1310$$
$$\rho = 0.544.$$

These results do not show a large change in the coefficient of income but the estimated variances are substantially bigger than before. The Durbin–Watson test as applied to the residuals of the transformed equation supports the hypothesis that there is no serial correlation (i.e. after the first-order correction has been made). We can carry out hypothesis tests on this equation by the normal 't' or F test procedure. For example to test the null hypothesis that the coefficient of income is zero we can use the t statistic approach—the t score is 20.28, while the critical value on a one-sided 5 per cent test with 19 degrees of freedom is 1.729. Clearly the null hypothesis is rejected and we continue to find evidence that income has a significant effect on consumption.

As a final testing procedure we consider the possibility that the autoregressive process is second order, i.e. $U_t = \rho_1 U_{t-1} + \rho_2 U_{t-2} + \varepsilon_t$, and apply a large sample LM test. First we obtain the residuals from the restricted equation (OLS with no serial correlation adjustment). These residuals ($\hat{U}$) are then regressed on intercept, income, $\hat{U}_{-1}$ and $\hat{U}_{-2}$ for observations 1963 to 1982. The value of R^2 from this equation is 0.269 so that the LM statistic is $TR^2 = 5.38$. In large samples under the null hypothesis of no serial correlation this statistic would be distributed as $\chi^2(2)$, which has a critical value with a 5 per cent test of 5.99. The hypothesis that the errors follow a second-order Markov process is not supported by the data. In practice it would be necessary to check for the presence of just first-order serial correlation (regressing $\hat{U}$ on the intercept, income, and $\hat{U}_{-1}$) but the DWS test, which is an exact test and not a large sample test, has already indicated that some serial correlation is present.

An important problem arises with the need to forecast from a model where the errors are serially correlated. The use of the GLS based equation for forecasting gives the optimal forecasts as usual. Consider the model $Y_t = \beta X_t + U_t$, $t = 1, \ldots, T$, where $U_t = \rho U_{t-1} + \varepsilon_t$. The parameter ρ is assumed known so that the optimal form for estimation is

$$Y_t^* = \beta X_t^* + U_t^*$$
$$Y_t^* = Y_t - \rho Y_{t-1} \quad \text{for } t \geq 2$$

etc. We have established that the BLUE of β comes from the regression of Y^* on X^*. Now if we require the best linear unbiased predictor of Y^*_{T+1} we can use the ideas of Chapter 3 on forecasting to suggest that the predictor is:

$$\hat{Y}^*_{T+1} = \hat{\hat{\beta}} X^*_{T+1} \tag{7.82}$$

where $\hat{\hat{\beta}}$ is the GLS estimator of β. This can be shown to have the same expectation as Y^*_{T+1} (the actual value of the weighted difference), is linear in Y^*_T, and has the smallest variance of all such predictors:

$$\text{Var } \hat{Y}^*_{T+1} = (X^*_{T+1})^2 \text{ Var } \hat{\hat{\beta}}. \tag{7.83}$$

These formulae need to be converted into the basic data forms:

$$\widehat{(Y_{T+1} - \rho Y_T)}^* = \hat{\hat{\beta}}(X_{T+1} - \rho X_T) \tag{7.84}$$

$$\text{Var } \widehat{Y^*_{T+1}} = (X_{T+1} - \rho X_T)^2 \text{ Var } \hat{\hat{\beta}}. \tag{7.85}$$

Now ρ, Y_T, and X_T are all known so that we can derive the predictor of Y_{T+1} itself:

$$\hat{Y}_{T+1} = \rho Y_T + \hat{\hat{\beta}}(X_{T+1} - \rho X_T) \tag{7.86}$$

$$= \hat{\hat{\beta}} X_{T+1} + \rho(Y_T - \hat{\hat{\beta}} X_T)$$

$$= \hat{\hat{\beta}} X_{T+1} + \rho \hat{U}_T. \tag{7.87}$$

Hence the optimal predictor of Y_{T+1} takes into account not only the actual value of the exogenous variable in the prediction period and the parameter estimate, but also the estimated error term that will occur in the prediction period—$\rho \hat{U}_T$ being an estimate of U_{T+1} by the logic of the autoregressive error process. If the value of ρ is not known it is then replaced by an estimate obtained from the data in the historical period. This process is illustrated in example 7.3.

Example 7.3: Forecasts with serially correlated errors

We wish to forecast from our basic single-variable equation for 1983 and 1984 using the hypothetical values of income of

$$\text{Income (1983)} = 160{,}000$$
$$\text{Income (1984)} = 165{,}000$$

If we had used the original estimation by OLS (not allowing for the first-order serial correlation) we would have obtained:

$$C(1983) = 139{,}750$$
$$C(1984) = 143{,}625$$

(based on $C = 15750 + 0.775Y$). However if we allow for first-order serial correlation the estimated relation was

$$C = 20495 + 0.743\,Y$$
$$U = 0.544\,U_{-1}.$$

The best forecast is obtained either by working with weighted first difference $(C - 0.544 C_{-1})$, etc. or equivalently by adding in the estimated serially correlated error in 1983 and 1984:

$$C_{83} = 20495 + 0.743\,Y_{83} + 0.544\,U_{82}$$

and

$$C_{84} = 20495 + 0.743\,Y_{84} + (0.544)^2 U_{82}.$$

Now in 1982 the estimated residual from the GLS parameters was

$$\hat{U}_{82} = C_{82} - 20495 - 0.743\,Y_{82}$$
$$= 138865 - 20495 - 0.743 \times 155627$$
$$= 2739$$

therefore

$$C_{83} = 140865$$
$$C_{84} = 143901.$$

The forecasts are in fact fairly similar in this case, but had the residual been large in 1982 there could have been a dramatic difference.

7.3 Seemingly unrelated regressions

A particularly useful application of the techniques for correcting the effects of serial correlation and heteroskedasticity occurs in what are known as seemingly unrelated regressions (SUR). Imagine two equations run over the same time-period but linking different equations with different parameter values: a typical example would be data from two different industries on investment and sales in each year. The equation could be written:

$$Y_{1t} = \beta_1 X_{1t} + U_{1t}$$
$$Y_{2t} = \beta_2 X_{2t} + U_{2t} \qquad t = 1 \ldots T. \qquad (7.88)$$

The equations are different if the *values* of the variables are different even though the concepts are the same. A new difficulty can come from the error terms—we can assume that within each equation there is serial independence and heteroskedasticity but we also need to make assumptions about the covariances of errors from the two equations. Often it is not possible or plausible to assume that all covariances are zero. The simplest case of non-zero contemporaneous covariance is

$$E(U_{it}U_{js}) = \sigma_{ij} \text{ for } t = s$$
$$= 0 \text{ for } t \neq s, \quad i = 1, 2 \quad (7.89)$$

where $\quad E(U_{it}) = 0$ all i, t.

It is very plausible that there could be shocks common to the two equations in the same time-period, particularly if the equations relate to similar concepts. The equations are now only seemingly unrelated because, despite a lack of a direct link (e.g. by an assumption of equal parameter values), they are in fact linked by the error covariance. The problem that we face is to derive the optimal estimator. As usual, this depends on the class of estimators that we are prepared to consider. If, for example, we wish to consider all estimators for β_1 of the form

$$\beta_1^* = \sum_1^T Y_{1t} w_t \quad (7.90)$$

then OLS of Y_{1T} on X_{1T} will be BLUE as usual (of estimators linear in Y_1). However, we have two relevant variables Y_1 and Y_2 so that we could consider a more general class of estimators linear in both:

$$\beta_1^* = \sum_1^T w_{1t} Y_{1t} + \sum_1^T w_{2t} Y_{2t}. \quad (7.91)$$

If we related the observations by the index s, $s = t$ for the first equation, $s = T + t$ for the second equation, then the equation (7.88) can be rewritten:

$$Y_s^* = \beta_1(X_{1s}^*) + \beta_2(X_{2s}^*) + U_s^* \quad (7.92)$$

where $s = 1 \ldots 2T$.

$$X_{1s}^* = X_{1s} \quad \text{for} \quad s \leq T$$
$$= 0 \quad \text{for} \quad s > T$$
$$X_{2s}^* = 0 \quad \text{for} \quad s \leq T$$
$$= X_{2s} \quad \text{for} \quad s > T$$
$$Y_s^* = Y_{1s} \quad \text{for} \quad s \leq T$$
$$= Y_{2s} \quad \text{for} \quad s > T$$
$$U_s^* = U_{1s} \quad \text{for} \quad s \leq T$$
$$= U_{2s} \quad \text{for} \quad s > T$$

This is known as 'stacking' the two equations—by introducing two (dummy) variables with zero values for the set of observations corresponding to the equation the variable does not appear in, we keep the exact relation of (7.87) at every point in time. We can now consider the estimation of (7.92). OLS of

Y^* on X_1^* and X_2^* (over $2T$ observations) would in one regression give estimates of both parameters and the estimators are clearly linear in Y^*. This in turn means that they are linear in Y_1 and Y_2 since Y^* is a linear form of the separate Y_s. The properties depend as usual on the error terms. We know that the estimators would be unbiased since $E(U^*)=0$ from the basic assumptions. However, the estimators cannot be BLUE in Y^* since we have in effect assumed that the U^* are not 'well-behaved'. We have two difficulties. First there is pairwise dependence:

$$E(U_s^* U_{s+T}^*) = \sigma_{12} \qquad s=1 \ldots T \qquad (7.93)$$

this being the contemporaneous covariance, arranged as if it were Tth-order serial correlation. Second there is heteroskedasticity:

$$E(U_s^2) = \sigma_{11} \qquad s \leq T \qquad (7.94)$$
$$= \sigma_{22} \qquad s > T$$

The optimality of OLS cannot be guaranteed by the Gauss–Markov theorem under these conditions and instead we need to use Aitken's version of the theorem to guarantee optimality of the resulting estimators. The essence of Aitken's method is that we need to find a linear transformation of the observations which makes the transformed error terms well behaved. The error terms, which are T observations apart in the 'stacked' equation, need to be made independent to get rid of the serial correlation and then transformed by a scale factor to adjust for homoskedasticity (if we adjust for homoskedasticity first, the subsequent transformation for pairwise independence would be likely to reintroduce heteroskedasticity). Since there is effectively Tth-order serial correlation we need to take the weighted Tth difference of the observations:

$$\tilde{U}_s^* = U_s \qquad \text{for} \quad s \leq T$$
$$\tilde{U}_s^* = U_s^* - A U_{s-T}^* \qquad \text{for} \quad s > T \qquad (7.95)$$

For independence we need

$$E(\tilde{U}_s^* \tilde{U}_{s-t}^*) = 0 \qquad \text{for} \quad s > T \qquad (7.96)$$

$$\therefore \quad E(U_s^* - A U_{s-T}^*)(U_{s-T}^*) = 0$$

$$\therefore \quad \sigma_{12} - A \sigma_{11} = 0$$

$$\therefore \quad A = \frac{\sigma_{12}}{\sigma_{11}} \qquad (7.97)$$

To correct for heteroskedasticity we need to bring these transformed errors $(s>T)$ to the same variance as those of the first T observations. We define

$$\check{U}_s^* = \tilde{U}_s^* = U_s^* \qquad s \leq T$$
$$\check{U}_s^* = B\tilde{U}_s = BU_s^* - BAU_{s-T}^* \qquad s > T. \qquad (7.97)$$

Now the variance of the first T error terms is σ_{11} so we require

$$E\{(BU_s^* - BAU_{s-T}^*)^2\} = \sigma_{11} \qquad s > T \qquad (7.98)$$

$$\therefore \quad B^2 \sigma_{22} + B^2 A^2 \sigma_{11} - 2B^2 A \sigma_{12} = \sigma_{11} \qquad (7.99)$$

$$\therefore \quad B^2 = \frac{\sigma_{11}^2}{\sigma_{22}\sigma_{11} - \sigma_{12}^2}. \qquad (7.100)$$

Hence the linear transform which gives an Aitken-form equation suitable for OLS estimation is

$$\left(\frac{Y_{1t}}{BY_{2t} - BAY_{1t}}\right) = \beta_1 \left(\frac{X_{1t}}{-BAX_{1t}}\right) + \beta_2 \left(\frac{0}{BX_{2t}}\right) + \left(\frac{U_{1t}}{BU_{2t} - BAU_{1t}}\right) \qquad (7.101)$$

The OLS regression of the $2T$ transformed observations on the two transformed variables is BLUE (in Y_1 and Y_2). Since OLS equation by equation is also in the class of estimators, for example

$$\beta_1^* = \sum_1^T w_{1t} Y_{1t} + \sum_1^T w_{2t} Y_{2t} \qquad (7.102)$$

with

$$w_{2t} = 0 \quad \text{all } t$$

and

$$w_{1t} = \frac{X_{1t}}{\sum_1^T X_{1t}^2}$$

it follows that our new estimator has in general a smaller variance than equation by equation OLS. As with most applications of the Aitken theorem we need to know the parameters of the error structure before we can apply GLS (or seemingly unrelated regressions estimation—SURE). It is unlikely that we will know the true parameters so in practice an approximation technique is used. First, OLS is applied equation by equation and the OLS residuals $\hat{U}_{it}$ are obtained. The error variances and covariances are estimated by:

$$\hat{\sigma}_{ij} = \frac{1}{T-K} \sum_1^T \hat{U}_{it} \hat{U}_{jt} \qquad (7.103)$$

where K is the number of parameters estimated in each equation. These estimated values are used to transform the variables as in (7.102) and OLS is then applied to the transformed equations. This technique is easily generalizable to the many-variable case and can be used in the many-equation case by an extension of the ideas involved in the two-equation case.

Testing hypothesis on the parameters β_1 and β_2 follows in the usual way from F tests using restricted and unrestricted estimates of (7.101). Forecasts also are based on the GLS form and then are transformed into the basic dependent variables. Finally testing for the presence of contemporaneous

Problems 7

7.1 In the model
$$Y_t = \beta X_t + U_t$$
$$E(U_t) = 0, \quad E(U_t U_s) = 0 \quad \text{for} \quad s \neq t$$

X_t is fixed in repeated samples $\Sigma X_t^2 > 0$, $E(U_t^2) = \sigma_t^2$. The equation is estimated by OLS and the variance of the estimator of β is estimated as if the equation were homoskedastic. Under what conditions will expectation of this variance equal the true variance of the OLS estimator? How does this variance relate to the true variance of the GLS estimator?

7.2 Data is generated by an equation
$$Y_t = \beta X_t + U_t$$
where
$$E(U_t) = 0$$
$$E(U_t^2) = \sigma^2$$
$$E(U_t U_s) = 0 \quad \text{for} \quad s \neq t$$
X_t is fixed in repeated samples
$$t = 1 \ldots T.$$

The experimenter does not have access to the original data but only to 'grouped' means. The data has been ordered by size $(X_1 < X_2 \ldots < X_T)$ and the average of the first N_1 Xs and associated Ys taken, i.e.

$$\bar{X}_1 = \frac{1}{N_1} \sum_1^{N_1} X_t$$

$$\bar{Y}_1 = \frac{1}{N_1} \sum_1^{N_1} X_t.$$

This is repeated over all groups, e.g.

$$\bar{X}_2 = \frac{1}{N_2} \sum_{N_1+1}^{N_1+N_2} X_t$$

There are J groups of sizes $N_j (j = 1 \ldots J)$. It is proposed to estimate the parameter β by the OLS regression of $\bar{Y}_j$ on $\bar{X}_j$. What are the means and variance of this estimator? What is the loss of efficiency due to grouping and under what circumstances will this loss be small?

7.3 Consider the model

$$Y_t = \beta X_t + U_t \qquad t = 1 \ldots J$$

where

$$E(U_t) = 0$$
$$E(U_t) = \sigma^2 \qquad \text{all } t$$
$$E(U_t U_s) = 0 \qquad s \ne t$$

X_t fixed in repeated samples.

Suppose that we have from another investigation of the same relationship with different data the results of the estimation of β (but that we do not have the data in the second case), i.e. $\bar{\beta}$ and Var $\bar{\beta}$ (the *true* variance). Assume that the errors in the two sets of data are independent and that we also know the value of σ^2.

(a) Derive the best linear (in Y_t and $\bar{\beta}$) unbiased estimator of β. (Hint: treat the relation between $\bar{\beta}$ and β as an extra observation.)
(b) Compare this estimator with OLS using the Y_t, X_t values.
(c) Suppose that the second set of data become available. What would be the optimum estimator of β? Interpret your results.
(d) Generalize the result to the case where a second independent estimate of β is available.

7.4 Consider the model

$$Y_t = \beta X_t + U_t$$

where

$$U_t = \rho U_{t-1} + \varepsilon_t$$

and

$$E(\varepsilon_t) = 0$$
$$E(\varepsilon_t \varepsilon_s) = 0 \qquad s \ne t$$
$$\qquad\qquad = \sigma_\varepsilon^2 \qquad s = t$$

X_t are fixed in repeated samples.

Also

$$\gamma_s^2 = \lim T \to \infty \left\{ \frac{\left(\frac{1}{T}\Sigma X_t X_{t-s}\right)^2}{\frac{1}{T}\Sigma X_t^2 \frac{1}{T}\Sigma X_{t-s}^2} \right\}$$

(a) Relate the variance of $\hat{\beta}$ (OLS estimator) to the serial correlation coefficient ρ and the lagged correlations in the X variables.
(b) Suppose that this variance is estimated by the conventional formula, i.e.

$$\widehat{\text{Var } \hat{\beta}} = s^2 / \Sigma X_t^2.$$

Discuss the bias involved under the special assumption that the γ_s are all positive.

(c) Compare the true variance of OLS with the variance of the GLS estimator in large samples (hint: the transformed first observation can be ignored in this case) for the case where $\gamma_i = \rho^i$

7.5 Consider the model
$$Y_t = \beta X_t + U_t$$
where
$$U_t = V_t + \lambda V_{t-1}$$
$$E(V_t) = 0$$
$$E(V_t V_s) = 0 \quad \text{for} \quad s \neq t$$
$$= \sigma^2 \quad \text{for} \quad s = t$$

(U_t is a first order moving average—MA(1)).

$$\gamma = \frac{(\Sigma X_t X_{t-1})^2}{\Sigma X_t^2 \Sigma X_{t-1}^2} \quad \text{over} \quad t = 2 \ldots T.$$

(a) Show that the true variance of the OLS estimator is given in large samples by
$$\text{Var } \hat{\beta} = (1 + \lambda^2 + 2\lambda\gamma) \frac{\sigma^2}{\Sigma X_t^2}$$

(b) Show that the expected value of the estimator of the OLS variance given by the conventional formula
$$\widehat{\text{Var } \hat{\beta}} = s^2 / \Sigma X_t^2$$
is in the large sample equal to
$$\frac{\sigma^2(1 + \lambda^2)}{\Sigma X_t^2} - \frac{2\lambda\gamma\sigma^2}{(T-1)\Sigma X_t^2}$$
and compare this to the correct variance of the OLS estimator.

7.6 Consider the model
$$Y_t = \beta X_t + U_t$$
where
$$U_t = \rho U_{t-1} + \varepsilon_t$$
and
$$E(\varepsilon_t) = 0$$
$$E(\varepsilon_t \varepsilon_s) = 0 \qquad s \neq t$$
$$= \sigma^2 \qquad s = t.$$

It is proposed to estimate the parameter β by regressing first differences $(\Delta Y, \Delta X)$ on each other.

(a) Show that the resulting estimator is linear and unbiased.

(b) Derive the variance of this estimator and show that the conventional estimator of this variance is biased.

7.7 Consider the two equations:

$$Y_{1t} = \beta_1 X_{1t} + U_{1t}$$
$$Y_{2t} = \beta_2 X_{2t} + U_{2t}$$
$$E(U_{it}) = 0 \quad \text{all } i \text{ and } t$$
$$E(U_{it} U_{js}) = 0 \quad t \neq s$$
$$= \sigma_{ij} \quad t = s$$

where the σ_{ij} are all known. Show that Seemingly Unrelated Regression estimation does not gain in efficiency over equation by equation OLS if $X_{1t} = X_{2t}$ all t.

7.8 Consider the model

$$Y_{1t} = \beta_1 X_{1t} + U_{1t}$$
$$Y_{2t} = \beta_2 X_{2t} + U_{2t}$$
$$E(U_{it}) = 0 \quad \text{all } i \text{ and } t$$
$$E(U_{it} U_{js}) = 0 \quad t \neq s;$$
$$= 0 \quad i \neq j.$$

Show that there is no gain in efficiency through using SUR, rather than equation by equation OLS, in the estimation of the β_i.

Answers 7

7.1 The estimator is $\hat{\beta} = \Sigma X_t Y_t / \Sigma X_t^2$ and the 'incorrect' variance is estimated by

$$\text{Var}_I \hat{\beta} = s^2 / \Sigma X_t^2$$

$$s^2 = \frac{1}{T-1} \Sigma (Y_t - \hat{\beta} X_t)^2.$$

Now

$$\Sigma \hat{U}_t^2 = \Sigma Y^2 - (\Sigma XY)^2 / \Sigma X^2$$

$$= 2\beta \Sigma XU + \Sigma U^2 - \frac{(\Sigma XU)^2}{\Sigma X^2}$$

∴

$$E(s^2) = \frac{1}{T-1} \left\{ \Sigma \sigma_t^2 - \frac{\Sigma X_t^2 \sigma_t^2}{\Sigma X_t^2} \right\}$$

∴

$$E(\text{Var}_I \hat{\beta}) = \frac{\Sigma X_t^2 \Sigma \sigma_t^2 - \Sigma X_t^2 \sigma_t^2}{(T-1)(\Sigma X_t^2)^2}.$$

From the chapter (7.16) we have for the true OLS variance

$$\text{Var } \hat{\beta} = \frac{\Sigma \sigma_t^2 X_t^2}{(\Sigma X_t^2)^2}. \tag{i}$$

These two expressions are equal if

$$\Sigma X_t^2 \Sigma \sigma_t^2 = T \Sigma X_t^2 \sigma_t^2$$

i.e. if the error variance is 'independent' of the size of the associated X_t^2.

Following (7.24) and (7.25) for the case of the single-variable model without intercept we have for GLS the regression of Y_t/σ_t on X_t/σ_t. Therefore,

$$\hat{\hat{\beta}} = \frac{\Sigma Y_t X_t / \sigma_t^2}{\Sigma X_t^2 / \sigma_t^2}$$

$$\text{Var } \hat{\hat{\beta}} = \frac{1}{\Sigma(X_t^2/\sigma_t^2)} \tag{ii}$$

(the error variances—$E[(U_t/\sigma_t)^2]$ are all unity). Now considering (i) and (ii) we need to show under what circumstances (i) is greater than (ii). i.e. $\Sigma \sigma_t^2 X_t^2 \cdot \Sigma X_t^2/\sigma_t^2 > (\Sigma X_t^2)^2$. Now call $\sigma_t X_t = A_t$, $X_t/\sigma_t = B_t$. The inequality becomes $\Sigma A^2 \Sigma B^2 > (\Sigma AB)^2$. But this is always true by the Cauchy–Schwarz inequality, unless $A_t = KB_t$ all t, in which case the two values are equal. Thus the variance of OLS is always greater than that of GLS unless $\sigma_t X_t = KX_t/\sigma_t$ all t, i.e. $\sigma_t^2 = K$. If there is homoskedasticity then the two variances are equal.

7.2. Writing the relationship between the grouped variables as $\bar{Y}_j = \beta \bar{X}_j + \bar{U}_j$ we need to establish the properties of $\bar{U}_j$ in order to evaluate the OLS estimator:

$$\hat{\beta}_G = \sum_{j=1}^{J} \bar{Y}_j \bar{X}_j / \Sigma \bar{X}_j^2.$$

Substituting in for the group means of $\bar{Y}_j$, $\bar{X}_j$, we have for example:

$$\frac{1}{N_1} \Sigma Y_t = \beta \frac{1}{N_1} \Sigma X_t + \bar{U}_1.$$

It is clear that for this equation to hold the $\bar{U}_j$ are defined by the same averaging $\bar{U}_j = \frac{1}{N_j} \Sigma U_t$. Now,

$$E(\hat{\beta}_G) = \beta + E\left(\frac{\Sigma \bar{X}_j \bar{U}_j}{\Sigma \bar{X}_j^2}\right)$$

and since $E(\bar{U}_j) = 0$, $E(\hat{\beta}_G) = \beta$, the estimator is clearly linear in $\bar{Y}_j$ (and hence is linear in Y_t). Similarly

$$\text{Var } \hat{\beta}_G = E\left\{\left(\frac{\Sigma \bar{X}_j \bar{U}_j}{\Sigma \bar{X}_j^2}\right)^2\right\}.$$

Now $E(\bar{U}_j \bar{U}_i) = 0$, since different group means contain no common U_t and all the U_t are pairwise independent. However,

$$E\{(\bar{U}_j)^2\} = E\left\{\left(\frac{1}{N_j}\Sigma U_t\right)^2\right\}$$

$$= \frac{1}{N_j^2} N_j \sigma^2$$

so that unless all groups are of equal size the grouped data is heteroskedastic.

$$\text{Var } \hat{\beta}_G = \frac{\sigma^2 \Sigma \bar{X}_j^2 / N_j}{(\Sigma \bar{X}_j^2)^2}$$

The lack of optimality in the grouped error properties suggests that a GLS approach should be utilized. Since the original data is not available we need the best linear (in $\bar{Y}_j$) unbiased estimator. Applying the Aitken transform to make the errors 'well behaved' we obtain

$$\sqrt{N_j}\, \bar{Y}_j = \beta \sqrt{N_j}\, \bar{X}_j + \sqrt{N_j}\, \bar{U}_j.$$

The resulting errors are homoskedastic and continue to have zero mean and be serially independent. Applying OLS to the transformed equation we have

$$\hat{\hat{\beta}}_G = \Sigma N_j \bar{Y}_j \bar{X}_j / \Sigma N_j \bar{X}_j^2$$

$$\text{Var } \hat{\hat{\beta}}_G = \frac{\sigma^2}{\Sigma N_j \bar{X}_j^2}.$$

This variance can be seen to be smaller than that of OLS (apply Cauchy–Schwarz inequality as in the previous question) with equality if N_j is constant all j.

To evaluate the loss of efficiency due to grouping we need to compare the variance of this best grouped estimator (given that it is unbiased) with the variance of the optimal estimator that could be obtained if the original data were available. The latter is clearly OLS on the raw data (since the original errors are 'well behaved') and has variance

$$\text{Var } \hat{\beta} = \frac{\sigma^2}{\sum_{1}^{T} X_t^2}.$$

To compare these two estimates we need to relate the two denominators (one summed from 1 to T and the other from 1 to J). If we partition the total data into groups we have:

$$\sum_{1}^{T} X_t^2 = \sum_{1}^{N_1} X_t^2 + \sum_{N_1+1}^{N_1+N_2} X_t^2 \ldots$$

$$= N_1 \bar{X}_1^2 + \left(\sum_{1}^{N_1} X_t^2 - N_1 \bar{X}_1^2 \right)$$

$$+ N_2 \bar{X}_2^2 + \left(\sum_{N_1+1}^{N_1+N_2} X_t^2 - N_2 \bar{X}_2^2 \right) \ldots$$

$$\therefore \quad \sum_{1}^{T} X_t^2 = \sum_{1}^{J} N_j \bar{X}_j^2 + \sum_{t=1}^{T} \sum_{j=1}^{J} (X_t - \bar{X}_j^2)$$

i.e. the total sum of squared values is the sum of squared means weighted by group sizes, plus the sum of the squared deviations around the group means for each group taken over all groups. The denominator of the ungrouped variance is larger by this second term than the variance of the grouped data. Hence if there are few observations to a group and if within each group the data is clustered around the group mean the loss of efficiency due to grouping is likely to be small.

7.3 (a) Write the data and the 'extraneous' information as

$$Y_t = \beta X_t + U_t \quad 1 \ldots T$$
$$\bar{\beta} = \beta 1 + V.$$

Treat this as an equation in $T+1$ observations.

$$Y_t^* = \beta X_t^* + U_t^* \qquad 1 \ldots T+1.$$

Checking the error properties we have

(i) $E(U_t^*) = 0$, because $E(U_t) = 0$ and $E(\bar{\beta}) = \beta$, so $E(V) = 0$.
(ii) $E(U_t^* U_s^*) = 0$, $s \neq t$, because $E(U_t U_s) = 0$, $s \neq t$, and $E(U_t V) = 0$, since the extraneous estimator is independent of the errors in the first set of data.
(iii) $E(U_t^{*2}) = \sigma^2$ for $t = 1 \ldots T$, $= \text{Var } \bar{\beta}$ for $T+1$.

The observations need to be transformed so that OLS applied to the new equation is optimal, because of the heteroskedasticity involved in the extra observation. Applying Aitken's technique we need only multiply the final observation by the factor R.

$$U_{T+1}^{**} = R U_{T+1}^*$$

where $$R = \sqrt{\sigma^2 / \text{Var } \bar{\beta}}.$$

The new errors are made homoskedastic and the equation is

$$Y_t = \beta X_t + U_t \quad 1 \ldots T$$
$$R\bar{\beta} = \beta R + RV.$$

Applying OLS to these $T+1$ observations we have

$$\hat{\hat{\beta}} = \frac{\sum_1^T Y_t X_t + R^2 \bar{\beta}}{\sum_1^T X_t^2 + R^2}$$

and

$$\text{Var } \hat{\hat{\beta}} = \frac{\sigma^2}{\sum_1^T X_t^2 + R^2}.$$

By the properties of the Gauss–Markov theorem this estimator is best unbiased in the class $\beta^+ = \Sigma W_t Y_t + W_0 \bar{\beta}$. This class clearly includes OLS on the basic data ($W_t = X_t / \Sigma X_t^2$, $W_0 = 0$) as a member so that the new estimator is better than OLS (see also (ii)). This technique is known as Mixed Regression or as the Theil–Goldberger estimator.

(b) We can rewrite $\hat{\hat{\beta}}$ by dividing all terms by ΣX_t^2:

$$\hat{\hat{\beta}} = \frac{\hat{\beta} + S\bar{\beta}}{1+S}$$

where $S = \text{Var } \hat{\beta} / \text{Var } \bar{\beta}$ ($\hat{\beta}$ is OLS on original data and Var $\hat{\beta}$ its true variance). The new estimator is a weighted average of OLS on the basic data and the extraneous information estimator $\bar{\beta}$. As S increases $\hat{\hat{\beta}}$ tends to the extraneous value—thus as the variance of the extraneous estimator (Var $\bar{\beta}$) is small relative to the variance of OLS, the information in the extraneous estimator dominates. Conversely when the extraneous estimator has a relatively large variance S tends to zero and the mixed estimator tends to OLS (the external information is of little value because of its variability). Similarly we have:

$$\text{Var } \hat{\hat{\beta}} = \frac{\text{Var } \hat{\beta}}{1+S}.$$

The variance of the mixed estimator is at largest equal to that of OLS (when S tends to zero) in the situation where the extraneous variance is very large. As the extraneous variance becomes very small then the variance of the mixed estimator becomes small. The variance of the mixed estimator is always less than that of the extraneous estimator since

$$\text{Var } \hat{\hat{\beta}} = \frac{\text{Var } \bar{\beta}}{1+(1/S)}$$

(c) Given the second set of data

$$Y_t = \beta X_t + U_t \qquad t = T+1 \ldots T+M$$

we see that

$$\bar{\beta} = \sum_{T+1}^{T+M} Y_t X_t \Big/ \sum_{T+1}^{T+M} X_t^2$$

$$\operatorname{Var} \bar{\beta} = \frac{\sigma^2}{\sum_{T+1}^{T+M} X_t^2}.$$

Substituting these values in the formula for $\hat{\hat{\beta}}$ we have on collecting terms and cancelling:

$$\hat{\hat{\beta}} = \sum_{1}^{T+M} Y_t X_t \Big/ \sum_{1}^{T+M} X_t^2$$

i.e. the OLS estimator of the regression of all values of Y on all the X. The variance of $\hat{\hat{\beta}}$ can similarly be shown to be identical to that of the OLS estimator on the full data set. The extraneous estimator (and its variance) in effect contain all the information used in a full OLS analysis.

(d) Imagine we have two extraneous estimators, $\bar{\beta}_1$ with Var $\bar{\beta}_1$ and $\bar{\beta}_2$ with Var $\bar{\beta}_2$, and that they are based on independent sets of data. Then the augmented equation is in $T+2$ observations:

$$Y_t = \beta X_t + U_t \qquad t = 1 \ldots T$$
$$\bar{\beta}_1 = \beta_1 + V_1$$
$$\bar{\beta}_2 = \beta_2 + V_2.$$

Applying the Aitken transform with factors

$$R_i = \sqrt{\sigma^2 / \operatorname{Var} \bar{\beta}_i}$$

to the last two observations, and then using OLS we obtain

$$\hat{\hat{\beta}} = \frac{\sum_{1}^{T} X_t Y_t + R_1^2 \bar{\beta}_1 + R_2^2 \bar{\beta}_2}{\Sigma X_t^2 + R_1^2 + R_2^2}$$

$$\operatorname{Var} \hat{\hat{\beta}} = \frac{\sigma^2}{\Sigma X_t^2 + R_1^2 + R_2^2}.$$

In practical terms this estimator cannot be directly used since we do not know R_i. The obvious technique is to use estimates of R_i. This is done by first carrying out OLS on the available data to obtain $\hat{\sigma}^2$, and by using the estimated variance of the extraneous estimator $\widehat{\operatorname{Var} \bar{\beta}_i}$. It can be shown that in large samples this estimator tends towards the true mixed regression estimator.

7.4 (a) From the standard formula for the OLS variance we have

$$\text{Var } \hat{\beta} = E\left\{\left(\frac{\Sigma X_t U_t}{\Sigma X_t^2}\right)^2\right\}.$$

Using the result that $E(U_t U_{t-\tau}) = \rho^\tau \sigma^2$ where $\sigma^2 = \sigma_\varepsilon^2/(1-\rho^2)$

$$\therefore \quad \text{Var } \hat{\beta} = \frac{1}{(\Sigma X_t^2)^2}\left(\sigma^2 \sum_1^T X_t^2 + 2\sigma^2 \sum_2^T \rho X_t X_{t-1} + 2\sigma^2 \sum_3^T \rho^2 X_t X_{t-2} \ldots\right)$$

$$= \frac{\sigma^2/T}{\Sigma X_t^2/T} + \frac{2\rho\sigma^2/T}{\Sigma X_t^2/T}\left(\frac{\sum_2^T X_t X_{t-1}}{\sum_1^T X_t^2} \ldots\right).$$

But rewriting

$$\frac{\sum_2^T X_t X_{t-1}}{\sum_1^T X_t^2} = \frac{\sum_2^T X_t X_{t-1}}{\sqrt{\left(\sum_2^T X_t^2 \sum_2^T X_{t-1}^2\right)}} \cdot \frac{\sqrt{\left(\sum_2^T X_t^2 \sum_2^T X_{t-1}^2\right)}}{\sum_1^T X_t^2}$$

and using the fact that in large samples the limit of

$$\left(\frac{1}{T-1}\right)\sum_2^T X_t^2$$

is equal to

$$\frac{1}{T}\sum_1^T X_t^2$$

we have

$$\text{Var } \hat{\beta} = \frac{\sigma^2/T}{m_{XX}}\left\{1 + 2\rho\left(\frac{T-1}{T}\right)\gamma_1 \ldots + 2\rho^2\left(\frac{T-2}{T}\right)\gamma_2 \ldots\right\}$$

where

$$m_{XX} = \lim \frac{1}{T}\Sigma X_t^2$$

Since the limit of terms $((T-m)/T)$ is always unity we have

$$\text{Var}\hat{\beta}_{\lim} = \frac{\sigma^2/T}{m_{XX}}(1 + 2\rho\gamma_1 + 2\rho^2\gamma_2 \ldots)$$

This tends to zero showing that the asymptotic variance of OLS is zero even when there is serial correlation. This coupled with the unbiasedness of OLS is sufficient to show that OLS is still consistent.

(b)
$$E\{\widehat{\text{Var }\hat{\beta}}\} = E\left\{\frac{s^2}{\Sigma X_t^2}\right\}$$

But
$$s^2 = \frac{1}{T-1}\Sigma(Y_t - \hat{\beta}X_t)^2.$$

Substituting in the true equation of Y_t and the OLS estimator value for $\hat{\beta}$

$$s^2 = \frac{1}{T-1}\Sigma U_t^2 - \frac{1}{T-1}\frac{(\Sigma X_t U_t)^2}{\Sigma X_t^2}.$$

Using
$$\widehat{\text{Var }\hat{\beta}} = E\{(\Sigma X_t U_t)^2/(\Sigma X_t^2)^2\}$$

we have:
$$E(\widehat{\text{Var }\hat{\beta}}) = \frac{T}{T-1}\cdot\frac{\sigma^2}{\Sigma X_t^2} - \frac{\text{Var }\hat{\beta}}{T-1}.$$

(Of course in the case of no serial correlation, Var $\hat{\beta} = \sigma^2/\Sigma X_t^2$, so that $E(\widehat{\text{Var }\hat{\beta}}) = \text{Var }\hat{\beta}$.) Using (a) we have for the bias in large samples of the estimator of the OLS variance

$$E(\widehat{\text{Var }\hat{\beta}}) - \text{Var }\hat{\beta} = \frac{T}{T-1}\left\{\frac{\sigma^2/T}{m_{XX}} - \frac{\sigma^2/T}{m_{XX}}(1 + 2\rho\gamma_1 + 2\rho^2\gamma_2 \ldots)\right\}.$$

The bias will be negative if ρ and all the γ_i are positive, i.e. the conventional formula for the OLS variance will underestimate the true variance. If ρ is negative and all the γ_i are positive then we can say nothing definite about the direction of bias.

(c) The variance of GLS is, by the standard formula applied to the Aitken transformed estimating equation,

$$\text{Var }\hat{\hat{\beta}} = \frac{\sigma_\varepsilon^2}{\Sigma(X_t - \rho X_{t-1})^2}$$

But
$$\sigma_\varepsilon^2 = \sigma^2(1 - \rho^2)$$

and
$$\lim \frac{1}{T}\Sigma(X_t - \rho X_{t-1})^2 = m_{XX}(1 - 2\rho\gamma_1 + \rho^2)$$

therefore
$$\lim \text{Var }\hat{\hat{\beta}} = \frac{\sigma^2}{T}(1 - \rho^2)/m_{XX}(1 - 2\rho\gamma_1 + \rho^2).$$

For the case where $\gamma_i = \rho^i$ we have:

$$\lim \operatorname{Var} \hat{\hat{\beta}} = \frac{\sigma^2(1-\rho^2)/T}{m_{xx}(1-\rho^2)} = \frac{\sigma^2/T}{m_{xx}}.$$

For OLS with $\gamma_i = \rho^i$:

$$\lim \operatorname{Var} \hat{\beta} = \frac{\sigma^2/T}{m_{xx}}(1 + 2\rho^2 + 2\rho^4 \ldots)$$

$$= \frac{\sigma^2/T}{m_{xx}}\left(\frac{1+\rho^2}{1-\rho^2}\right).$$

The magnitude of ρ gives a measure of the reduction in variance attainable by using GLS.

7.5 (*a*) As usual the variance of OLS (being unbiased in this model) is given by

$$\operatorname{Var} \hat{\beta} = E\left\{\frac{(\Sigma X_t U_t)^2}{(\Sigma X_t^2)^2}\right\}$$

$$= \frac{\sum_{1}^{T} X_t^2 E(U_t^2)}{\left(\sum_{1}^{T} X_t^2\right)^2} + \frac{2\sum_{2}^{T} X_t X_{t-1} E(U_t U_{t-1})}{\left(\sum_{1}^{T} X_t^2\right)^2} \ldots$$

Now

$$E(U_t^2) = E\{(V_t + \lambda V_{t-1})^2\}$$
$$= \sigma^2(1 + \lambda^2)$$
$$E(U_t U_{t-1}) = E\{(V_t + \lambda V_{t-1})(V_{t-1} + \lambda V_{t-2})\} = \lambda \sigma^2$$
$$E(U_t U_{t-s}) = 0 \quad \text{for } s > 1.$$

$$\therefore \quad \operatorname{Var} \hat{\beta} = \frac{\sigma^2}{\Sigma X_t^2}(1 + \lambda^2) + 2\lambda \sigma^2 \frac{\Sigma X_t X_{t-1}}{(\Sigma X_t^2)^2}.$$

In large samples this becomes

$$\operatorname{Var} \hat{\beta} = \frac{\sigma^2}{\Sigma X_t^2}(1 + \lambda^2 + 2\lambda \gamma)$$

(even for small samples this approximation is very close with the last term including a correction factor $(T-1)/T$ which tends to unity as the sample increases).

(*b*) (see answer 7.4)

$$E(s^2) = \frac{1}{T-1} E\left\{\Sigma U_t^2 - \frac{(\Sigma X_t U_t)^2}{\Sigma X_t^2}\right\}$$

$$= \frac{1}{T-1}\{T(1+\lambda^2)\sigma^2 - \Sigma X_t^2 \operatorname{Var} \hat{\beta}\}.$$

Using (a)

$$E(\widehat{\operatorname{Var} \hat{\beta}}) = \frac{T(1+\lambda^2)}{T-1} \frac{\sigma^2}{\Sigma X_t^2} - \frac{\operatorname{Var} \hat{\beta}}{T-1}$$

$$= \frac{\sigma^2}{\Sigma X_t^2}\left(1 + \lambda^2 - \frac{2\lambda\gamma}{T-1}\right).$$

If
$$2\lambda\gamma > \frac{-2\lambda\gamma}{T-1}$$

then the true variance is on average underestimated by the conventional formula. This inequality holds if $2\lambda\gamma(T-1) > -2\lambda\gamma$ or $2\lambda\gamma T > 0$. For λ and γ both positive or both negative there is an underestimate, while if they are of opposite signs there is an overestimate on average of the true OLS variance.

7.6 The estimator is based on

$$\Delta Y_t = \beta \Delta X_t + \Delta U_t \qquad (2 \ldots T)$$

$$\bar{\beta} = \Sigma \Delta Y \Delta X / \Sigma \Delta X^2.$$

(a) This is immediately seen to be linear in Y_t ($\Sigma W_t Y_t$):

$$W_T = \Delta X_T / \Sigma \Delta X^2$$

$$W_{T-1} = (-\Delta X_T + \Delta X_{T-1})/\Sigma \Delta X^2$$

etc. Substituting in the transformed equation $\bar{\beta} = \beta + \Sigma \Delta X \Delta U / E \Delta X^2$, since $E(\Delta U) = 0$, therefore $E(\bar{\beta}) = \beta$.

(b) The variance of this estimator is

$$\operatorname{Var} \bar{\beta} = E\left\{\frac{(\Sigma \Delta X \Delta U)^2}{(\Sigma \Delta X^2)^2}\right\}.$$

To evaluate terms of the form $E\{\Delta U_t \Delta U_s\}$ we use the result that (7.61)

$$E(U_t U_{t-s}) = \frac{\rho^s \sigma^2}{1-\rho^2} = \rho^s \theta$$

where
$$\theta = \sigma^2/(1-\rho^2).$$

$\therefore$
$$\operatorname{Var} \bar{\beta} = \frac{1}{(\Sigma \Delta X^2)^2}\left\{\sum_2^T \Delta X_t^2 E(\Delta U_t^2)\right.$$

$$\left. + 2\sum_3^T \Delta X_t \Delta X_{t-1} E(\Delta U_t \Delta U_{t-1}) \cdots \right\}$$

$$= \frac{1}{(\Sigma \Delta X^2)^2}\left\{2\theta(1-\rho)\sum_2 \Delta X_t^2\right.$$

$$+2\theta(1-\rho)(\rho-1)\sum_3 \Delta X_t \Delta X_{t-1}$$

$$+2\theta\rho(1-\rho)(\rho-1)\sum_4 \Delta X_t \Delta X_{t-2} \ldots \bigg\}.$$

The conventional formula for the estimated variance is

$$\widehat{\text{Var }\bar{\beta}} = E(s^2)/\Sigma \Delta X_t^2$$

where

$$s^2 = \frac{1}{T-2}\sum_2^T (\Delta Y - \bar{\beta}\Delta X)^2.$$

But

$$E(s^2) = \frac{1}{T-2}\left\{E(\Sigma \Delta U)^2 - \frac{E(\Sigma \Delta X \Delta U)^2}{\Sigma \Delta X^2}\right\}$$

$$\therefore \quad E\left\{\frac{s^2}{\Sigma \Delta X^2}\right\} = \frac{1}{T-2}\left\{\frac{(T-1)2\theta(1-\rho)}{\Sigma \Delta X^2} - \text{Var }\bar{\beta}\right\}.$$

If

$$\frac{2\theta(1-\rho)}{\Sigma \Delta X^2} = \text{Var }\bar{\beta}$$

then the estimator will be unbiased. However, we see that the first term in the bracket in effect omits all but the first term for Var $\bar{\beta}$. If these omitted terms are all positive then we underestimate the true variance of $\bar{\beta}$.

7.7 For equation by equation OLS we have

$$\text{Var }\hat{\beta}_1 = \frac{\sigma_{11}}{\Sigma X_{1t}^2}.$$

For SUR we have the standard multiple regression variance formula applied to (7.101)

$$\text{Var }\hat{\hat{\beta}}_1 = \frac{\sigma_{11}\sum_1^{2T} X_{2t}^{*2}}{\Sigma X_{1t}^{*2}\Sigma X_{2t}^{*2} - (\Sigma X_{1t}^* X_{2t}^*)^2}.$$

But

$$\sum_1^{2T} X_{2t}^{*2} = B^2 \sum_1^T X_{2t}^2$$

$$\sum_1^{2T} X_{1t}^{*2} = \sum_1^T X_{1t}^2 + B^2 A^2 \sum_1^T X_{1t}^2$$

$$\sum_1^{2T} X_{1t}^* X_{2t}^* = -B^2 A \sum_1^T X_{1t} X_{2t}$$

$$\therefore \quad \text{Var }\hat{\hat{\beta}} = \frac{\sigma_{11}\Sigma X_{2t}^2}{(1+B^2 A^2)\Sigma X_1^2 \Sigma X_2^2 - B^2 A^2 (\Sigma X_1 X_2)^2}.$$

If $\quad X_{1t} = X_{2t}(\text{or } KX_{2t})$

$\therefore \quad \text{Var } \hat{\hat{\beta}}_1 = \text{Var } \hat{\beta}_1$

and there is no gain in efficiency from using SURE.

7.8 The variance of the equation by equation OLS estimator is

$$\text{Var } \hat{\beta}_1 = \sigma_{11} \Big/ \sum_1^T X_{1t}^2$$

while that for SURE is given (as in the answer to question 7.7) by

$$\text{Var } \hat{\hat{\beta}}_1 = \frac{\sigma_{11} \Sigma X_{2t}^2}{(1+B^2 A^2)\Sigma X_1^2 \Sigma X_2^2 - B^2 A^2 (\Sigma X_1 X_2)^2}.$$

Now from (7.97) and (7.100):

$$A = \sigma_{12}/\sigma_{11}$$

$$B^2 = \frac{\sigma_{11}^2}{\sigma_{22}\sigma_{11} - \sigma_{12}^2}.$$

Hence if $\quad \sigma_{12} = 0$

$$A = 0$$

$$B = \left(\frac{\sigma_{11}}{\sigma_{22}}\right)^{\frac{1}{2}}$$

$\therefore \quad \text{Var } \hat{\hat{\beta}}_1 = \frac{\sigma_{11}}{\Sigma X_1^2} = \text{Var } \hat{\beta}_1$

and there is no gain in efficiency from using SURE.

8
Specification and Measurement Errors

In the preceding chapters we have analysed situations where the econometrician knew the true model and was concerned with its estimation or with inference about the size of its parameters. We must now stand back and ask what are the consequences for the use of standard estimation techniques if the model has been incorrectly specified. This can happen in many ways:

(a) an explanatory variable is omitted completely from an equation that is being estimated;
(b) a variable (either dependent or independent) is included in the wrong functional form—e.g. linear instead of in logarithms;
(c) a variable (either dependent or independent) is incorrectly measured;
(d) the errors do not have the properties assumed for the estimation method used to be appropriate.

Case (d) has already arisen with the problems of serial correlation and heteroskedasticity where the use of OLS led to inefficient estimators. The other cases reflect more directly the failure of the behavioural part of the model to be correctly specified and will therefore be of great importance for cases where the equation does not follow from some very tightly specified model.

8.1 Omitted variables

The topic of omitted variables can be used to analyse cases where a variable has been completely ignored and cases where variables have been mismeasured in some way, e.g. a model should be

$$\log Y_t = \alpha + \beta \log X_t + U_t \tag{8.1}$$

but the experimenter writes

$$Y_t = \alpha + \beta X_t + V_t \tag{8.2}$$

then we can see that V_t is in fact a composite term

$$V_t = U_t + \beta(\log X_t - X_t) - (\log Y_t - Y_t). \tag{8.3}$$

Hence if U_t were truly 'well behaved' then V_t could not in general be 'well behaved' and OLS applied to (8.2) would not have the optimal properties.

We analyse in detail the case of a two-variable model where one variable is omitted because of a specification error.

True: $$Y_t = \beta_0 + \beta_1 X_{1t} + \beta_2 X_{2t} + U_t \tag{8.4}$$

where
$$E(U_t)=0$$
$$E(U_t^2)=\sigma^2 \text{ all } t$$
$$E(U_t U_s)=0 \quad s\neq t.$$

The model which is estimated is written as

False:
$$Y_t = \beta_0 + \beta_1 X_{1t} + V_t. \tag{8.5}$$

We can see that this mistake in fact sets:

$$V_t = \beta_2 X_{2t} + U_t. \tag{8.6}$$

However, we analyse the consequences of estimating (8.5) as if V_t were well behaved, i.e. with zero mean, constant variance, and serial independence. The OLS estimates of β_0 and β_1 would be

$$\check{\beta}_1 = \frac{\Sigma y_t x_{1t}}{\Sigma x_{1t}^2}$$

$$\check{\beta}_0 = \bar{Y} - \check{\beta}_1 \bar{X}_1. \tag{8.7}$$

These are clearly still linear. To check for unbiasedness we substitute in the value of Y_t from the true equation.

Therefore:
$$\check{\beta}_1 = \beta_1 + \beta_2 \frac{\Sigma x_1 x_2}{\Sigma x_1^2} + \frac{\Sigma x_1 u}{\Sigma x_1^2} \tag{8.8}$$

$$\check{\beta}_0 = \beta_0 + \beta_2 \bar{X}_2 - \beta_2 \frac{\Sigma x_1 x_2}{\Sigma x_1^2} \cdot \bar{X}_1 - \frac{\Sigma x_1 u}{\Sigma x_1^2}. \tag{8.9}$$

If we take expectations we can see that in general both $\check{\beta}_0$ and $\check{\beta}_1$ are biased as estimators of the parameters in the true equation (8.4). The size of the bias for β_1 depends on the magnitude of β_2 and on the covariance between X_1 and X_2. This second term is in fact the regression coefficient of a regression of X_1 on a constant and X_2 and measures the extent to which the variables are associated. Only if there is no correlation between X_1 and X_2 does the omission of X_2 leave the coefficient of X_1 unbiased—otherwise the estimation $\check{\beta}_1$ reflects not only the 'direct' influence of X_1 on Y but also the extent to which X_2 affects Y while being reflected in a co-movement in X_1, i.e. a one-unit change in X_2 affects Y by β_2 units while X_1 tends to change at the same time—as measured by the regression coefficient of X_1 and X_2—and this fraction of β_2 is falsely allocated to the genuine marginal impact of X_1 on Y.

The constant is more likely still to be biased, since for lack of bias it requires not only that the two variables X_1 and X_2 be uncorrelated but also that the mean of the omitted variable be zero over the sample period. This first result, that OLS estimators are in general biased, is of great importance because it is clear that when data is highly correlated, as is usually the case with economic time series, the bias can be very substantial. It is also helpful to derive the variance of $\check{\beta}_1$, which as can be seen from (8.8) is:

$$\operatorname{Var} \check{\beta}_1 = \sigma^2 / \Sigma x_1^2. \tag{8.10}$$

This has to be compared with the variance of OLS applied to (8.4) which can be written

$$\operatorname{Var} \hat{\beta}_1 = \frac{\sigma^2}{\Sigma x_1^2 (1 - r^2)} \tag{8.11}$$

where r is the correlation between X_1 and X_2. What is striking is that the true variance for the mis-specified estimation is smaller than that for the estimation of the correct equation, and that this difference is greater the higher is the correlation between the two variables.

At first sight it would seem possible, if we were willing to accept bias in estimation in return for a reduction in variance, to decide which estimation would be preferable. Hence if we were particularly interested (say) in the coefficient of β_1 we could deliberately accept a specification error by dropping X_2 when the collinearity between X_1 and X_2 was very high. There are two important practical difficulties in the way of this procedure. The most obvious is that we will not be able to know the size of the bias since it depends on β_2—only if we have very clear ideas on the possible size of β_2 can we assess the likely degree of bias. The second difficulty concerns the variance term: in practice we need to estimate this term, and to do so we use the estimated residuals. If we use the mis-specified equation we have

$$\check{\sigma}^2 = \frac{1}{T-2} \Sigma \check{u}_t^2 \tag{8.12}$$

and by substituting in (8.7), and using the assumptions on the true error terms, this estimator can be shown in general to be biased.

The implications of this are that the estimated variance will often be larger than the true variance, so that a correct assessment of this aspect of a potential trade-off will not be correctly made. The effect is smaller when the correlation between the variables is larger.

8.2 Wrongly included variables

The second type of specification error that we consider is the wrongful inclusion of some variable that in fact has no impact on the dependent variable. Let us consider the following simple case

True: $\qquad Y_t = \beta_0 + \beta_1 X_{1t} + U_t \tag{8.13}$

with U_t well behaved.

False: $\qquad Y_t = \beta_0 + \beta_1 X_{1t} + \beta_2 X_{2t} + V_t. \tag{8.14}$

We can see that the error term in the incorrectly specified equation can be written:

$$V_t = U_t - \beta_2 X_{2t}. \tag{8.15}$$

At first sight the inclusion in the error term of a variable which is correlated with the variables used in the regression might lead us to expect OLS applied to (8.14) to be biased, but in fact the special structure of the equation means that this does not happen. We take the OLS estimates for (8.14)

$$\breve{\beta}_1 = \frac{\Sigma y x_1 \Sigma x_2^2 - \Sigma y x_2 \Sigma x_1 x_2}{\Sigma x_1^2 \Sigma x_2^2 - (\Sigma x_1 x_2)^2} \tag{8.16}$$

and substitute in for Y_t from the true equation (8.14).

$$\breve{\beta}_1 = \beta_1 + \frac{\Sigma x_1 u \Sigma x_2^2 - \Sigma x_2 u \Sigma x_1 x_2}{\Sigma x_1^2 \Sigma x_2^2 - (\Sigma x_1 x_2)^2}. \tag{8.17}$$

Taking expectations we can see that this is a linear unbiased estimator. Similarly we can show that

$$E(\breve{\beta}_0) = \beta_0$$
$$E(\breve{\beta}_2) = 0 \tag{8.18}$$

It also follows that $\breve{\beta}_1$ etc. are not BLUE because the error terms given by (8.15) cannot have the properties which are sufficient for OLS to be BLUE (except for very special sets of data). It is no surprise that the variance of (8.16) is greater than necessary. We have as usual for a multiple regression:

$$\text{Var } \breve{\beta}_1 = \frac{\sigma^2}{\Sigma x_1^2 (1 - r^2)} \tag{8.19}$$

while the variance of the OLS estimation applied to (8.13) will be as usual

$$\text{Var } \hat{\beta}_1 = \frac{\sigma^2}{\Sigma x_1^2} \tag{8.20}$$

which is always smaller unless the correlation between the two X variables is zero. Hence wrongful inclusion of a variable leads to a more imprecise (large-variance) estimation than necessary. Finally, we need to check whether the variance (8.19) is correctly estimated if we use the residuals from the mis-specified equation. By substitution and use of the properties of U_t it can be shown that

$$E(\breve{\sigma})^2 = \sigma^2 \tag{8.21}$$

so that the variance will on average be correctly estimated by the standard technique.

It is clear that the inclusion of extra variables, which in fact do not affect Y_t, is a much less serious problem than the omission of genuinely important variables. The inclusion of such variables does not create bias, it increases the variance by a factor which is in principle calculable, while the error variance is correctly estimated by standard methods. Statistical inferences based on such a model would be correct but would have unnecessarily low power because of the higher than necessary variance of the estimator. On the other hand, incorrect exclusion leads to a bias of unknown amount and overestimation of the variance by an unknown amount, with the result that significance tests are not of the size stated.

8.3 Measurement error

The problem of measurement error arises not from the failure to specify the model correctly but from the failure to measure the genuine variables in the relationship correctly. This may occur because

(i) the data provided is not actually reporting what it purports to claim, e.g. households in a survey not declaring their true income;
(ii) the wrong concept of a variable is measured, e.g. gross income rather than net income;
(iii) the variable is not directly observable, e.g. an expected rate of inflation.

In all such cases it is useful to distinguish between the 'true' level of the variable's and the 'measured' level (the difference between the two being the 'measurement error'). The problem is caused by the fact that the relationship is between 'true' values while the econometrician attempts to estimate the parameters of the relationship from the measured data. To develop these ideas we introduce the notation X_t^*, $Y_t^* =$ true values, X_t, $Y_t =$ measured values, X_t^{**}, $Y_t^{**} =$ measurement errors, with a simple basic relationship:

$$Y_t^* = \beta X_t^* + U_t. \qquad (8.23)$$

Now the econometrician knows that the relation is of the form (8.23) but has to work with measured values. If we now write

$$Y_t^* = Y_t + Y_t^{**}$$
$$X_t^* = X_t + X_t^{**} \qquad (8.24)$$

we can substitute these into (8.23) to yield:

$$Y_t = \beta X_t + V_t \qquad (8.25)$$

where
$$V_t = U_t + \beta X_t^{**} - Y_t^{**}. \qquad (8.26)$$

The properties of the use of OLS in (8.25) using the measured values will, as always, depend on the properties of the error term V_t as expressed in (8.26). It may well be that the equation error U_t is well behaved, and in particular $E(U_t) = 0$. The properties of the measurement error terms are equally crucial but would depend on the actual economic model being discussed. To keep the discussion at the simplest level we shall assume that there are no measurement errors in the dependent variable and that both the true and measured values of the independent variable are fixed in repeated samples. This is a more specialized case than one in which the measurement error is seen as a random variable that would inevitably vary over repeated samples at the same level of the true variable, but it does allow us to obtain some easy results which are not misleading for more general cases.

It is conventional to make two further assumptions about the error processes:

$$\Sigma X_t^* = 0$$

$$\Sigma X_t^* X_t^{**} = 0. \tag{8.27}$$

The first assumes that the average measurement error is zero while the second assumes that the true values and measurement errors are uncorrelated. If we now consider the OLS estimator based on (8.25)

$$\hat{\beta} = \Sigma X_t Y_t / \Sigma X_t^2 \tag{8.28}$$

and substitute in for Y using (8.25) and (8.26) we obtain

$$\hat{\beta} = \beta + \frac{\Sigma(U_t - \beta X_t^{**})X_t}{\Sigma X_t^2}. \tag{8.29}$$

Using the error properties assumed, the expected value is then

$$E(\hat{\beta}) = \beta - \beta \frac{\Sigma X_t^{**2}}{\Sigma X_t^{**2} + \Sigma X_t^{*2}}. \tag{8.30}$$

Hence we see that, under the assumptions made about the error term, the OLS estimator is biased towards zero. The size of the bias is dependent on the relative variation of the measurement errors and of the true value. We can see also from (8.29) why the estimation is biased: the observed values X_t are built up of the true value and measurement error. If the measurement error is uncorrelated with the true variable then it follows that the composite error V_t and the measured value are correlated because of their common component X_t^{**}. Equally we can see that if the measurement error were uncorrelated with the measured value (but therefore correlated with the true value) then OLS would be unbiased.

Faced with this serious and widespread problem it is not surprising that attention has been given to deriving a technique of estimation that would be unbiased when there are measurement errors present. This technique is known as 'instrumental variables estimation' and to introduce it we first present a different approach to obtaining the ordinary least squares estimator. Consider the equation without errors in variables:

$$Y_t = \beta X_t + U_t \tag{8.31}$$

where U_t is well behaved. It follows that

$$\Sigma Y_t X_t = \beta \Sigma X_t^2 + \Sigma X_t U_t \tag{8.32}$$

$$\frac{\Sigma Y_t X_t}{\Sigma X_t^2} = \beta + \frac{\Sigma X_t U_t}{\Sigma X_t^2}. \tag{8.33}$$

If the expected value of the second part of the term on the right-hand side is zero then the left-hand side is an unbiased estimator for β. Given that $E(U_t) = 0$ and X_t is fixed (and that $\Sigma X_t^2 > 0$) it follows that this estimator, which is in fact OLS, is unbiased. We can generalize by multiplying both sides of (8.31) by any variable Z_t to give

$$\Sigma Y_t Z_t = \beta \Sigma Z_t X_t + \Sigma Z_t U_t \tag{8.34}$$

so that

$$\frac{\Sigma YZ}{\Sigma XZ} = \beta + \frac{\Sigma ZU}{\Sigma ZX}. \tag{8.35}$$

Again, provided that $E(\Sigma ZU) = 0$ and $\Sigma ZX \neq 0$, the left-hand side provides an unbiased estimator for β. The estimator

$$\tilde{\beta} = \frac{\Sigma YZ}{\Sigma XZ} \tag{8.36}$$

is known as an 'instrumental variable estimator'—Z_t being the instrument. It is, in this set up, a linear unbiased estimator and hence inferior to OLS. Its variance, from (8.35), is seen to be

$$\operatorname{Var} \tilde{\beta} = \frac{\sigma^2 \Sigma Z^2}{(\Sigma ZX)^2} \tag{8.37}$$

if the error terms U_t are independent and homoskedastic. The ratio of the OLS variance to IV variance has a particularly simple form when U is well behaved:

$$\frac{\operatorname{Var} \hat{\beta}}{\operatorname{Var} \tilde{\beta}} = \frac{(\Sigma ZX)^2}{\Sigma Z^2 \Sigma X^2} \tag{8.38}$$

which is the analogue to the correlation coefficient in raw data between Z and X. From the Cauchy–Schwarz inequality it lies between zero and unity with the upper bound being attained when Z is an exact multiple of X for all t. We can see that in OLS the variable X can be interpreted as acting as its own instrument.

However, although we would not wish therefore to use IV techniques when U is well behaved, once we turn to the errors-in-variables framework, we can see how the method could be applied. Turning back to equation (8.25) we can see that for a general instrument Z the estimator (8.36) will be unbiased if the

term
$$\frac{\Sigma ZV}{\Sigma ZX} = \frac{\Sigma Z(U+\beta X^{**})}{\Sigma ZX} \tag{8.39}$$

has zero expectation. Given that Z is also a variable fixed in repeated samples we see that Z must obey two conditions for the IV to be unbiased:

(i) $\qquad\qquad\qquad \Sigma ZX^{**} = 0$

(ii) $\qquad\qquad\qquad \Sigma ZX \neq 0 \quad \text{i.e. } \Sigma ZX^* \neq 0. \tag{8.40}$

If the instrument is not correlated with the measurement error but is correlated with the true values, then IV is unbiased. The variance of the estimator is again

$$\text{Var } \tilde{\beta} = \frac{\sigma^2 \Sigma Z^2}{(\Sigma ZX)^2}. \tag{8.41}$$

The error variance σ^2 is estimated using residuals from the equation:

$$\tilde{U}_t = Y_t - \tilde{\beta} X_t. \tag{8.42}$$

We can see that for any instrument to yield unbiased estimates we must be sure that it is uncorrelated with the measurement error while being correlated with the true values. The variance formula (8.41) shows that the stronger this correlation the smaller will be the variance for a given value of ΣX^2. We can check the latter condition from the actual correlation between Z and X, but the former cannot of course be checked empirically (since the essence of the problem is that we do not know X^{**}). Hence we have to rely on a priori reasoning to satisfy ourselves that it is reasonable to use a particular instrument.

There is an alternative method of carrying out the same procedure which uses OLS twice and is often therefore called two-stage least squares (TSLS). The first stage is to regress the measured variable X_t on the instrument (call the resulting coefficient γ)

$$\hat{\gamma} = \frac{\Sigma XZ}{\Sigma Z^2} \tag{8.43}$$

and then form the fitted value for X_t (call this $\hat{X}_t$):

$$\hat{X}_t = Z_t \frac{\Sigma XZ}{\Sigma Z^2}. \tag{8.44}$$

The second stage is to replace X in the basic equation by $\hat{X}$ and then to regress Y on $\hat{X}$:

$$\beta^* = \Sigma Y\hat{X}/\Sigma \hat{X}^2. \tag{8.45}$$

By simple substitution we can see that

$$\beta^* = \tilde{\beta}$$

so that IV and TSLS are equivalent in this case (a third way of obtaining the same estimator would be to use $\hat{X}$ as the instrument). What this approach does is to scale the Z variable by the regression coefficient in the first stage so that it is brought to a scale where movements in the fitted value have the same impact on Y as movements in X. It is important *not* to use residuals from the second stage ($\tilde{U}_t = Y_t - \tilde{\beta}\hat{X}_t$) to estimate the error variance since these include the difference between X and its fitted value.

Before we turn to various important generalizations of this technique it is helpful to consider some instruments that have been suggested for fairly general use:

(i) $Z_t = 1$ all t
(ii) $Z_t = t$
(iii) $Z_t = r(X_t)$ where r is the rank order of the tth observation.
(iv) $Z_t = 1$, $X_t >$ median X
  $= -1$, $X_t <$ median X.

If the model does not include an intercept term then the unit variable is not being used and would be available as an instrument. It is likely to be a suitable candidate since if the average measurement error is zero, as is usually assumed, then the unit variable (*a*) is uncorrelated with the measurement error and, providing that the average true value is not zero as well, the second condition will also be obeyed. However, the correlation between this Z and X (in raw form) is likely to be substantially lower than that of other potential instruments and hence its variance unnecessarily large. For time series data the trend (*b*) is likely to be much more strongly correlated with the measured (and true) values while still being uncorrelated with the measurement errors. The use of a rank variable (*c*) or rank related variable (*d*) attempts to pick up the dominant part of variations in X_t^* while not being sensitive to the effects of small measurement errors, i.e. the assumption is that the ranking of the true and of the measured values will be the same if the measurement errors are relatively small. The larger these errors become the more likely are the complete rankings to be different, so that a simple dichotomy, which will be substantially different only for relatively large measurement errors, will be preferred.

There are two important generalizations to this technique. The first is to the case where there is more than one instrument available and the second looks at the problems raised in the multiple explanatory variable case. We begin with the case where there is a single explanatory variable measured with error and two potential instruments, Z_1 and Z_2. The simple way of choosing between them is to find which Z is more strongly correlated with X and use it. However, the alternative approach of constructing the instrument from the fitted value of the regression of X on Z suggests that, since the multiple regression of X on Z_1 and Z_2 will have at least as high a correlation as either simple correlation, we should construct an instrument which is the 'best' average of Z_1 and Z_2 in terms of correlation of X and this instrument. A two-stage procedure is needed in this case. The first stage is to regress X on Z_1 and Z_2 giving multiple regression coefficients $\hat{\gamma}_1$ and $\hat{\gamma}_2$ and hence the fitted value:

$$\hat{X}_t = \hat{\gamma}_1 Z_{1t} + \hat{\gamma}_2 Z_{2t}. \tag{8.46}$$

This fitted value is used as an instrument in the second stage, or replaces X in a second regression of Y on $\hat{X}$. Such a technique could be extended to any number of instruments which obeyed the conditions.

The extention to a multi-variable equation is also important even in cases where only one of the explanatory variables is measured with error. We start, however, with a two-variable equation in which both explanatory variables are measured with error. This yields an equation with observed variables X_1 and X_2, both of which are correlated with the composite error term.

$$Y_t = \beta_1 X_{1t} + \beta_2 X_{2t} + V_t. \tag{8.47}$$

Suppose also that we have two instruments, Z_1 and Z_2, both of which are uncorrelated with the errors of measurement. We can as before form the so called 'normal' equations.

$$\Sigma Z_1 Y = \beta_1 \Sigma Z_1 X_1 + \beta_2 \Sigma Z_1 X_2 + \Sigma Z_1 V$$
$$\Sigma Z_2 Y = \beta_1 \Sigma Z_2 X_1 + \beta_2 \Sigma Z_2 X_2 + \Sigma Z_2 V \tag{8.48}$$

Now we can see that, providing Z_{1t} is not an exact linear multiple of Z_{2t}, we can express these two equations as solutions for the two true parameters.

$$\beta_1 = \frac{\Sigma Z_1 Y \Sigma Z_2 X_2 - \Sigma Z_2 Y \Sigma Z_1 X_2}{\Sigma Z_1 X_1 \Sigma Z_2 X_2 - \Sigma Z_1 X_2 \Sigma Z_2 X_1}$$
$$+ \frac{\Sigma Z_1 X_2 \Sigma Z_2 V - \Sigma Z_2 X_2 \Sigma Z_1 V}{\Sigma Z_1 X_1 \Sigma Z_2 X_2 - \Sigma Z_1 X_2 \Sigma Z_2 X_1}. \tag{8.49}$$

Now if Z_1 and Z_2 are independent of V then using the first term of (8.49) will yield an estimator which is unbiased for β_1

$$\tilde{\beta}_1 = \frac{\Sigma Z_1 Y \Sigma Z_2 X_2 - \Sigma Z_2 Y \Sigma Z_1 X_2}{\Sigma Z_1 X_1 \Sigma Z_2 X_2 - \Sigma Z_1 X_2 \Sigma Z_2 X_1}. \tag{8.50}$$

The estimator $\tilde{\beta}_2$ can be written immediately by noting the complete symmetry between X_1 and X_2 and between Z_1 and Z_2. The variance of $\tilde{\beta}_1$ is obtained from, as usual, taking the expectation of the square of the second term in (8.49). Assuming V is homoskedastic and not serially correlated we have:

$$\operatorname{Var} \tilde{\beta}_1 = \sigma \{ \Sigma Z_2^2 (\Sigma Z_1 X_2)^2 + \Sigma Z_1^2 (\Sigma Z_2 X_2)^2 - 2 \Sigma Z_1 Z_2 \Sigma Z_1 X_2 \Sigma Z_2 X_2 \} / D^2 \tag{8.51}$$

where D is the denominator of (8.50). We can see that were Z_{1t} to be an exact linear multiple of Z_{2t} or X_{1t} of X_{2t} then the variance would be infinite and no estimator would exist. Hence in the multiple instrument case we add the extra assumption that the instruments, as well as the variables, must be linearly independent.

Were there to be more instruments than parameters to be estimated, then we should again have to revert to the two-stage procedure. The first stage would be to regress in turn each X variable on all the Z variables to obtain the fitted values $\hat{X}_{1t}$, $\hat{X}_{2t}$ etc. These could then be substituted into the structural equation for OLS to be used, or equivalently the fitted values could then be used as the instruments for the basic equation.

Having given the general approach in the multivariable case we can now see how to apply it when, say, one variable is measured with error (X_2) and the other (X_1) is not. In such a case the variable X_2 must be 'replaced' by an instrument Z which is uncorrelated with the measurement error. The variable X_1 also needs an instrument (we must have two normal equations in order to obtain a solution), but the value of X_1 is the optimal instrument. It is uncorrelated with the error term and is perfectly correlated with X_1—thus yielding the smallest variance. A particular application of this idea is the single variable plus intercept model where the variable is measured with error:

$$Y_t = \alpha + \beta X_t + V_t. \tag{8.52}$$

We use as instruments variable Z (to act for X) and the unit variable to 'act as its own instrument'. The normal equations would be

$$\Sigma Y = T\alpha + \beta \Sigma X + \Sigma V \tag{8.53}$$

and
$$\Sigma Z Y = \alpha \Sigma Z + \beta \Sigma Z X + \Sigma Z V; \tag{8.54}$$

or
$$\bar{Y} = \alpha + \beta \bar{X} + \bar{V} \tag{8.55}$$

$$\therefore \quad \bar{Y}\Sigma Z = \alpha \Sigma Z + \beta \bar{X} \Sigma Z + \bar{V} \Sigma Z$$

and subtracting this from (8.54) and using the properties of means we obtain

$$\Sigma zy = \beta \Sigma zx + \Sigma zv \tag{8.56}$$

$$\therefore \quad \frac{\Sigma zy}{\Sigma zx} = \beta + \frac{\Sigma zv}{\Sigma zx} \tag{8.57}$$

and hence the left-hand side is an unbiased estimator:

$$\tilde{\beta} = \Sigma zy/\Sigma zx; \; (\tilde{\alpha} = \bar{Y} - \tilde{\beta}\bar{X}) \tag{8.58}$$

with variance

$$\operatorname{Var} \tilde{\beta} = \frac{\sigma^2 \Sigma z^2}{(\Sigma zx)^2} \tag{8.59}$$

which is the same form as for the model without an intercept, but has the variables measured in derivations from their means.

8.4 Random measurement errors and probability limits

The model we have considered for the analysis of measurement error is restrictive in the assumptions it makes about the behaviour of the measurement errors. It assumes that within the data sample, however big or small, the average measurement error is zero and the covariance between 'true value' and measurement error is also zero. Since the size of the sample is purely arbitrary those assumptions are not tenable in exact form, even though they may be good approximations for large samples. In order to obtain results which are applicable to a wider range of situations we clearly need to lessen the restrictiveness of the assumptions. We start by recognizing that the measurement error itself is likely to be a random variable—for a given value of the true variable X_t^*, the measurement error X_t^{**} would vary in repeated samples. Given this relaxation of the earlier assumption that X_t^{**} was fixed for each t, we need to make assumptions about its sampling behaviour. To do this in any useful way it is necessary to use the idea of a probability limit (see Chapter 6). This we can only do in an informal and unrigorous fashion.

This can be seen to be necessary by considering again the model (8.27) where

$$\hat{\beta} = \beta + \frac{\Sigma XV}{\Sigma X^2} \tag{8.27}$$

where V contains the measurement error, which is a random variable. The expectation of the OLS estimator is

$$E(\hat{\beta}) = \beta + E\left(\frac{\Sigma XV}{\Sigma X^2}\right). \tag{8.60}$$

However, since X and V are now both random variables and not independent of each other (both contain the random measurement error) we cannot, for a given finite sample size, set the expectation of a ratio equal to the ratio of the expectations (such an operation is correct only for independent variables). Hence we cannot guarantee that OLS is unbiased. However, the results on probability limits suggest a way to evaluate OLS in this situation. Provided that

$$\text{plim}\left(\frac{1}{T}\Sigma XV\right) = \text{plim}\left(\frac{1}{T}\Sigma X^{**2}\right), \text{ and exists;}$$

$$\text{plim}\left(\frac{1}{T}\Sigma X^2\right) = \text{plim}\left(\frac{1}{T}\Sigma X^{*2}\right), \text{ and exists;} \tag{8.61}$$

then we have:

$$\operatorname{plim} \hat{\beta} = \beta - \beta \frac{\operatorname{plim}\left(\frac{1}{T}\Sigma X^{**2}\right)}{\operatorname{plim}\left(\frac{1}{T}\Sigma X^{*2}\right)} \tag{8.62}$$

and we see that OLS is inconsistent (the probability limit of the estimator does not equal the parameter). The assumptions (8.61) introduce the factor $1/T$ in front in order to keep the limits finite. The probability limit of a sum is automatically the sum of the probability limits so that we are implicitly assuming

$$\operatorname{plim}\left(\frac{1}{T}\Sigma X^* X^{**}\right) = 0$$

$$\operatorname{plim}\left(\frac{1}{T}\Sigma XU\right) = 0 \tag{8.63}$$

which is effectively assuming that for large samples the true values and measurement errors are independent of each other and of the equation error.

The instrumental variable estimator can also be brought within this framework. We can accept that the instrument is a random variable and would vary in repeated samples; we consider the estimator

$$\tilde{\beta} = \frac{\Sigma ZY}{\Sigma ZX} = \beta + \frac{\Sigma ZV}{\Sigma ZX}. \tag{8.64}$$

If we add the assumptions that

$$\operatorname{plim}\left(\frac{1}{T}\Sigma ZV\right) = 0$$

$$\operatorname{plim}\left(\frac{1}{T}\Sigma ZX\right) \neq 0 \tag{8.65}$$

than we see that IV based upon instrument Z is consistent. Large sample values of the estimator will all be close to the true value of the parameter. The probability limit of the variance is also obtained from (8.64):

$$\operatorname{plim}(\operatorname{Var}\tilde{\beta}) = \left\{\operatorname{plim}\left(\frac{\frac{1}{T}\Sigma ZV}{\frac{1}{T}\Sigma ZX}\right)^2\right\} \tag{8.66}$$

(given that the variance is evaluated *around the large sample mean* β);

$$\text{plim Var }\tilde{\beta} = \frac{\sigma_v^2 \frac{1}{T}\text{plim}\left(\frac{1}{T}\Sigma Z^2\right)}{\left\{\text{plim}\left(\frac{1}{T}\Sigma XZ\right)\right\}^2} \tag{8.67}$$

Given the assumption that plim $(1/T \Sigma Z^2)$ also exists we see that this is the analogue to the variance for the fixed-error case. However, it is unclear as to how useful the formula (8.67) really is since it is clearly equal to zero. This is of course no surprise because if IV is to be a consistent estimator the variance of the repeated samples (when the sample size is sufficiently large) must be zero. It is true that the finite sample version (using values estimated from the data instead of the probability limits) is often used to express the finite sample variance of IV but this approximation is of dubious validity for two reasons. First, the finite sample mean is not the true parameter as (8.67) assumes, but is in general biased even for IV, and second, the derivation of (8.67) from (8.66) also relies on independence arguments for plims which do not generally hold for finite samples. Hence the attempt to use an IV to estimate the true parameter is indeed consistent for the true parameter under quite general conditions, but we cannot give full weight to attempts to base significance tests or construct confidence intervals based on the asymptotic variance formula for the IV estimator. The price paid for the wider applicability of the estimator is that of a less specific knowledge of its sampling properties.

8.5 Testing for specification and measurement errors

So far we have analysed, from a position of omniscience, the effects of the mistaken assumptions made by an econometrician in using OLS in the presence of various errors. In practical terms this knowledge is useful to us in analysing the potential effects of errors when we suspect that they might be present. It is clear, in the face of the suspicion, but not certainty, that such problems might be present, that tests to check for the presence of these difficulties would be very useful.

We begin with tests for specific errors in which the list of variables included in the regression is incorrect. The simpler problem is obviously to test for the incorrect inclusion of some variables. Suppose that a variable is to be regressed on two variables, one of which X_2 is of uncertain validity. The standard test for the marginal impact of (X_2) on Y, given that X_1 is already in the model, is the $F(t)$ test developed in Chapter 5. The null hypothesis is

$$H_0: \beta_2 = 0$$
$$H_1: \beta_2 \neq 0$$

(given that X_1 is included in the model). Hypotheses on subsets of variables can also be handled via the F-test procedure.

The problem of incorrect exclusion is also able to be handled in the same way if we know the variable that may have been incorrectly excluded. In practical terms this is much less likely than being able to specify a variable which is included but is of doubtful validity. By its nature it will tend to include all those factors which we have forgotten, or for which our theory is incomplete. Therefore, as well as being able to test against specified variables, it is desirable to see whether some more general tests for omitted variables can be constructed. The key to such tests is the observation that if a variable has been omitted it will have been in effect included in the error term. Now the biased OLS estimator will of course have absorbed some of the influence of this variable (depending on the correlation between the included and the excluded variables) but some will in effect still be left in the residuals derived from the incorrect OLS estimation. If these residuals show any pattern this is evidence that some systematic factor has been omitted. In time-series analysis the most likely pattern to emerge is some form of serial correlation since most economic variables are themselves autocorrelated. Hence it is very common to use the Durbin–Watson test for serial correlation as a test of specification. A significant value of $\hat{d}$ is taken to indicate that some variable has been omitted, but of course it does not tell us which variable has been omitted and the response to the presence of serially correlated errors is often just to re-estimate using the Cochrane–Orcutt or a related procedure. As we saw in Chapter 7, the Durbin–Watson test is perhaps most usefully applied if the alternative hypothesis is that there is first-order serial correlation. We can see what type of specification error would indeed correspond to this situation. Suppose that the true model is

$$Y = \alpha + \beta X + \gamma X_{-1} + \delta Y_{-1} + \varepsilon \quad (8.68)$$

where the ε are well behaved. This can be compared to a simpler equation with an AR(1) error process:

$$Y = A + \beta X + U \quad (8.69)$$

where
$$U = \rho U_{-1} + V. \quad (8.70)$$

The GLS form for estimating (8.69) is

$$Y - \rho Y_{-1} = A(1 - \rho) + B(X - \rho X_{-1}) + V. \quad (8.71)$$

This can be rewritten

$$Y = A(1 - \rho) + BX - B\rho X_{-1} + \rho Y_{-1} + V. \quad (8.72)$$

The equation (8.72) has exactly the same form as (8.68) except that there are restrictions between the various coefficients. The coefficient on X_{-1} divided by that on X is equal to minus that on Y_{-1}. Hence estimating (8.68) subject to the restriction

$$\hat{\gamma}/\hat{\beta} = -\hat{\delta} \tag{8.73}$$

is equivalent to estimating (8.69) subject to first-order serial correlation. If (8.69) and (8.70) are correct then this model sees serial correlation as a 'convenient simplification' of (8.68), i.e. omitting the variables X_{-1} and Y_{-1} but using GLS may be an easy way to impose the restriction on (8.68). However, if (8.68) is correct in unrestricted form, i.e. (8.73) does not hold, then attempts to treat it as if it is a single equation (8.70) subject to first-order serial correlation will be incorrect. Thus when we find evidence that the residuals from OLS are not serially independent (e.g. by the Durbin–Watson test) we should not automatically conclude that the specification error of omitting variables (here X_{-1} and Y_{-1}) will be exactly solved by using a technique designed to correct for first-order serial correlation. The variables X_{-1} and Y_{-1} can be genuinely omitted and produce serially correlated residuals without being fully incorporated by the serial correlation correction. If we suspect that these are the omitted variables then we could estimate (8.68) in unrestricted form and also subject to the restriction (8.73) using ML techniques to impose the restriction, and use an LR test to check whether the non-linear restriction is obeyed by the data.

A second important type of check is to allow for the possibility of higher-order serial correlation by using the LM test suggested in Chapter 7. Evidence on higher-order serial correlation suggests longer lags in omitted variables. In all cases where there is evidence of specification error as shown by serial correlation tests we must also think about the possibility that some other variable, not included in the model in any form, has been omitted.

Not all omitted variables or specification errors will produce serially correlated residuals from OLS on the incorrect equation, so that other tests can be used to advantage. If the omitted variable is rather cyclical and the magnitude of the cycle varies then the residuals may show evidence of heteroskedasticity. The tests discussed in Chapter 7 (F tests and the White LM test) could be used if something could be specified about the nature of the possible heteroskedasticity. A final approach is that of checking the stability of the equation. If there is an omitted variable then its correlation with included variables might be different for different time periods and this would be reflected in substantially different estimators of the same parameter when the model was split into sub-periods (reflecting the different degrees of bias). Hence an F test for structural stability can be used to check for possible misspecification. It is common practice to estimate the model for a sub-period (say $1 \ldots \tau$) reserving the final observations ($\tau + 1 \ldots T$) for checking. Once the equation has been finalized on the initial data it is then re-estimated on the new data and a standard F test for equality of *all* parameters is carried out (see Chapter 5). Rejection of the null hypothesis of no structural change is taken to suggest there may have been a specification error of a larger type, rather than that the model was basically correct but that the value of one of the parameters shifted.

In the case of measurement error we may be in a very different position if we can specify a suitable IV to use if there is such error, and tests for measurement error are based on analysing the magnitude of the difference between an OLS estimator and an IV estimator for the same model. Consider the simple case:

$$Y_t = \beta X_t + U_t \tag{8.74}$$

where there may be measurement error in X_t. The data is assumed to be fixed in repeated samples.

There is also available an instrumental variable Z (fixed in repeated samples) which is independent of the measurement error if present. Now under the null hypothesis of no measurement error we have two possible estimators for the parameter—OLS and IV:

$$\hat{\beta} = \Sigma X Y / \Sigma X^2 \tag{8.75}$$

$$\tilde{\beta} = \Sigma Z Y / \Sigma Z X. \tag{8.76}$$

Under the null hypothesis both estimators will be unbiased but of course OLS will be efficient (having smaller variance). However, if the alternative hypothesis is true, that there is measurement error present, the OLS will be biased while IV will continue to be unbiased. Hence the difference between the two estimators should form the basis of a test with good power since this difference will be small usually under the NH and large usually under the AH. A test, due to Hausman, uses the mean and variance of the difference. Under the NH it is clear that:

$$\hat{\beta} - \tilde{\beta} = \frac{\Sigma X_t U_t}{\Sigma X_t^2} - \frac{\Sigma Z_t U_t}{\Sigma Z_t X_t}. \tag{8.77}$$

If the X and Z are fixed and if the U are independently identically normally distributed then this statistic also has a normal distribution with zero mean. The variance of (8.77) is easily shown to be

$$\text{Var}(\hat{\beta} - \tilde{\beta}) = \frac{\sigma^2 \Sigma Z^2}{(\Sigma Z X)^2} - \frac{\sigma^2}{\Sigma X^2}. \tag{8.78}$$

However this is the difference between the IV variance and the OLS variance. To make this operational we can estimate σ^2 from the OLS residuals and derive the associated t (or F) statistic:

$$\frac{\hat{\beta} - \tilde{\beta}}{(\text{Var}\, \tilde{\beta} - \text{Var}\, \hat{\beta})^{\frac{1}{2}}} \sim t(\text{d.f.}). \tag{8.79}$$

Under the NH this statistic follows a t distribution with $T-1$ degrees of freedom, while if the AH is true the statistic will be pulled away from zero

236 Specification and Measurement Errors

(IV ≠ OLS) and the test will have high power in picking up the presence of the measurement errors. Clearly the nearer is the IV variance to the OLS variance the higher will be the value of the 't' statistic and the greater the power of the test, i.e. the more sensitive it will be to a given divergence between the values of the two estimates.

The use of IV estimation and a test for possible measurement error are shown in example 8.1.

Example 8.1 A test for measurement error

We consider the basic model of consumption related to income, but without an intercept. The basic OLS equation is

$$C = 0.894 Y$$
$$(0.00468) \quad SEE = 2879$$

This is re-estimated by instrumental variables using the unit variable as instrument (i.e. assuming that the true average error is zero and thus is uncorrelated with the instrument):

$$C = 0.897 Y$$
$$(0.00478) \quad SEE = 2904$$

The instrumental variable makes very little difference either to the coefficient or the standard error of the estimated coefficient. We test for a significant difference between the two estimators (two-tailed test) using Hausman's test. The statistic is

$$H = \frac{(0.897 - 0.894)}{\sqrt{(0.00478^2 - 0.00468^2)}}$$
$$= 0.003/0.00097$$
$$= 3.09.$$

This statistic, if the null hypothesis is true, is distributed as a t statistic with 21 degrees of freedom. The critical value of t with a 95 per cent significance level and two-tailed test is 2.08. We conclude that the two estimators are significantly different. In general this suggests that the error term is correlated with income (provided that we are sure that the errors have mean zero). One case of this is measurement error for the explanatory variable, as explained in the text, but another cause could be the presence of an omitted variable. It is noticeable but even a very small difference in the estimated coefficients is seen as significant because of the very small variance of the difference between estimators.

Problems 8

8.1 Consider the true model

$$Y_t = \beta_1 X_{1t} + \beta_2 X_{2t} + U_t$$
$$E(U_t) = 0$$
$$E(U_t U_s) = 0 \quad s \neq t$$
$$\quad\quad\quad\quad = \sigma^2 \quad s = t.$$

Let $\hat{\beta}_1$ be the OLS estimator obtained from regressing Y on X_1 and X_2, with β_1^* that from regressing Y just on X_1.

(a) Show that

$$\frac{\text{MSE}(\beta_1^*)}{\text{MSE}(\hat{\beta}_1)} = 1 + r_{12}^2(t_2^2 - 1)$$

where
$$r_{12}^2 = (\Sigma X_1 X_2)^2 / \Sigma X_1^2 \, \Sigma X_2^2$$
$$t_2^2 = \beta_2^2 / \text{Var } \hat{\beta}_2.$$

(b) Consider the weighted estimator

$$\check{\beta}_1 = \lambda \hat{\beta}_1 + (1 - \lambda)\beta_1^*.$$

For what value of λ is the MSE $(\check{\beta}_1)$ minimized?

(c) Comment on the usefulness of these results.

8.2 There is a model:

$$Y_t = \beta X_t^* + U_t$$

when
$$E(U_t X_t^*) = 0$$
$$E(U_t U_s) = 0 \quad s \neq t$$
$$\quad\quad\quad\quad = \sigma^2 \quad s = t.$$

X_t^* is unobservable but is measured by P_t (a 'proxy' variable) where

$$P_t = \gamma X_t^* + e_t$$

as
$$E(e_t X_t^*) = 0$$
$$E(e_t U_t) = 0$$

(a) It is proposed to regress Y on P. Discuss the properties of the resulting coefficient as an estimator for β.

238 Specification and Measurement Errors

(b) There is an instrumental variable Z_t available such that

$$\text{plim} \frac{1}{T}\Sigma Z_t U_t = \text{plim} \frac{1}{T}\Sigma Z_t e_t = 0$$

$$\text{plim} \frac{1}{T}\Sigma Z_t P_t \neq 0.$$

Discuss the properties of the coefficient obtained from the use of instrumental variables to estimate the relationship between Y and P.

(c) Suppose that we change the assumption that $E(e_t X_t^*) = 0$ to $E(e_t P_t) = 0$. What will be the effect on the properties of the estimator from regressing Y on P?

(d) How would the results in (c) be modified if we could not assume $E(e_t U_t) = 0$?

8.3 Consider the model

$$Y_t = \beta X_t^* + \gamma Z_t + U_t$$

where
$$E(U_t) = 0$$
$$E(U_t U_s) = 0 \qquad s \neq t$$
$$= \sigma^2 \qquad s = t$$

X^* is not observed but could be replaced by a proxy variable P from the relationship

$$P_t = X_t^* + V_t.$$

Assume
$$E(VU) = E(VZ) = E(VX^*)$$
$$= E(ZU) = E(X^*U) = 0$$
$$E(V) = 0$$
$$E(V_s V_t) = 0 \qquad s \neq t$$
$$E(V_t^2) = \sigma^2.$$

We estimate the equation first by omitting X^* altogether and regressing Y on Z, and second by using the proxy and regressing Y on P and Z. Show that the bias in estimating γ is smaller when the proxy is included.

8.4 There is an equation

$$Y_t = \beta X_t + U_t$$

where
$$E(U_t) = 0$$
$$E(U_t U_s) = 0 \qquad s \neq t$$
$$= \sigma^2 \qquad s = t.$$

There is a variable Z_t such that $\frac{1}{T}\Sigma Z_t X_t \neq 0$. Show that the following estimation procedures give identical values for the estimator of β:
 (a) Use Z_t as an instrumental variable;
 (b) Regress X_t on Z_t (without an intercept) and obtain the fitted value $\hat{X}_t$— then regress Y on $\hat{X}$;
 (c) Use the $\hat{X}$ as an instrumental variable.

A second instrument W_t is available. Show how two-stage least squares generally has a smaller variance than an IV estimator using just Z_t as the instrument.

8.5 Consider the model
$$Y = \beta_1 X_1^* + \beta_2 X_2^* + U$$
where X_1^* and X_2^* are measured with error by X_1 and X_2 respectively. The measurement errors are uncorrelated with the true values. There are two instrumental variables Z_1 and Z_2 available such that they are uncorrelated with the measurement errors or the equation error.

 (a) It is proposed to use the standard IV formula (8.50) to estimate β_1. Under what conditions will this estimator not exist (will the variance be infinite)?
 (b) Use the results of (a) to discuss the following case:
$$Y = \beta_1 X_1^* + \beta_2 X_2 + U$$
(i.e. only X_1^* is measured with error). It is proposed to estimate this by IV but the instrument Z_t that is proposed to replace X_1 is in fact a multiple of X_{2t}. What are the properties of such an estimator?

8.6 Show that in the model
$$Y_1 = \beta_1 X_1 + \beta_2 X_2 + U$$
where
$$E(U_t) = 0$$
$$E(U_t U_s) = 0 \quad s \neq t$$
$$= \sigma^2 \quad s = t$$
when we regress Y solely on X_1, and obtain the residuals $\hat{U}_t = (Y - \hat{\beta} X_1)$, the value of the estimated error variance ($\hat{S}^2$) based on these residuals is biased upwards.

8.7 There is a model:
$$Y_t = \beta X_t + U_t$$
where
$$E(U_t) = E(X_t) = E(X_t U_{t-j}) = 0 \text{ all } j.$$
$$E(U_t U_{t-s}) = 0 \quad \text{for } s \neq 0$$
$$= \sigma^2 \quad s = 0.$$

Variable X is measured with error by Z such that $Z_t = X_t - V_t$, where $E(V_t) = E(X_t V_{t-j}) = E(U_t V_{t-j}) = 0$ all j. The variable Y is regressed on Z and the residuals are denoted $\hat{U}_t$. Show that the probability limit of the estimated first-order serial correlation coefficient $\hat{\rho}$, where $\hat{\rho} = \Sigma \hat{U}_t \hat{U}_{t-1} / \Sigma \hat{U}_{t-1}^2$, is given by

$$\text{plim } \hat{\rho} = \frac{\beta^2 r^2 \sigma_{XX}(-1) + \beta^2(1-r)^2 \sigma_{VV}(-1)}{\sigma_{UU} - \beta^2 r^2 \sigma_{XX} + \beta^2(1-r^2)\sigma_{VV}}$$

where

$$\sigma_{KK} = \text{plim } \frac{1}{T} \Sigma K_t^2$$

$$\sigma_{KK}(-1) = \text{plim } \frac{1}{T} \Sigma K_t K_{t-1}$$

(K represents U, V, and X), and

$$r = \frac{\text{plim } \frac{1}{T} \Sigma V^2}{\text{plim } \frac{1}{T} \Sigma Z^2}$$

Answers 8

8.1 (a) See the answer to question 4.15 using the value $C = 0$.

(b) Using the result that $\text{MSE}(a\theta) = a^2 \text{MSE}(\theta)$ and (4.15) we have

$$\text{MSE}(\breve{\beta}_1) = \lambda^2 \text{MSE}(\hat{\beta}_1) + (1-\lambda)^2 \text{MSE}(\beta_1^*)$$

Minimizing with respect to λ:

$$\frac{\partial \text{MSE}(\breve{\beta}_1)}{\partial \lambda} = 2\lambda \text{MSE}(\hat{\beta}_1) - 2(1-\lambda) \text{MSE}(\beta_1^*) = 0$$

$$\therefore \quad \lambda^* = \frac{2 \text{MSE}(\beta_1^*)}{2 \text{MSE}(\hat{\beta}_1) + 2 \text{MSE}(\beta_1^*)}$$

$$\lambda^* = \frac{\text{MSE}(\beta_1^*)/\text{MSE}(\hat{\beta}_1)}{1 + \text{MSE}(\beta_1^*)/\text{MSE}(\hat{\beta}_1)}$$

and using the result of (a) we have for the optimal weight:

$$\lambda^* = \frac{1 + r_{12}^2(t_2^2 - 1)}{2 + r_{12}^2(t_2^2 - 1)}.$$

(c) The first result shows that, under certain circumstances, if we omit a variable the MSE for the parameter on the other is smaller than had we included both in the regression (this corresponds to letting $\lambda = 0$). The decision rule to choose between $\lambda = 0$ and $\lambda = 1$ (keeping both in) is that if the 'true' 't' statistic ($\beta_2/(\text{Var } \hat{\beta}_2)^{\frac{1}{2}}$) is less than unity then X_2 should be omitted. This cannot however be used as an operational rule to suggest that if the actual 't' statistic—$\hat{\beta}_2/(\widehat{\text{Var } \hat{\beta}_2})^{\frac{1}{2}}$—is less than unity then omitting X_2 yields the smaller MSE.

The result is suggestive however, in that when the estimated variance of the second variable is high and the correlation between X_1 and X_2 is high, then omitting X_2 may lead to a more reliable estimator of β_1.

The second part of the question clearly generalizes the first and instead of choosing $\lambda = 0$ or $\lambda = 1$ it finds the optimal average. Again the optimal value of λ is a function of the 'true' t and cannot be used operationally. It shows that at $t_2 = 1$ the optimal estimator is an equal weighted average, while as t increases above 1 the value of λ^* runs towards unity (OLS on both variables). Again if the value of r_{12}^2 is zero an equal weight average should be used, while as r_{12}^2 increases towards unity the optimal weight tends to $t^2/(1+t^2)$. The only circumstance in which pure OLS should be preferred to the weighted average estimator is when t^2 is very large. This is not a surprising result—we have shown that in the single-variable case the minimum means square error estimator for a parameter is (problem 3.13)

$$\beta^* = \frac{\Sigma X_t Y_t}{\sigma^2/\beta^2 + \Sigma X_t^2}$$

$$= \frac{\beta}{(1/t^2) + 1}$$

where $t^2 = \beta^2/\text{Var } \hat{\beta}$, so that here OLS coincides with MMSE only as t grows large.

8.2 The way of writing the proxy variable equation with a coefficient conceals the question of dimensionality. Since X^* is unobservable it can have any unit we choose and γ will just take the inverse scaling so that the dimension of γX^* is the same as P. It is conventional to 'normalize' γ to the value units so that the model is measured in units of X^* (a unit change in X^* is associated with a unit change in P). We shall assume that this is done for the rest of the question.

(a) Substituting the proxy variable equation into the main equation we have

$$Y = \beta P_t + U_t - \beta e_t.$$

Carrying out proxy variable estimation

$$\tilde{\beta} = \Sigma YP/\Sigma P^2$$

and substituting in we have:

$$\tilde{\beta} = \beta + \frac{\Sigma(U - \beta e)P}{\Sigma P^2}$$

To evaluate the properties of this estimator we must recognize that we cannot generally assume P to be fixed (it varies with e) so that we cannot use expectations. Accordingly we need to take probability limits:

$$\text{plim } \check{\beta} = \beta + \frac{\text{plim}\left(\frac{1}{T}\Sigma(U-\beta e)P\right)}{\text{plim}\left(\frac{1}{T}\Sigma P^2\right)}.$$

Now
$$E(UP) = E\{U(X^*+e)\} = 0$$

and
$$E(eP) = E\{e(X^*+e)\} = E(e^2)$$

$$\therefore \quad \text{plim } \check{\beta} = \beta - \beta\frac{\text{plim }\frac{1}{T}\Sigma e^2}{\text{plim }\frac{1}{T}\Sigma P^2}.$$

As we have already shown for measurement error models OLS is biased downwards.

(b) The IV estimator is $\tilde{\beta} = \Sigma ZY/\Sigma ZP$, therefore

$$\tilde{\beta} = \beta + \frac{\Sigma Z(U-\beta e)}{\Sigma ZP}.$$

Again taking plims we have plim $\tilde{\beta} = \beta$ (since plim$(\Sigma Ze)/T) = 0$ and plim$(\Sigma ZP)/T \neq 0$). The variance is given by

$$\text{Var } \tilde{\beta} = \frac{\frac{1}{T}\Sigma Z^2 \frac{1}{T}\Sigma V^2}{\left(\frac{1}{T}\Sigma ZP\right)^2}$$

where $V = U - \beta e$. This can be estimated by using the residuals from the IV estimation.

$$\tilde{V}_t = Y_t - \tilde{\beta}P_t.$$

(c) If $E(e_t P_t) = 0$ then we see from (a) that plim $\check{\beta} = \beta$ (because plim$(\Sigma UP)/T = 0$ still). This result is important because it shows that the measurement error does not necessarily lead to bias. The existence (and direction) of bias depends on the error properties.

(d) If we alter the assumption so that plim$(\Sigma eU)/T \neq 0$ then it follows that plim$(\Sigma UP)/T \neq 0$, since $E(UP) = E\{U(X^*+e)\}$ and hence OLS is again biased, but the sign of the bias depends on the sign of the covariance between the measurement error (e) and the equation error (U).

8.3 The relationship can be written $Y = \beta P + \gamma Z + W$, where $W = U - \beta V$.

(a) Regressing Y on Z we have:

$$\check{\gamma} = \Sigma YZ / \Sigma Z^2.$$

Substituting in for Y and taking probability limits we have (using the assumptions on the errors)

$$\text{plim } \check{\gamma} = \gamma + \beta \text{ plim} \left(\frac{1}{T} \Sigma PZ \bigg/ \frac{1}{T} \Sigma Z^2 \right) = \gamma + \beta \theta.$$

(b) Regressing Y on P and Z we have

$$\gamma^0 = \frac{\Sigma YZ \Sigma P^2 - \Sigma YP \Sigma ZP}{\Sigma Z^2 \Sigma P^2 - (\Sigma ZP)^2}$$

$$= \gamma + \frac{\Sigma WZ \Sigma P^2 - \Sigma WP \Sigma ZP}{D}$$

where $\qquad D = \Sigma Z^2 \Sigma P^2 - (\Sigma ZP)^2.$

Replacing W by $U - \beta V$, taking plims and using our assumptions we have:

$$\text{plim } \gamma^0 = \gamma + \frac{\beta \text{ plim } \frac{1}{T} \Sigma V^2 \text{ plim } \frac{1}{T} \Sigma ZP}{\text{plim } \frac{1}{T} \Sigma Z^2 \text{ plim } \frac{1}{T} \Sigma P^2 - \text{plim} \left[\left\{ \frac{\frac{1}{T}(\Sigma ZP)^2}{\frac{1}{T} \Sigma Z^2} \right\} \right]}$$

Now

$$\text{plim } \frac{1}{T} \Sigma P^2 = \text{plim } \frac{1}{T} \Sigma X^{*2} + \text{plim } \frac{1}{T} \Sigma V^2$$

$$\text{plim } \frac{1}{T} \Sigma ZP = \text{plim } \frac{1}{T} \Sigma ZX^*$$

$$\text{plim } \frac{1}{T} \Sigma X^{*2} = \sigma_x^2$$

$\therefore \qquad \text{plim } \gamma^0 = \gamma + \beta \theta \left\{ \frac{\sigma^2}{\sigma^2 + \sigma_x^2(1 - r_{ZX}^2)} \right\}$

where r_{ZX}^2 is the probability limit of the squared correlation between X and Z. Hence the factor in square brackets is definitely less than unity and the bias in γ^0 is less than the bias in $\check{\gamma}$. As the correlation between Z and X increases the bias of the two estimators tends to equality. This model is sometimes taken to imply that in cases of measurement error it is better to include even a 'poor' proxy rather than none at all. Such an inference depends critically on the error assumptions—different assumptions can reverse the effect.

8.4 (a) Using IV solution we have $\tilde{\beta} = \Sigma ZY/\Sigma ZX$ (standard result).

(b) Regressing X on Z we have for the fitted value $\hat{X}_t = Z_t \Sigma ZX/\Sigma Z^2$ and regressing Y on $\hat{X}$,

$$\beta^* = \Sigma Y\hat{X}/\Sigma \hat{X}^2$$

$$= \frac{\Sigma YZ\left(\frac{\Sigma ZX}{\Sigma Z^2}\right)}{\Sigma Z^2\left(\frac{\Sigma ZX}{\Sigma Z^2}\right)^2}$$

$$= \Sigma YZ/\Sigma ZX = \tilde{\beta}.$$

(c) Using $\hat{X}$ as instrument:

$$\beta^0 = \Sigma \hat{X} Y/\Sigma \hat{X} X$$

$$= \frac{\Sigma ZY \frac{(\Sigma ZX)}{(\Sigma Z^2)}}{\Sigma ZX \frac{(\Sigma ZX)}{(\Sigma Z^2)}} = \frac{\Sigma ZY}{\Sigma ZX} = \tilde{\beta}.$$

(d) The variance of an instrumental variable (M_t) estimator is

$$\operatorname{Var} \tilde{\beta} = \frac{\sigma^2 \Sigma M^2}{(\Sigma MX)^2}.$$

We could argue that the stronger the correlation of M with X the nearer is the variance of IV to that of OLS (approaching from above). To obtain the instrument M we regress X on the list of variables Z_t and W_t and use the fitted value

$$M_t = \hat{X}_t = \hat{\pi}_1 W_t + \hat{\pi}_2 Z_t.$$

Since the correlation of X with W and Z is always at least as great as that of X on Z alone (see chapter on multiple regression) the instrument M has a higher correlation with X than the single instrument. This leads to a variance nearer to OLS and hence lower than that of IV based on a single variable.

8.5 (a) The estimator is derived from the normal equations as usual to give:

$$\tilde{\beta}_1 = \frac{\Sigma Z_1 Y \Sigma Z_2 X_2 - \Sigma Z_2 Y \Sigma Z_1 X_2}{\Sigma Z_1 X_1 \Sigma Z_2 X_2 - \Sigma Z_1 X_2 \Sigma Z_2 X_1}$$

$$\operatorname{Var} \tilde{\beta}_1 = \sigma^2 \{\Sigma Z_2^2 (\Sigma Z_1 X_2)^2 + \Sigma Z_1^2 (\Sigma Z_2 X_2)^2 - 2\Sigma Z_1 Z_2 \Sigma Z_1 X_2 \Sigma Z_2 X_1\}/D^2$$

where D is the same denominator as for $\tilde{\beta}$. No estimator will exist if $D=0$. This condition can fail in three separate ways:

(i) $\quad\quad\quad\quad\quad\quad\quad\quad X_{1t} = KX_{2t}$ all t

(ii) $\quad\quad\quad\quad\quad\quad\quad\quad Z_{1t} = MZ_{2t}$ all t

(iii) $$\frac{\Sigma Z_1 X_1}{\Sigma Z_1 X_2} = \frac{\Sigma Z_2 X_1}{\Sigma Z_2 X_2}$$

Now condition (iii) is implied by (i) or (ii) but could occur even if neither held. It means that the relative strength of correlation of Z_1 to X_1 or X_2 is the same as that of Z_2 to X_1 or X_2, i.e. the two instruments have the same *relative* correlations to the two measured variables.

(b) The instruments are $Z_{1t} = KX_{2t}$ and $Z_{2t} = X_{2t}$; they are clearly perfectly collinear and so the estimator does not exist.

8.6 $$E(\hat{S}^2) = E\left\{\frac{1}{T-1}\Sigma(Y_t - \hat{\beta} X_{1t})^2\right\}$$

where $$\hat{\beta}_1 = \Sigma YX_1/\Sigma X_1^2.$$

Substituting in for the true value of Y_t we have

$$E(\hat{S}^2) = E\left[\frac{1}{T-1}\Sigma\left\{\beta_1 X_1 + \beta_2 X_2 + U - X_1 \frac{\Sigma(\beta_1 X_1 + \beta_2 X_2 + U)X_1}{\Sigma X_1^2}\right\}^2\right]$$

$$= E\left\{\frac{1}{T-1}\Sigma\left(\beta_2 X_2 - \beta_2 X_1 \frac{\Sigma X_1 X_2}{\Sigma X_1^2} + U - X_1 \frac{\Sigma U X_1}{\Sigma X_1^2}\right)^2\right\}$$

$$= \frac{\beta_2^2 \Sigma\left(X_2 - X_1 \frac{\Sigma X_1 X_2}{\Sigma X_1^2}\right)^2 + T\sigma^2 - \sigma^2}{T-1}$$

$$= \frac{\beta_2^2 \Sigma X_2^2(1-r_{12}^2)}{T-1} + \sigma^2$$

where r_{12} is the correlation (in raw data form) between X_1 and X_2. The estimated error variance is biased upwards and this bias does not disappear as the sample size increases since both ΣX_2^2 and T increase. If the variables are strongly correlated then the effects of omitting one has less effect on the estimated variance (but of course does increase the bias).

8.7 Now $$\Sigma \hat{U}^2 = \Sigma\left(Y - \frac{\Sigma YZ}{\Sigma Z^2}\cdot Z\right)^2$$

$$= \Sigma W^2 - \frac{(\Sigma WZ)^2}{\Sigma Z^2}$$

where $$W = U + \beta V$$

But $$\text{plim}\frac{1}{T}\Sigma W^2 = \sigma_{UU} + \beta^2 \sigma_{VV}$$

$$\frac{\text{plim}\frac{1}{T}\Sigma WZ}{\text{plim}\frac{1}{T}\Sigma Z^2} = -\beta r$$

and
$$\text{plim}\frac{1}{T}\Sigma Z^2 = \sigma_{XX} + \sigma_{VV}$$

$\therefore$
$$\text{plim}\frac{1}{T}\Sigma \hat{U}^2 = \sigma_{UU} + \beta^2\sigma_{VV} - (\beta r)^2(\sigma_{XX} + \sigma_{VV})$$
$$= \sigma_{UU} + \beta^2(1-r^2)\sigma_{VV} - \beta^2 r^2 \sigma_{XX}.$$

$$\Sigma \hat{U}\hat{U}_{-1} = \Sigma\left(W - Z\frac{\Sigma WZ}{\Sigma Z^2}\right)\left(W_{-1} - Z_{-1}\frac{\Sigma WZ}{\Sigma Z^2}\right)$$

But
$$\text{plim}\frac{1}{T}\Sigma WW_{-1} = \beta^2 \sigma_{VV}(-1)$$

$$\text{plim}\frac{1}{T}\Sigma ZZ_{-1} = \sigma_{XX}(-1) + \sigma_{VV}(-1)$$

$$\text{plim}\frac{1}{T}\Sigma W_{-1}Z = \text{plim } 1\ \Sigma WZ_{-1} = -\beta\sigma_{VV}(-1)$$

Using these and the value for
$$\left(\frac{\text{plim}\frac{1}{T}\Sigma WZ}{\text{plim}\frac{1}{T}\Sigma Z^2}\right)$$

we have
$$\text{plim}\frac{1}{T}\Sigma \hat{U}\hat{U}_{-1} = \beta^2\sigma_{VV}(-1) + \beta^2 r^2[\sigma_{XX}(-1) + \sigma_{VV}(-1)] - 2\beta(\beta r)\sigma_{VV}(-1)$$

$\therefore$
$$\text{plim } \tilde{\rho} = \frac{\beta^2 r^2 \sigma_{XX}(-1) + \beta^2(1-r)^2\sigma_{VV}(-1)}{\sigma_{UU} + \beta^2(1-r^2)\sigma_{VV} - \beta^2 r^2 \sigma_{XX}}.$$

The importance of this result is obvious. With errors in variables the use of OLS on the measured values (i.e. ignoring the measurement error) can produce the appearance of Serial Correlation in the residuals even though the equation errors (U) are not serially correlated. Serial correlation in the measured value and/or the measurement error is enough to create this effect. As we have pointed out before, the solution to this problem is not to look automatically to a serial correlation type correction (e.g. Cochrane–Orcutt procedure) but instead to look for the specification error that produced the measurement error.

9
Simultaneous Equations

The discussion in the previous chapters has centred on models which are single equations only—that is, where there is a single link from the explanatory variable (X) to the explained variable (Y). Many economic models recognize that within the same time period there will be a 'feedback' also from Y to X. The simplest example is the basic Keynesian model:

$$C = a + bY \qquad (9.1)$$

$$Y = C + I. \qquad (9.2)$$

The first equation tells us that changes in income will lead to changes in the aggregate demand for consumer goods, while the second tells us that aggregate income, which equals aggregate output, is adjusted to equal aggregate demand within the period. Thus, if for any reason consumption were to increase independently of the income level then, unless aggregate demand for investment fell in a compensating fashion, output would rise to meet this demand within the same period. This would be translated into a rise in income which would have a further effect on consumption within the period. Hence income affects consumption and at the same time consumption affects income. The two variables are simultaneously determined. There are numerous other examples of such systems of simultaneous interaction between variables and, as we shall see, this interaction creates special problems for the econometrician.

We can illustrate the problem by letting the first equation of our system become stochastic by adding an error term,

$$C_t = a + bY_t + U_t \qquad (9.3)$$

thus recognizing that consumption is affected by other factors apart from income. If we also add the assumption that investment demand is not affected by consumption or income within the period, we can see that income and the error term must be correlated. A larger error causes C to rise and this in turn causes Y to rise etc. This covariance between movements in a right-hand side variable and the error term has the same consequences for OLS as it did in the cases of measurement error or omitted variables—OLS will be biased.

In order to analyse this problem systematically we need some general notation. Variables determined (simultaneously) within the system are called endogenous and are represented by $Y_{it} (i = 1 \ldots I)$, while those unaffected by the system are called exogenous and are denoted by $X_{jt} (j = 1 \ldots J)$. A simple system could be represented by the following equations.

$$Y_1 = \gamma_{11} X_1 + \beta_{12} Y_2 + U_1 \qquad (9.4)$$

$$Y_2 = \beta_{21} Y_1 + U_2 \qquad (9.5)$$

(the first suffix refers to the equation number while the second refers to the variable number). Both equations are stochastic. It is usual to write the equations each with a different endogenous variable on the left-hand side even though the system as a whole does not give priority to one or the other. In a supply/demand model the demand equation can equally well be thought of as having price or quantity on the left-hand side since they are in fact jointly determined.

We assume that the error terms are not serially correlated or heteroskedastic.

For this very simple system we can obtain the OLS bias for the coefficient, although even for the simplest systems the exact evaluation of the bias is cumbersome. Consider the estimator for γ_{11}:

$$\hat{\gamma}_{11} = \frac{\Sigma Y_1 X_1 \Sigma Y_2^2 - \Sigma Y_1 Y_2 \Sigma X_1 Y_2}{\Sigma X_1^2 \Sigma Y_2^2 - (\Sigma X_1 Y_2)^2}. \qquad (9.6)$$

The presence of the non-independent random variables (Y_1 and Y_2) in the numerator and denominator mean that we cannot evaluate piecemeal the expectations of the expression. It will clearly be necessary to use probability limits. To obtain an expression in terms of the parameters of the structural equations (9.4) and (9.5) it is necessary to substitute out the endogenous variables. We give the solutions to these before substituting into (9.6).

$$Y_1 = \frac{\gamma_{11}}{1 - \beta_{12}\beta_{21}} X_1 + \frac{U_1 + \beta_{12} U_2}{1 - \beta_{12}\beta_{21}} \qquad (9.7)$$

$$Y_2 = \frac{\gamma_{11}\beta_{21}}{1 - \beta_{12}\beta_{21}} X_1 + \frac{\beta_{21} U_1 + U_2}{1 - \beta_{12}\beta_{21}} \qquad (9.8)$$

These equations, known as the reduced form of the system, are often written in the compact notation:

$$Y_1 = \pi_{11} X_1 + V_1 \qquad (9.9)$$

$$Y_2 = \pi_{21} X_1 + V_2 \qquad (9.10)$$

where the π_{ij} (reduced-form parameters) are known functions of the structural parameters, and the V_i (reduced-form disturbances) are known functions of the structural errors. For our system we can substitute directly from (9.9) and (9.10) into (9.6).

$$\hat{\gamma}_{11} = \frac{\Sigma(\pi_{11} X_1 + V_1) X_1 \Sigma(\pi_{21} X_1 + V_2)^2 - \Sigma(\pi_{11} X_1 + V_1)(\pi_{21} X_1 + V_2) \Sigma X_1 (\pi_{21} X_1 + V_2)}{\Sigma X_1^2 \Sigma(\pi_{21} X_1 + V_2)^2 - \{\Sigma X_1 (\pi_{21} X_1 + V_2)\}^2} \qquad (9.11)$$

If we multiply numerator and denominator by $(1/T)^2$ and then take probability limits (assuming that all the relevant limits exist), we have after cancelling:

$$\lim \hat{\gamma}_{11} = \frac{\pi_{11} \operatorname{plim} \frac{1}{T} \Sigma V_2^2 - \pi_{21} \operatorname{plim} \frac{1}{T} \Sigma V_1 V_2}{\operatorname{plim} \frac{1}{T} \Sigma V_2^2}. \qquad (9.12)$$

Now expressing the V_i in terms of components we can obtain

$$\lim \hat{\gamma}_{11} = \pi_{11} - \pi_{21} \left\{ \frac{\gamma_{21}\sigma_{11} + (1+\gamma_{12}\gamma_{21})\sigma_{12} + \gamma_{12}\sigma_{22}}{\gamma_{21}^2 \sigma_{11} + 2\gamma_{12}\sigma_{12} + \sigma_{22}} \right\} \qquad (9.13)$$

where

$$\operatorname{plim} \frac{1}{T} \Sigma U_{it} U_{jt} = \sigma_{ij}. \qquad (9.14)$$

Here we allow for the possibility that errors in the *same period* from different equations have a non-zero and unknown covariance σ_{12}.

$$\therefore \qquad \operatorname{plim} \hat{\gamma}_{11} = \gamma_{11} \left(\frac{1 - \beta_{21}\theta}{1 - \beta_{21}\beta_{12}} \right) \qquad (9.15)$$

where the term θ is the term in curly brackets from (9.13). There are clearly two sufficient conditions for $\hat{\gamma}_{11}$ to be unbiased—

(i) $\qquad\qquad\qquad\qquad\qquad \beta_{21} = 0$

(ii) $\qquad\qquad\qquad\qquad\qquad \theta = \beta_{12}.$

The former condition means that there is no feedback from Y_1 to Y_2 so that the system is not simultaneous. The second condition could be met by a series of special relations between the parameters. The simplest condition for $\theta = \beta_{12}$ is $\sigma_{12} = 0$; and $\beta_{21} = 0$ or $\sigma_{11} = 0$. This would occur again if there were no simultaneity between the equations or if the first equation were non-stochastic. Hence we can see that in general the use of OLS will be biased when applied to one of a set of simultaneous equations. We also see that we can anticipate under what conditions the bias is likely to be large or small.

The fact that OLS is biased raises the question of whether any estimation method exists that will obtain consistent estimates of the structural parameters, given the information specified by the system.

The easiest way to approach estimation is to ask whether we can find (at least) as many independent 'normal' equations based on a given structural equation as there are parameters to be estimated from that equation. If it is possible to do so then we can solve for the unknown structural parameters. This approach, which effectively underlies the technique of instrumental variables, can be simply illustrated for our small system (9.4), (9.5).

250 Simultaneous Equations

We start with the second equation,

$$Y_2 = \beta_{21} Y_1 + U_2.$$

We need to find a variable, in fact to act as an instrument, that is uncorrelated with the error term U_2, but which is correlated with Y_1. Suppose that we have such a variable Z, then

$$\frac{\Sigma Z Y_2}{\Sigma X Y_1} = \beta_{21} + \frac{\Sigma Z U_2}{\Sigma Z Y_1}. \tag{9.16}$$

If the variable Z has the properties:

$$\text{plim} \frac{1}{T} \Sigma Z U_2 = 0$$

$$\text{plim} \frac{1}{T} \Sigma Z Y_1 \neq 0$$

then we have

$$\text{plim} \frac{\left(\frac{1}{T} \Sigma Z Y_2\right)}{\left(\frac{1}{T} \Sigma Z Y_1\right)} = \beta_{21}. \tag{9.17}$$

Thus we have the IV estimator $\tilde{\beta}_{21} = \Sigma Z Y_2 / \Sigma Z Y_1$. Further the variance would be:

$$\text{Var } \tilde{\beta}_{21} = \frac{\sigma_{22} \frac{1}{T} \text{plim} \frac{1}{T} \Sigma Z^2}{\text{plim} \left(\frac{1}{T} \Sigma Z Y_1\right)^2} \tag{9.18}$$

and this will tend to zero as T becomes large, so that the estimator is consistent. Clearly the problem is simply that of a choice of Z. If we consider the second requirement, that Z be correlated with Y_1, we can see that the system (9.4) (9.5) provides the list of variables that are relevant. Only Y_2 and X are correlated with Y_1 directly—if any other variable were to have a significant relation with Y_1 we should expect it to be in the equation itself. However Y_2 is not a suitable instrument since it is certainly correlated with U_2, so that we are just left with the single variable X as an instrument. Providing that X is independent not merely of the error in its own equation (U_1) but also of U_2, it is a valid instrument and we have:

$$\tilde{\beta}_{21} = \frac{\Sigma Y_2 X}{\Sigma Y_1 X} \tag{9.19}$$

whose variance we can approximate by:

$$\widehat{\text{Var}\,\tilde{\beta}_{21}} = \tilde{\sigma}_{22} \frac{\Sigma X^2}{(\Sigma Y_1 X)^2}. \tag{9.20}$$

The error variance σ_{22} can be estimated from the structural residuals:

$$\hat{U}_{2t} = Y_{2t} - \tilde{\beta}_{21} Y_{1t}. \tag{9.21}$$

Hence the second equation can be satisfactorily estimated solely from the structure of the system.

We look next at the first equation

$$Y_1 = \beta_{12} Y_2 + \gamma_{11} X + U_1.$$

Here we need two independent normal equations, i.e. we must find two separate variables that are correlated with one or both of the explanatory variables but which are not correlated with U_1. Obviously X can act as its own instrument since it obeys both conditions, so we have

$$\Sigma Y_1 X = \beta_{12} \Sigma Y_2 X + \gamma_{11} \Sigma X^2 + \Sigma U_1 X \tag{9.22}$$

which gives a first equation in the two unknowns (the term $1/T(\Sigma U_1 X)$ having a plim of zero). However the second variable Y_2 cannot act as its own instrument—being correlated with U_1 (which would introduce another unknown into the normal equation). We need an instrument for Y_2. The system tells us that equation (9.5) describes Y_2 and this equation tells us that only Y_1 has a direct systematic influence on Y_2. However Y_1 cannot act as an instrument here, being correlated with U_1. We could thus look at the determinants of Y_1 to find that part which affects Y_2 (via Y_1) but which is uncorrelated with U_1. This is the variable X—but we cannot use it a second time as an instrument since the two normal equations would be identical (the instruments perfectly correlated) and no unique solution could be found. This shows us that from within the system there are not enough exogenous variables to act as instruments to allow the parameters of the first equation to be estimated. The first equation is said to be *underidentified*, while the second equation is *just identified* (there being exactly as many possible normal equations as there are parameters to be estimated).

We have shown that a simultaneous system of equations cannot in general be consistently estimated by OLS because the feedback produces a correlation between the error term and the endogenous variables appearing on the right-hand side of the equations, and that (without some other information) it may not be possible to estimate the parameters of an equation by any method.

We can see that the problems of estimation and identification are completely interdependent for, if an equation is identified then an estimator exists for the parameters, and, in showing that it is identified, this in fact will suggest a method of estimation.

9.1 Identification and prior information

Our brief discussion has shown that the key to being able to obtain consistent estimators of an equation lies in the prior information that is available. Even in the standard single equation case it is the information that X is fixed (or is independent of the error term) which tells us that OLS is unbiased and consistent. As we saw in the case of measurement error, this condition can fail and the structural parameter is then underidentified (estimated with bias) unless there is additional information that some other variable Z is uncorrelated with the error term.

In the context of a set of simultaneous equations there are three sources of prior information which may be available to the econometrician and which can aid in identifying the parameters:

(a) 'zero' or 'exclusion' restrictions on parameters;
(b) non-zero parameter restrictions both within and between equations;
(c) error variance-covariance restrictions.

The first of these is by far the most important since it corresponds to the economic information of the theory which generates the model being estimated, while the other sources of prior information are much less likely to be generated by the theory.

(a) Zero restrictions and identification

We begin with a general system of equations with two endogenous and two exogenous variables:

$$Y_1 = \gamma_{11} X_1 + \gamma_{12} X_2 + \beta_{12} Y_2 + U_1 \tag{9.23}$$

$$Y_2 = \gamma_{21} X_1 + \gamma_{22} X_2 + \beta_{21} Y_1 + U_2. \tag{9.24}$$

In everything which follows we treat the constant variable $X_t = 1$ all t as if it were any other variable. It is important to notice that it should only appear in an equation if economic theory specifies its presence. As we shall see later the knowledge that it appears in some equations but not in others can be valuable for identification.

The sole information we begin with is that the X variables are exogenous i.e.

$$\lim \frac{1}{T} \Sigma X_{it} U_{jt} = 0 \qquad i, j = 1 \ldots 2. \tag{9.25}$$

We have already seen that both equations of such a system are clearly underidentified. Each equation needs three independent instruments or normal equations which involve solely the parameters of that equation, while there are only two available (X_1 and X_2) in each case. We can consider the impact of excluding X_2 from the first equation, i.e. the prior restriction that $\gamma_{12}=0$. This immediately changes the identifiability of (9.23). The equation requires two independent instruments—both uncorrelated with the error term but being correlated with the right-hand side variables. X_1 can act as its own instrument since it satisfies both conditions—now X_2 is also available, not for itself but for Y_2, as an instrument which is independent of other instruments used in the equation. This is because X_2 is assumed uncorrelated with the error term (being an exogenous variable of the system) but is correlated with Y_2 because it appears in the system. Therefore, unless X_1 and X_2 are accidently perfectly correlated we can base our estimates on the normal equations:

$$\Sigma Y_1 X_1 = \tilde{\gamma}_{11} \Sigma X_1^2 + \tilde{\beta}_{21} \Sigma Y_2 X_1 \qquad (9.26)$$

$$\Sigma Y_1 X_2 = \tilde{\gamma}_{11} \Sigma X_1 X_2 + \tilde{\beta}_{21} \Sigma Y_2 X_2 \qquad (9.27)$$

yielding

$$\tilde{\gamma}_{11} = \frac{\Sigma Y_1 X_1 \Sigma Y_2 X_2 - \Sigma Y_2 X_1 \Sigma Y_1 X_2}{\Sigma Y_2 X_2 \Sigma X_1^2 - \Sigma Y_2 X_1 \Sigma X_1 X_2}. \qquad (9.28)$$

Hence the exclusion of X_2 from the first equation (while it was included in the second equation) served to allow the first equation to be identified.

We can see that a (necessary) condition for identification, when the only information available (in addition to the list of exogenous variables present in the system) is the list of zero restrictions, can be formulated by a simple rule which counts the number of instruments required and compares this with the number that are available. Every variable in the equation (apart from the one on the left-hand side) requires an instrument so that the minimum number of instruments required for an equation is the number of variables in the equation less one. The number of variables which can serve as instruments are the exogenous variables of the system. Hence if the number of exogenous variables in the system is less than the number of all variables (both endogenous and exogenous) in the equation less one, then the equation cannot be identified without further information. Thus it is necessary (but not in fact sufficient) for identification of an equation that the number of exogenous variables in the system should be at least as many as the number of variables in the equation less one. This equation, known as the *order condition*, can be stated in an equivalent fashion. Since all exogenous variables in the equation can serve as their own instrument we need to have other exogenous variables to act as instruments for the endogenous variables appearing on the right-hand side of the equation. Thus we can state that a necessary (but not sufficient) condition for identification is that the number of excluded exogenous variables must be at least as great as the number of included endogenous variables less one.

Necessary conditions for identification of an equation when there are exclusion restrictions: the order condition

(a) The number of exogenous variables in the system ≥ number of variables in the equation less one;

or

(b) The number of exogenous variables excluded from the equation ≥ number of endogenous variables included in the equation less one.

We can see immediately that if the order condition fails there will be no estimator—the set of normal equations would be too few in number and no solution could be obtained. As we shall see when considering estimation techniques, this generates an automatic failure whatever the sample size because the estimator would always suffer from perfect multicollinearity so that in practice it is not necessary to check the order condition—the failure of the computer programme to yield a solution would indicate the problem.

However there is a much more subtle problem involved with the condition which is required for sufficiency (as well as necessity). There are special cases where the order condition is satisfied but there is no consistent estimator. As we shall show, this condition is not evident from the finite set of data alone but rather from the logic of the system as a whole, so it has to be checked before estimation is carried out.

Consider the system:

$$Y_1 = \beta_{12} Y_2 + \beta_{13} Y_3 + U_1 \qquad (9.29)$$

$$Y_2 = \gamma_{21} X_1 + \gamma_{22} X_2 + \beta_{21} Y_1 + U_2 \qquad (9.30)$$

$$Y_3 = \beta_{32} Y_2 + U_3. \qquad (9.31)$$

Now by the order condition the first equation is identified—there are two right-hand side endogenous variables (Y_2, Y_3) and two excluded exogenous (X_1 and X_2) which could serve as instruments.

$$\Sigma Y_1 X_1 = \tilde{\beta}_{12} \Sigma Y_2 X_1 + \tilde{\beta}_{13} \Sigma Y_3 X_1 \qquad (9.32)$$

$$\Sigma Y_1 X_2 = \tilde{\beta}_{12} \Sigma Y_2 X_2 + \tilde{\beta}_{13} \Sigma Y_3 X_2. \qquad (9.33)$$

We can solve for the equations to give:

$$\tilde{\beta}_{12} = \frac{\Sigma Y_1 X_1 \Sigma Y_3 X_2 - \Sigma Y_1 X_2 \Sigma Y_3 X_1}{\Sigma Y_2 X_1 \Sigma Y_3 X_2 - \Sigma Y_2 X_2 \Sigma Y_3 X_1}. \qquad (9.34)$$

This estimator will exist unless of course the denominator is zero. There are three cases where this can happen:

(i) X_{1t} and X_{2t} are perfectly collinear;

(ii) Y_{2t} and Y_{3t} are perfectly collinear;
(iii) the relative magnitude of the correlations of X_1 with Y_2 and Y_3 is the same as the relative magnitude of the correlations of X_2 with Y_2 and Y_3.

These cases were noted in the many-variable instrumental variable analysis. The first condition insists that the instruments are not perfectly collinear while the second insists that the structural variables are not collinear, while the third was treated as a curiosity. It is not unlikely that there will be perfect correlation between the exogenous variables, unless by mistake the same variable has in effect been counted as two separate variables. However, the logic of the system can imply that the endogenous variables will be perfectly correlated. In our system here we can see that Y_2 and Y_3 would be exact multiples of each other except for the influence of U_3. In the limit, since the expected value of U_3 is zero, Y_3 will be an exact multiple of Y_2. This implies that the value we obtain in (9.34) is not a valid estimator for $\tilde{\beta}_{12}$ since it does not tend towards the true value β_{12}. This type of failure is said to be a failure of the *rank* condition—the interrelationship between the variables in the system as a whole (here the exclusion of all other variables than Y_2 from the third equation) means that the estimator will not tend towards the true value. It is important to recognize that if we failed to notice a priori that the logic of the system implies such a failure, and instead attempted to use formula (9.34) we should obtain a finite value from the computer programme; the denominator will not always be zero because Y_3 is not necessarily an exact multiple of Y_2 (U_3 is not always zero). This number cannot be claimed as an estimator for β_{12} since its limit does not equal the true value. However, the fact that there is not an automatic computer failure when the rank condition fails (but the order condition does not) shows how important it is in practical terms to check the identifiability of the system before attempting to estimate it. To summarize, this example demonstrates the following.

1. The order condition is necessary but is not sufficient for identification.

2. As is the case with all failures of this type, the breakdown will not normally be spotted just from using the standard formula for an estimator (9.34) and seeing whether the computer can solve for it. The failure of the estimator is defined at the limit, so that for a finite sample of date, the use of (9.34) can yield a finite number and the econometrician may thus believe that it is a reliable estimate of β_{12}. The upshot of this is that it is essential to check for sufficiency before estimation.

3. The failure was caused not, as is possible, by a peculiar combination of non-zero parameter values which just happen to produce a zero denominator and which would, with a minor perturbation, allow the equation to be identified.[1] Rather, the failure was intrinsic to the model because of the role of

---

[1] In such cases it is extremely unlikely that the econometrician would know that the true parameters in question took exactly the critical values for identification to fail.

the X variables. The exclusion (zero restriction) of both from the third equation meant that Y_2 and Y_3 in the first equation were effectively collinear. No change in the value of the β_{ij} or non-zero γ_{ij} would alter this lack of identifiability. This case can therefore be recognized in principle a priori solely from knowledge of the location of the zero restrictions (which is intrinsic to the economics of the model).

4. The breakdown for the identification of a particular equation could be spotted only by analysis of the *whole system*. It is clearly not enough to write down just one equation of a system and list the exogenous variables that would appear elsewhere in the system. The ability of this set to give enough independent information to obtain a consistent IV estimator from the first equation will depend crucially on the exact way in which these instruments appear in the system.

The general statement of the necessary and sufficient *rank* condition for an equation to be identified is too complex to state here for the general case, and it requires the use of matrix algebra.

(b) Non-zero restrictions

We have seen that zero restrictions on parameters occur naturally in economic theory since they correspond to the assumption that a particular variable does not influence the endogenous variable in question. Occasionally economic theory may give us some other information on the parameters. The first case is where parameters within an equation are linked. In the system (9.23) (9.24) this could be illustrated by:

$$\gamma_{11} = \gamma_{12} \tag{9.35}$$

We see that this does not tell us the value of any individual parameter but it does reduce the number of unknowns.

The second type of restriction is where we know the actual value of a parameter, e.g.

$$\gamma_{11} = r \tag{9.36}$$

where r is known. This type of restriction clearly also reduces the number of unknowns in an equation.

The third type of restriction is one linking parameters between equations: e.g.,

$$\gamma_{12} = \gamma_{21}. \tag{9.37}$$

Clearly the total number of unknowns is reduced by such a restriction but there is an ambiguity about which equation has its number of unknowns reduced by the restriction.

Just as in the case where underidentification, because of perfect multicollinearity, can be rectified by a parameter restriction (see Chapter 4) so also the lack of identification based solely on zero restrictions can be rectified by the (correct) addition of non-zero restrictions. There are generalizations of the necessary (order) and of the necessary and sufficient (rank) conditions for an equation to be identified, given the zero and non-zero restrictions. The generalization of the order conditions also effectively gives the key to how the equation should be estimated.

The first step is to substitute the restriction(s) into the equation (for the within-equation case). This will reduce the number of right-hand side exogenous or endogenous variables by the number of restrictions. The order condition is then applied to this restricted equation. Suppose for the system (9.23) (9.24) we have the non-zero restriction (9.35). Now by the order condition on the basic equation (9.23) is underidentified (by one). However substituting (9.35) into (9.23) and collecting terms we have

$$Y_1 = \gamma_{11}(X_1 + X_2) + \beta_{12} Y_2 + U_1. \tag{9.23a}$$

This, when taken with (9.24), is identified. There are two right-hand side variables in the equation ($X_1 + X_2$ and Y_2) and there are two (independent) exogenous variables in the system—($X_1 + X_2$) and X_1 (or X_2). We note that we cannot count the sum of the two X variables and *both* individual X variables as three separate exogenous because of the exact linear relation linking the sum to the components. The instruments for estimation would be ($X_1 + X_2$) and X_1 and the equation is now clearly identified.

The second type of restriction can similarly be handled. Suppose we have

$$\gamma_{11} = r \tag{9.38}$$

where r is a known value. Substituting into (9.23) and taking all known coefficients to the left-hand side we have:

$$Y_1 - rX_1 = \gamma_{12} X_2 + \beta_{12} Y_2 + U_1. \tag{9.23b}$$

Again applying the order condition we see that the equation becomes just identified, with X_1 and X_2 being the instruments for the two right-hand-side variables.

The case of inter-equation restrictions is not so straightforward since there is a choice of equations in which to substitute it (the restriction cannot be substituted into both equations thus saving two unknowns because it contains only one piece of information). If one equation is identified and the other is underidentified by one, then the choice is clear. The parameter must be estimated (consistently) from the identified equation and then this consistent value used in the restriction is substituted into the second equation. Consider the system:

$$Y_1 = \gamma_{11} X_1 + \gamma_{12} X_2 + \beta_{12} Y_2 + U_1 \tag{9.23}$$

$$Y_2 = \gamma_{21} X_1 + \beta_{21} Y_1 + U_2. \tag{9.24a}$$

If we apply the standard order condition the first equation is underidentified (by one) while the second equation is just identified. However if there is also the restriction

$$\gamma_{12} = \gamma_{21} \tag{9.39}$$

we can estimate the second equation by our instrumental variable technique to yield a consistent estimator $\tilde{\gamma}_{21}$. This is then substituted into (9.23):

$$Y_1 - \tilde{\gamma}_2 X_2 = \gamma_{11} X_1 + \beta_{12} Y_2 + W_1 \tag{9.23c}$$

where W_1 includes U_1 and the estimation error from $\tilde{\gamma}_{21}$. However W_1 will not be correlated with either X_1 or X_2 in the limit, so that again the number of exogenous variables in the system will be equal to the number of right-hand-side variables in (9.23c). Hence the equation is identified and can be estimated by instrumental variables using X_1 and X_2.

In general a necessary condition for the identification of an equation is that the number of restrictions on an equation plus the number of excluded exogenous variables should be at least as great as the number of right-hand-side endogenous variables (*generalized order condition*). This is easily shown not to be a sufficient condition by the example:

$$Y_1 = \gamma_{11} X_1 + \beta_{12} Y_2 + U_1 \tag{9.23d}$$

$$Y_2 = \gamma_{21} X_1 + \beta_{21} Y_1 + U_2. \tag{9.24b}$$

Each equation is underidentified by one, applying the standard order condition. Now we suppose that there is the extra restriction:

$$\gamma_{11} = \gamma_{21} \tag{9.40}$$

This would serve to identify both by the generalized order condition, but it is clear that neither can be identified—we cannot first estimate either γ_{11} or γ_{21} consistently; so that we cannot substitute it into the other equation via (9.40).

The generalized rank condition (which is both necessary and sufficient for identification) can again be formulated with matrix algebra and in practice is tested before estimation is carried out.

(c) Reduced forms and variance-covariance restrictions

The joint problems of identification and estimation can also be approached from analysis of the reduced forms of the system. We can illustrate this by considering again the structural equations (9.4) and (9.5) which had reduced forms:

$$Y_1 = \frac{\gamma_{11}}{1 - \beta_{12}\beta_{21}} X_1 + \frac{U_1 + \beta_{12} U_2}{1 - \beta_{12}\beta_{21}} \tag{9.7}$$

$$Y_2 = \frac{\gamma_{11}\beta_{21}}{1 - \beta_{12}\beta_{21}} X_1 + \frac{\beta_{21} U_1 + U_2}{1 - \beta_{12}\beta_{21}} \tag{9.8}$$

which we denoted
$$Y_1 = \pi_{11} X_1 + V_2 \tag{9.9}$$
$$Y_2 = \pi_{21} X_1 + V_2. \tag{9.10}$$

These equations, by construction, have only exogenous variables on the right-hand side and so can be estimated satisfactorily. The application of ordinary least squares (or equivalently instrumental variables estimation with X_1 as instrument) yields unbiased and consistent estimators of the π_{ij}:

$$E(\hat{\pi}_{11}) = E\left(\frac{\Sigma Y_1 X_1}{\Sigma X_1^2}\right) = \pi_{11} \tag{9.41}$$

$$E(\hat{\pi}_{21}) = E\left(\frac{\Sigma Y_2 X_1}{\Sigma X_1^2}\right) = \pi_{21} \tag{9.42}$$

so that we could estimate the single parameter of the second structural equation from the two estimated reduced form coefficients:

$$\tilde{\beta}_{21} = \hat{\pi}_{21}/\hat{\pi}_{11}. \tag{9.43}$$

It can be proved that this procedure is in general consistent (but we cannot prove that it is unbiased since the expectation of a ratio is not generally equal to the ratio of the expectations). Given this value of $\tilde{\beta}_{21}$, there are infinitely many pairs of $\tilde{\beta}_{12}$ and $\tilde{\gamma}_{11}$ that would satisfy the equations:

$$\hat{\pi}_{11} = \tilde{\gamma}_{11}/(1 - \tilde{\beta}_{12}\tilde{\beta}_{21}) \tag{9.44}$$

$$\hat{\pi}_{21} = \tilde{\gamma}_{11}\tilde{\beta}_{21}/(1 - \tilde{\beta}_{12}\tilde{\beta}_{21}) \tag{9.45}$$

so that the first equation is underidentified while the second is just identified. The technique of estimating the reduced forms by OLS and then, when possible, solving back for the structural parameters from the estimated reduced form coefficients is known as *indirect least squares*.

However, we can see that information on the structural parameters is also contained in the reduced form error terms V_1 and V_2. Assuming that the errors have the following properties.

$$E(U_{it} U_{js}) = \sigma_{ij} \qquad t = s \tag{9.46}$$
$$\qquad\qquad\quad = 0 \qquad t \neq s$$

i.e. that there is no serial correlation (either within or between equations), that there is homoskedasticity but there is a possibly non-zero contemporous covariance between errors from different equations (i.e. that part of the current shocks are common to both equations), then we can see that

$$E\left(\frac{1}{T}\Sigma V_{1t}^2\right) = \frac{\sigma_{11} + 2\beta_{12}\sigma_{12} + \beta_{12}^2\sigma_{22}}{(1 - \beta_{12}\beta_{21})^2} \tag{9.47}$$

$$E\left(\frac{1}{T}\Sigma V_{2t}^2\right) = \frac{\beta_{21}^2 \sigma_{11} + 2\beta_{21}\sigma_{12} + \sigma_{22}}{(1-\beta_{12}\beta_{21})^2} \qquad (9.48)$$

$$E\left(\frac{1}{T}\Sigma V_{1t} V_{2t}\right) = \frac{\beta_{21}\sigma_{11} + (\beta_{21}\beta_{12}+1)\sigma_{12} + \beta_{12}\sigma_{22}}{(1-\beta_{12}\beta_{21})^2}. \qquad (9.49)$$

The three reduced-form error variances and covariances are functions of the structural parameters (β_{21} and β_{12}) as well as of the error parameters (σ_{11}, σ_{12}, and σ_{22}). We can see that even if we could estimate the reduced-form variances and covariances consistently from the reduced form residuals, the three new equations introduced three more unknowns so that we could not hope to solve for all six parameters from the five estimated pieces of information.

In certain circumstances we may however have prior information on the structural errors which can be utilized. The most common form of restriction is one of the value of the error covariance σ_{12}. In certain models we may be in a position to specify, for example, that there is no contemporaneous correlation between the errors from the two structural equations i.e. $\sigma_{12}=0$. This in effect means that the equations (9.49), (9.48), and (9.49) introduce only two unknowns and are also a function of β_{12} which is unknown, and β_{21} which could be estimated via ILS. In principle we can proceed firstly to estimate β_{21} by ILS (or IV), then use the estimated residuals from the reduced forms to substitute into (9.47), (9.48) and (9.49) and thus to solve for $\tilde{\sigma}_{11}$, $\tilde{\sigma}_{22}$, and $\tilde{\beta}_{12}$. Given $\tilde{\beta}_{12}$ we could then return to the estimated reduced-form coefficients to solve for $\tilde{\gamma}_{11}$.

Another variation on this theme, although very rare in practice, is where the ratio of the structural error variances is known

$$\frac{\sigma_{11}}{\sigma_{22}} = k \qquad (9.50)$$

where k is a known value. A similar solution procedure could be used as with the zero covariance restriction.

We can see that even where a system is heavily underidentified using just the exclusion restriction criteria, the presence of enough restrictions on the error variances and covariances could identify all the parameters of the system. We can see that a further weakening of the (necessary) order condition for the identification on an equation is possible. There is also again a further generalization of the rank condition which is necessary and sufficient for identification.

9.2 The estimation of simultaneous equations

Our discussion of identification has repeatedly made the point that if an equation is identified then there is a method of estimating it. However, there may be more than one consistent estimator available so there is a second-order problem of choosing the best estimator. We shall concentrate for the present on systems where the only prior information is the zero restrictions on the exclusion of certain variables from an equation.

We have already exhibited two methods of estimation—instrumental variables and indirect least squares. In the case where an equation is just identified the values of these estimators are in fact exactly the same. Consider, for example the system (9.4) and (9.5) where the second equation is just identified. The IV estimator is given by:

$$\tilde{\beta}_{21} = \frac{\Sigma Y_2 X_1}{\Sigma Y_1 X_1} \tag{9.51}$$

(it should be noted that as usual with IV estimators the same value would be obtained by a two-stage procedure or regressing Y_2 on X and substituting the fitted value $\hat{Y}_1$ into (9.5) and regressing Y_2 on $\hat{Y}_1$). The indirect least squares estimator is (see 9.48)):

$$\beta_{21}^* = \hat{\pi}_{21}/\hat{\pi}_{11}$$

$$= \frac{\Sigma Y_2 X_1 / \Sigma X_1^2}{\Sigma Y_1 X_1 / \Sigma X_1^2} = \tilde{\beta}_{21}. \tag{9.52}$$

The case of an *overidentified* equation is not one that has concerned us so far, but in practice this is the norm. The presence of more instruments than required (more excluded exogenous than right-hand-side endogenous) indicates obviously that there is a choice of which subset to use and there is a similar plurality with indirect least squares. Consider the system

$$Y_1 = \gamma_{11} X_1 + \gamma_{12} X_2 + \beta_{12} Y_2 + U_1 \tag{9.53}$$

$$Y_2 = \gamma_{23} X_3 + \beta_{21} Y_1 + U_2. \tag{9.54}$$

By the order condition the first equation is just identified; three instruments are needed for the normal equations and we have X_1, X_2, and X_3 available. However, the second equation is overidentified—the same three instruments are available while there are only two parameters to be estimated. By the properties of IV estimation using X_3 and X_1, or X_3 and X_2 (assuming that we naturally use X_3 as its own instrument) will both give consistent estimates of γ_{23} and β_{21}. In any finite sample the two sets of estimated values would be different, but they would tend to equality for a sufficiently large sample. If we look at the reduced forms we have:

$$Y_1 = \pi_{11} X_1 + \pi_{12} X_2 + \pi_{13} X_3 + V_1 \tag{9.55}$$

$$Y_2 = \pi_{21} X_1 + \pi_{22} X_2 + \pi_{23} X_3 + V_2 \tag{9.56}$$

or

$$Y_1 = \frac{\gamma_{11} X_1}{D} + \frac{\gamma_{12} X_2}{D} + \frac{\beta_{12} \gamma_{23} X_3}{D} + V_1 \tag{9.57}$$

$$Y_2 = \frac{\beta_{21}\gamma_{11}X_1}{D} + \frac{\beta_{21}\gamma_{12}X_2}{D} + \frac{\gamma_{23}X_3}{D} + V_2 \qquad (9.58)$$

where
$$D = (1 - \beta_{12}\beta_{21}). \qquad (9.59)$$

The presence of six reduced-form coefficients while there are only five structural equation parameters indicates the difficulty. There is more than one way of recovering the structural parameters from the estimated reduced-form parameters. For example, once we have the $\hat{\pi}_{ij}$ from the regressions of each Y on all the exogenous variables, we can form:

$$\tilde{\beta}_{12} = \hat{\pi}_{13}/\hat{\pi}_{23} \qquad (9.60)$$

$$\tilde{\beta}_{12} = \hat{\pi}_{22}/\hat{\pi}_{12} \qquad (9.61)$$

$$\tilde{\gamma}_{13} = \hat{\pi}_{23}(1 - \tilde{\beta}_{12}\tilde{\beta}_{21}) \qquad (9.62)$$

$$\tilde{\gamma}_{12} = \hat{\pi}_{12}(1 - \tilde{\beta}_{12}\tilde{\beta}_{21}) \qquad (9.63)$$

and
$$\tilde{\gamma}_{11} = \hat{\pi}_{11}(1 - \tilde{\beta}_{12}\tilde{\beta}_{21}) \qquad (9.64a)$$

or
$$\tilde{\gamma}_{11} = \hat{\pi}_{21}(1 - \tilde{\beta}_{12}\tilde{\beta}_{21})/\tilde{\beta}_{21}. \qquad (9.64b)$$

In the large sample the alternative estimators (9.64a) and (9.64b) will tend to equality, but they will be different in the small sample because we have not imposed any restriction on the reduced-form coefficients. A *non-linear* restriction of the form:

$$\frac{\hat{\pi}_{11}}{\hat{\pi}_{21}} = \frac{\hat{\pi}_{12}}{\hat{\pi}_{22}} \qquad (9.65)$$

would, if the estimation was carried out subject to it, guarantee that the two solutions for $\tilde{\gamma}_{11}$ were equal. Techniques do exist to impose inter-equation non-linear restrictions but these are considerably more advanced and we shall not consider them here.

The key to estimation in the overidentified case was given in Chapter 8 when considering how an optimal instrument is to be chosen from a list of variables all uncorrelated with the error term. The criteria for a good instrument (in terms of the variance of the estimator) is the degree of correlation between the instrument and the variable it is to replace—the higher the correlation the lower the variance. We suggested that where a variable could be replaced by more than one instrument we should find the linear combination of the instruments which are most strongly correlated with the variable to be replaced. This 'average' instrument would also be independent of the error terms and would minimize the variance of the estimated parameter. The idea is easily extended to systems of simultaneous equations by the procedure known as *two-stage least squares* (*TSLS or 2SLS*). The first stage is to regress (in turn) each endogenous variable on all the exogenous variables in the system and form the fitted value (i.e. to estimate the reduced forms without any inter-equation restrictions). The second stage is to replace in the structural equations any right-hand-side endogenous variable by its fitted values and carry out a second regression on the exogenous and fitted endogenous variables. This procedure, as we saw in Chapter 8, is equivalent to using the fitted values as the instruments for the right-hand-side endogenous variables (and the exogenous variables as their own instruments) when estimating the structural equations by instrumental variables. We can illustrate the procedure for (9.53) and (9.54).

Step 1: regress Y_1 on X_1, X_2, X_3; Y_2 on X_1, X_2, X_3 and obtain fitted values:

$$\hat{Y}_{1t} = \hat{\pi}_{11} X_{1t} + \hat{\pi}_{12} X_{2t} + \hat{\pi}_{13} X_{3t}$$

$$\hat{Y}_{2t} = \hat{\pi}_{21} X_{1t} + \hat{\pi}_{22} X_{2t} + \hat{\pi}_{23} X_{3t}$$

where the $\hat{\pi}_{ij}$ are the estimated reduced-form coefficients from the two ordinary least squares estimations.

Either *Step 2(a)*: regress Y_1 on X_1, X_2 and $\hat{Y}_2$, Y_2 on X_3 and $\hat{Y}_1$; or *Step 2(b)*: do instrumental variables on equation (9.53) using X_1, X_2, and $\hat{Y}_2$ as instruments and equation (9.54) using X_3 and $\hat{Y}_1$ as instruments.

The identity of the two approaches is easily demonstrated: from (2a)

$$\tilde{\gamma}_{23} = \frac{\Sigma Y_2 X_3 \Sigma \hat{Y}_1^2 - \Sigma Y_2 \hat{Y}_1 \Sigma \hat{Y}_1 X_3}{\Sigma \hat{Y}_1^2 \Sigma X_3^2 - \Sigma (\hat{Y}_1 X_3)^2} \tag{9.66}$$

while from (2b) using (8.50):

$$\gamma_{23}^* = \frac{\Sigma Y_2 X_3 \Sigma Y_1 \hat{Y}_1 - \Sigma Y_2 \hat{Y}_1 \Sigma X_3 Y_1}{\Sigma X_3^2 \Sigma \hat{Y}_1 Y_1 - \Sigma X_3 Y_1 \Sigma X_3 \hat{Y}_1}. \tag{9.67}$$

These two formulae are identical since using the properties of least squares fitting we have

$$\hat{Y}_1 + e = Y_1 \tag{9.68}$$

and $\qquad\qquad\qquad\Sigma e \hat{Y}_1 = 0 \tag{9.69}$

and $\qquad\qquad \Sigma e X_1 = \Sigma e X_2 = \Sigma e X_3 = 0 \tag{9.70}$

which implies that $\qquad\Sigma X_3 Y_1 = \Sigma X_3 \hat{Y}_1 \tag{9.71}$

and $\qquad\qquad\qquad \Sigma Y_1 \hat{Y}_1 = \Sigma \hat{Y}_1^2. \tag{9.72}$

The demonstration of the coincidence between these two methods of calculating the estimator is purely an expository step, since in practical terms computer programmes can obtain the estimator from a single formula—there is no actual need to do two separate regressions since the standard formula for a fitted value can be substituted into (9.66) or (9.67) to give a one-step estimator.

Once we have a consistent estimator it is useful to have an estimate of its variance so that we can obtain an idea of the precision of estimation and even carry out significance tests. Since we have seen that TSLS is a particular case of instrumental variables, we can calculate the variances of the parameters by the generalizations of the formulae developed in Chapter 8. Again we need to be cautious about interpreting these formulae, which are only exact as the sample size becomes large (at which point the variance of the consistent estimators tends to zero). For fairly large samples it is conventional to estimate these variances by sample values, with the error variance estimated from the residuals of the consistently estimated structural equations. These are then utilized in conventional F or t test formulae as if they followed such a distribution exactly.

We next turn to the problem of estimation when there are other than zero restrictions available to the econometrician. We saw in fact that within and between equation linear restrictions on parameters present no new problems. Once the restrictions are substituted into the appropriate equations they can then, if identified, be estimated by TSLS. The case of estimation where there are restrictions on the variance–covariance matrix of residuals for the system, can be solved by the techniques used to demonstrate identifiability. However in the case of zero covariance restrictions there is a more elegant and simpler approach. Consider the system:

$$Y_1 = \gamma_{11} X_1 + \beta_{12} Y_2 + U_1 \qquad (9.4)$$

$$Y_2 = \beta_{21} Y_1 + U_2 \qquad (9.5)$$

where:
$$E(U_{1t} U_{2t}) = 0 \quad \text{all} \quad t. \qquad (9.73)$$

Without the error covariance restriction (9.73) the second equation is just identified while the first is underidentified (by one). The problem with the first equation is that we require another IV apart from X_1, i.e. a variable which is correlated with Y_2 but which is not correlated with U_1. The restriction (9.73) makes it clear that the error term U_{2t} has both these properties—it is of course correlated with Y_2 as specified by (9.5) and it is not correlated with U_1 by (9.70). We cannot of course observe U_{2t}, but because the second equation can be consistently estimated we can obtain reliable error terms $\hat{U}_{2t}$:

$$\hat{U}_{2t} = Y_{2t} - \tilde{\beta}_{21} Y_{1t} \qquad (9.74)$$

where $\tilde{\beta}_{21}$ is estimated by some consistent technique, e.g. TSLS. The instruments for estimating the first equation are then X_{1t} and $\hat{U}_{2t}$—the resulting estimators can be shown to be consistent.

In one case the zero covariance restriction even breaks the apparent simultaneity between equations, and hence removes the need to use TSLS. Consider the system

$$Y_1 = \gamma_{11} X_1 + \beta_1 Y_2 + U_1 \qquad (9.4)$$

$$Y_2 = \gamma_{21} X_1 + U_2 \qquad (9.75)$$

This is a very special form where the 'causality' of the model flows one way—

into Y_2 and from Y_2 to Y_1 but does not feed back from Y_1 to Y_2—such a system is said to be *recursive*. The second equation here is clearly identified and can in effect be estimated by OLS (X_1 being the instrument in an IV interpretation). However the first equation is more problematical. A straightforward application of the order condition shows that it is not identified—two instruments are needed and only one (X) is available. It may seem tempting at first sight to argue that Y_2 does not need an instrument since it appears not to be correlated with U_1—a rise in U_1 causes Y_1 to rise but there is no feedback from Y_1 to Y_2 to create a correlation between U_1 and Y_2. However, unless we know a priori that $E(U_1 U_2) = 0$, there is an indirect link between U_1 and Y_2: if U_1 shifts then U_2 will respond. The first equation is indeed then underidentified. Consider now the impact of the zero covariance restriction $E(U_1 U_2) = 0$. This implies that in the recursive structure Y_2 is uncorrelated with U_1 and hence Y_2 can act as its own instrument—i.e. we can consistently estimate the equation by ordinary least squares.

This discussion of estimation shows that for most purposes TSLS, applied in the appropriate fashion, will be an acceptable estimator. Clearly we need never apply ILS, since TSLS gives identical results when the system is just identified and is superior for overidentified systems. There are a few situations when TSLS does not use all the information in the system and when a more general estimator is required:

(i) when there are a priori restrictions on the error variance–covariance matrix other than the zero covariance restriction;
(ii) when there are non-linear parameter restrictions;
(iii) when the error covariances are non-zero, but are not known *a priori*.

The first two problems require techniques which can handle non-linearities and which can estimate all the parameters at the same time (rather than equation by equation). Such techniques, generally referred to as Full Information Maximum Likelihood (FIML) are beyond the scope of a introductory course. The third problem, which is quite common in practice, can be seen to be an analogue to the SUR problem discussed in Chapter 7—we saw that two equations related through the contemporaneous covariance of the error term could be estimated more efficiently by applying GLS to the equations as a system. The same is true for a set of simultaneous equations. Consider

$$Y_1 = \gamma_{11} X_1 + \gamma_{12} X_2 \beta_{12} Y_2 + U_1 \tag{9.76}$$

$$Y_2 = \gamma_{23} X_3 + \beta_{21} Y_1 + U_2 \tag{9.77}$$

where

$$E(U_1 U_2) = \sigma_{12} \tag{9.78}$$

and σ_{12} is unknown. At the second stage of TSLS we have as usual the fitted values $\hat{Y}_1$ and $\hat{Y}_2$ to be substituted into the structural equations. However, these equations as a set:

$$Y_1 = \gamma_{11} X_1 + \gamma_{12} X_2 + \beta_{12} \hat{Y}_2 + V_1$$
$$Y_2 = \gamma_{23} X_3 + \beta_{21} \hat{Y}_1 + V_2 \qquad (9.79)$$

will have a correlated error. As in the SUR case we can stack the equations to obtain a single form in $2T$ observations

$$\begin{pmatrix} Y_1 \\ Y_2 \end{pmatrix} = \gamma_{11} \begin{pmatrix} X_1 \\ 0 \end{pmatrix} + \gamma_{12} \begin{pmatrix} X_2 \\ 0 \end{pmatrix} + \gamma_{13} \begin{pmatrix} 0 \\ X_3 \end{pmatrix} + \beta_{12} \begin{pmatrix} \hat{Y}_2 \\ 0 \end{pmatrix}$$
$$+ \beta_{21} \begin{pmatrix} 0 \\ \hat{Y}_1 \end{pmatrix} + \begin{pmatrix} V_1 \\ V_2 \end{pmatrix} \qquad (9.80)$$

(the 'long' variables put in zeros for equations where the basic variables did not appear). Regarded as a single equation ready for OLS estimation at the second stage we can see that the error terms would not be well behaved since the 'stacked' error has pairwise correlation—every pair of observations T units apart are correlated (by (9.78)). In the presence of what is in effect pure Tth-order serial correlation we need to find the Aitken transformation that will make the errors well behaved. As in the SUR case of Chapter 7 the transformation involves purging the serial independence first and then correcting for heteroskedasticity. The transformed equation is:

$$\begin{pmatrix} Y_{1t} \\ AY_{2t} + BY_{2t} \end{pmatrix} = \gamma_{11} \begin{pmatrix} X_{1t} \\ BX_{1t} \end{pmatrix} + \gamma_{12} \begin{pmatrix} X_{2t} \\ BX_{2t} \end{pmatrix} + \gamma_{13} \begin{pmatrix} 0 \\ AX_{3t} \end{pmatrix}$$
$$+ \beta_{12} \begin{pmatrix} \hat{Y}_{2t} \\ B\hat{Y}_{2t} \end{pmatrix} + \beta_{21} \begin{pmatrix} 0 \\ A\hat{Y}_{1t} \end{pmatrix} + \begin{pmatrix} V_{1t} \\ AV_{2t} + BV_{1t} \end{pmatrix} \qquad (9.81)$$

where

$$A = \sigma_{11}^2/(\sigma_{22}\sigma_{11} - \sigma_{12}^2)^{1/2} \qquad (9.82)$$
$$B = \sigma_{12} A/\sigma_{11}. \qquad (9.83)$$

If the terms σ_{11}, σ_{22}, and σ_{12} were known then we could estimate (9.81) by ordinary least squares and obtain consistent estimators with a smaller variance than TSLS—such a technique is known as *three-stage least squares* (3SLS) and is said to be a *system estimator* rather than a *single equation estimator*. As with all the Aitken transformations it is extremely unlikely that we will know the σ_{ij}, and instead we have to estimate them. To do so we obtain consistent estimates of the coefficients by TSLS and then obtain an estimate of the structural residuals. From these we can estimate the variance and covariances:

$$\hat{\sigma}_{ij} = \frac{1}{T} \Sigma \hat{U}_{it} \hat{U}_{jt}. \qquad (9.84)$$

The gain from applying approximate 3SLS is likely to be at its greatest when

σ_{12} is large (relatively to σ_{11} and σ_{22}) so that examination of the residuals derived from the TSLS estimators is a sensible check on whether it is worth proceeding to 3SLS.

9.3 Simultaneous equations and forecasting

An important use of sets of simultaneous equations is in forecasting, where the simultaneity raises a problem not met in the single-equation analogue. Consider the system:

$$Y_1 = \gamma_{11} X_1 + \beta_{12} Y_2 + U_1 \tag{9.85}$$

$$Y_2 = \gamma_{21} X_2 + \beta_{21} Y_1 + U_2 \tag{9.86}$$

where both equations are (just) identified. If we wish to forecast the level of Y_1 (say) in the next period then we must know or guess the levels of the explanatory variables X_1 and Y_2. However, Y_2 will itself be determined by Y_1, etc. The simultaneity between the endogenous variables means in effect that we must be able to specify all the values for the exogenous variables in the next period and this will allow the reduced forms to be solved. From (9.85) and (9.86) we have

$$Y_1 = \pi_{11} X_1 + \pi_{12} X_2 + V_1 \tag{9.87}$$

$$Y_2 = \pi_{21} X_1 + \pi_{22} X_2 + V_2 \tag{9.88}$$

where

$$\pi_{11} = \frac{\gamma_{11}}{1 - \beta_{12}\beta_{21}} \text{ etc.} \tag{9.89}$$

If we know the reduced-form parameters, as well as the values of the exogenous variables for the forecast period, we can construct a forecast value for the endogenous variables, e.g.

$$\hat{Y}_{1F} = \hat{\pi}_{11} X_{1F} + \hat{\pi}_{12} X_{2F}. \tag{9.90}$$

Clearly we can obtain consistent estimates of the π_{ij} even for a system including some underidentified equations (see (9.41) and (9.42)). This is done by simply carrying out the first step of ILS (TSLS) which regresses each endogenous variable on all the exogenous variables of the system. The resulting (unrestricted) estimates of the reduced-form coefficients are unbiased and consistent (as estimators of the true reduced-form coefficients) and when substituted into (9.90) yield consistent and unbiased estimates of the endogenous variable. Such a procedure can indeed be used even when there are some overidentified equations in the system, but as we have argued the use of unrestricted reduced forms (ILS) ignores some information available from the exclusion restrictions. Hence the more efficient technique is to estimate the structural parameters by TSLS and then construct estimates of the reduced-form parameters using the known relation between the two. The resulting estimates of the reduced-form parameters will obey the restrictions implicit in the system. Consider the system:

$$Y_1 = \gamma_{11} X_1 + \gamma_{12} X_2 + \beta_{12} Y_2 + U_1 \tag{9.53}$$

$$Y_2 = \gamma_{23} X_3 + \beta_{21} Y_1 + U_2 \tag{9.54}$$

where we wish to forecast Y_1 (and Y_2) given knowledge of X_1, X_2, and X_3 for the forecast period. We know that the true reduced-form coefficient is:

$$\pi_{11} = \frac{\gamma_{11}}{1 - \beta_{12} \beta_{21}}. \tag{9.91}$$

We estimate first by TSLS to obtain $\tilde{\gamma}_{11}$, $\tilde{\beta}_{12}$, and $\tilde{\beta}_{21}$ and then we estimate the restricted reduced-form coefficient:

$$\tilde{\pi}_{11} = \tilde{\beta}_{11}/(1 - \tilde{\gamma}_{12} \tilde{\gamma}_{21}). \tag{9.92}$$

It is easy to see that the restriction is obeyed exactly

$$\frac{\tilde{\pi}_{11}}{\tilde{\pi}_{21}} = \frac{\tilde{\pi}_{12}}{\tilde{\pi}_{22}}, \tag{9.93}$$

unlike the unrestricted coefficients where in general

$$\frac{\hat{\pi}_{11}}{\hat{\pi}_{21}} \neq \frac{\hat{\pi}_{12}}{\hat{\pi}_{22}}. \tag{9.94}$$

The restricted reduced form is used for the forecast:

$$Y_{1F} = \tilde{\pi}_{11} X_{1F} + \tilde{\pi}_{12} X_{2F} + \tilde{\pi}_{13} X_{3F}. \tag{9.95}$$

The coefficient π_{ij} is known as the 'multiplier' of the exogenous variable; for the endogenous variable i—it measures the total effect on Y_i, within the observation period, of a one-unit change in X_j (allowing for all the instantaneous feedbacks in the system). It is compared with the first round multiplier (or impact multiplier) γ_{ij} which measures the direct effect of one unit change in X_j on Y_i before the feedbacks are allowed for.

The principal results on estimation are summarized in the following points, and a numerical illustration is then given of the estimation and use of a simple simultaneous system in example 9.1.

1. A consistent method of estimation of an identified equation in a set of simultaneous equations is two-stage least squares (TSLS). The first step is to regress each endogenous variable in the system in turn on all the exogenous variables (unrestricted reduced form) and construct the fitted values. The second step either (a) replaces right-hand-side endogenous variables by their fitted values from the first stage and then does OLS on this equation or (b) uses the fitted values as instruments together with the included exogenous variables in an IV estimation of the equation.

2. For a just identified equation TSLS = ILS, and for an overidentified

equation TSLS is more efficient than ILS, so in practice indirect least squares (recovering structural parameters from the estimated unrestricted reduced forms) is not utilized.

3. For recursive systems where the inter-equation error covariances are zero it is not necessary to use TSLS because there is no 'simultaneous feedback' and so OLS can be used instead.

4. To construct the residuals from TSLS, for use in estimating parameter variances to be used in significance tests, the structural equation must be used and not the second stage (where the right-hand-side endogenous were replaced by their fitted values).

5. Significance tests are conventionally carried out as if the statistics were exactly distributed as 't', 'F', or DWS forms. The approximations to these distributions improve in large samples so that tests eventually are of the size claimed.

6. When there is inter-equation non-zero error covariance then we can improve upon TSLS by using the Aitken estimator of the complete system of equations—3SLS. In practice this is approximated, since the parameters of the Aitken transformation have to be estimated first.

7. Linear restrictions are handled just as in the single-equation case. The restrictions are substituted into the equation, variables with known coefficients collected on the left-hand side, variables with common coefficients gathered together on the right-hand side, and the appropriate instruments are used to estimate the equation.

8. Forecasts from a system of simultaneous equations are based (if possible) on the estimated restricted reduced form rather than the unrestricted reduced form. The restricted reduced form is constructed from the estimates of the structural parameters.

Example 9.1: The estimation of a system of equations

The use of simultaneous estimation techniques is illustrated with our simple consumption function model to which we add the aggregate output equation: $C = \alpha + \beta Y + U$, $C + A = Y$. We specify that A (which is all expenditure except consumption) is exogenous—it does not depend on the level of current income. We continue to assume that U is well behaved and normally distributed.

First the identifiability of the system must be checked. The two endogenous variables are C and Y and the exogenous variables are the unit variable (attached to α) and A. There is no other information on the system except for the exclusion (zero) restrictions. The second equation, as is the case with all identities, is identified—all its coefficients are known exactly so there is no problem of estimation. The first equation has one right-hand-side endogenous variable (Y) and there is one excluded exogenous variable (A) so that the order condition for identification is just satisfied. This is necessary for the equation to be identified but not sufficient, and in reality the rank condition would also be checked—this system in fact satisfies the rank condition so that the equation is indeed identified. For a just identified equation TSLS and ILS give identified results (and since there is only one excluded exogenous variable there is no choice as between potential IV estimators—the only one is indeed TSLS).

$$C = 18888 + 0.751\,Y$$
$$(2394)\quad(0.018)$$

These values are quite close to those obtained by OLS in Chapter 2 and may suggest that the degree of bias is not very large. We must be careful with such an interpretation since it rests upon the assumption that the so-called 'autonomous' expenditure is genuinely independent of current income. If this were not so then TSLS estimation would also be biased.

Significance tests can be carried out on this model using the estimated parameter variances and residual variance (based on the TSLS residuals). For example, to test that the coefficient on income is not significantly less than unity we use a one-tailed t test. The appropriate t statistic is

$$\frac{1 - 0.751}{0.018} = 13.8.$$

If the null hypothesis is correct then a statistic calculated on this basis will approximately follow a t distribution (with the approximation improving in large samples). We treat the test as if the distribution were exact—the critical value for a one-tailed test of size 5 per cent with 20 degrees of freedom is 1.725 so that the null hypothesis is rejected and we accept that the m.p.c. is less than unity.

We next use the equation to construct a forecast. In order to obtain the forecast with the smallest variance we need first to obtain the restricted reduced-form estimates by substituting into:

$$C = \frac{\alpha}{1-\beta} + \frac{\beta A}{1-\beta}$$

therefore
$$C = 75{,}855 + 3.016 A.$$

Given the assumption that $A = £20{,}000$ million in 1983 we have:

$$C(83) = £136{,}176 \text{ million}$$

and hence we can also forecast total income:

$$Y(83) = £156{,}176 \text{ million}.$$

Finally we can consider the goodness of fit of the equation. The best measure is the standard error of estimate, based on the residuals from the structural equation using TSLS estimates, i.e.

$$\tilde{U}_t = C_t - \tilde{\alpha} - \tilde{\beta} Y_t$$

(and not the residuals from the second stage if TSLS has been carried out as a regression: i.e. do not use:

$$U_t^* = C_t - \tilde{\alpha} - \tilde{\beta} \hat{Y}_t$$

where $\hat{Y}_t$ is the fitted value from the first stage);

$$\text{SEE} = £1,625 \text{ million}$$

(as opposed to the OLS value of £1,544 million). An R^2 type measure could also be used, provided it is taken as the correlation between actual consumption and the estimated value—$\hat{C} = \hat{\alpha} + \hat{\beta} Y$—so that it lies between zero and unity. Finally we can check for serial independence by applying the Durbin–Watson test to the statistic based on the structural residuals. In this case we have $\hat{d} = 0.99$. Now the values of the bounds are no longer exactly valid in small samples but it is conventional to treat them as if they were valid (again the approximation improves for large samples). For twenty-two observations and one explanatory variable apart from the constant, with a two-sided 5 per cent test the critical values are $d_L = 1.24$ and $d_u = 1.43$. Hence we reject the hypothesis that the residuals are serially independent.

·Problems 9

9.1 Consider the system

$$Y_{1t} = \gamma_{11} X_{1t} + \beta_{12} Y_{2t} + U_{1t}$$
$$Y_{2t} = \gamma_{20} + \beta_{21} Y_{1t} + U_{2t}$$
$$E(U_{it} X_{it}) = 0 \qquad i = 1, 2$$
$$E(U_{it}) = 0 \qquad i = 1, 2$$

(a) Are the equations identified and if so how can their parameters be estimated?

(b) The first equation is modified to become:

$$Y_{1t} = \gamma_{10} + \gamma_{11} X_{1t} + \beta_{12} Y_{2t} + U_{1t}.$$

Are the equations identified now?

(c) The extra information that

$$E(U_{1t} U_{2t}) = 0$$

is available. How does this affect identifiability under (b)? How could the parameters of the equations be estimated?

9.2 Consider the system:

$$Y_{1t} = \beta Y_{2t} + U_{1t}$$
$$Y_{2t} = \alpha Y_{1t} + U_{2t}$$
$$E(U_{it}) = 0 \quad \text{for } i = 1, 2 \quad \text{all } t$$
$$E(U_{it} U_{js}) = \sigma_{ij} \quad \text{for } t = s$$
$$E(U_{it} U_{js}) = 0 \quad \quad t \neq s$$

(a) Show that the probability limit of the OLS estimator of β from the first equation is

$$\beta + (1 - \alpha\beta)(\alpha\sigma_{11} + \sigma_{12})/(\alpha^2 \sigma_{11} + 2\alpha\sigma_{12} + \sigma_{22}).$$

(b) Discuss the possibility of evaluating the direction of bias in relation to a priori knowledge of the signs of the parameters.

(c) Show how knowledge of the error variance ratio and the value of the error covariance would allow both α and β to be consistently estimated.

9.3 Consider the equations

$$Y_1 = \beta_{12} Y_2 + \gamma_{11} X_1 + U_1$$
$$Y_2 = \beta_{21} Y_1 + \gamma_{22} X_2 + \gamma_{23} X_3 + U_2.$$

Because of a specification error the variable X_3 is omitted from the second equation:

$$Y_2 = \beta_{21} Y_1 + \gamma_{22} X_2 + V$$

(where $V = U_2 + \gamma_{23} X_3$). The equations are to be estimated by TSLS. The technique used is to estimate the fitted values of the endogenous variable from the reduced forms and (a) replace right-hand-side endogenous variables with fitted values and carry out OLS again, or (b) use the fitted value as an instrument as well as the included exogenous variable. Show that for the first equation technique (a) is not consistent in the presence of specification error, while technique (b) is still consistent.

9.4 Consider the system

$$Y_1 = \beta_{12} Y_2 + \gamma_{11} X_1 + \gamma_{12} X_2 + U_1$$
$$Y_2 = \beta_{21} Y_1 + U_2$$

where $\quad E(U_i X_j) = 0, \quad \text{all } i, j$

(a) With no other information available on the system what is the state of identification of the equations?

(b) How is identifiability altered by the information that:

(i) $\gamma_{11} = \gamma_{12}$

(ii) $\beta_{21} = 2\beta_{12}$

(iii) $\sigma_{12} = 0$

(where $E(U_i U_j) = \sigma_{ij}$).

In cases where equations are identified describe how they could be consistently estimated?

9.5 Consider the equations

$$Y_1 = \beta_{12} Y_2 + \gamma_{11} X_1^* + \gamma_{12} X_2 + U_1$$
$$Y_2 = \beta_{21} Y_1 + \gamma_{23} X_3 + U_2$$

where $E(X_2 U_i) = E(X_3 U_i) = 0$, $i = 1, 2$. X_1^* is measured with error by X_1 where: $X_1^* + X_1^{**} = X_1$.

$$E(X_1^* U_i) = 0 \qquad E(X_1^{**} U_1) = 0.$$

(a) Discuss the identifiability of the equations.

(b) Suppose that X_3 were also measured with error. How would this affect your answer to (a)? Would the information that $E(X_1^{**} U_2) = 0$ be helpful in achieving identifiability for either of the equations?

Answers 9

9.1(a) Given that the only information on the system is on zero restrictions we start with the order conditions. For the first equation there is one right-hand-side endogenous (Y_2) and one excluded exogenous (the unit variable)—the equation just satisfies the order condition. For the second equation there is one right-hand side endogenous variable (Y_1) so that it too just satisfies the order condition. We finally need to check for perfect collinearity. Provided (i) the instruments X_{1t} and 1 are not perfectly correlated, (ii) the right-hand side endogenous are not perfectly correlated in the limit, (iii) the relative correlation of instruments to endogenous are not equal, then there is no danger that the estimator will not exist. In a two equation system there is only a single right-hand-side endogenous variable so that conditions (ii) and (iii) are inoperative. Hence, provided $X_{1t} \neq K$ all t, the system is identified. Estimation is by TSLS (or ILS because both equations are just identified). Regress Y_1 on 1 and X_1 to obtain $\hat{Y}_{1t}$, and similarly obtain $\hat{Y}_2$. Next regress Y_1 on X_1 and $\hat{Y}_2$, and regress Y_2 on 1 and $\hat{Y}_1$. The resulting estimators will be consistent.

(b) Changing the first equation means that there is now one right-hand side endogenous and no excluded exogenous—the equation is underidentified since it fails the order condition. The second equation continues to be just identified. As before the parameters of the second equation can be consistently estimated by TSLS. However, if we apply the technique to the first equation we will find that there is perfect multicollinearity at the second stage (the instruments being 1, X_{1t}, and $\hat{\pi}_{21}.1 + \hat{\pi}_{22} X_{1t}$).

(c) The extra information that $E(U_{1t} U_{2t}) = 0$ means that we cannot test for identifiability by the simple order condition. This piece of information effectively supplies one more 'exogenous' variable to be used as an instrument, providing that U_{2t} can itself be estimated. Since the second equation is identified we can obtain

$$\hat{U}_{2t} = Y_{2t} - \tilde{\gamma}_{20} - \tilde{\beta}_{21} Y_{1t}$$

The first equation is now just identified: the one right-hand-side endogenous variable is matched by the one excluded 'exogenous' variable $\hat{U}_{2t}$. Instrumental variables for the second stage are 1, X_{1t}, and $\hat{U}_{2t}$.

9.2. (a) We need to evaluate the plim of $\hat{\beta}$

$$\text{plim } \hat{\beta} = \frac{\text{plim } \frac{1}{T} \Sigma Y_1 Y_2}{\text{plim } \frac{1}{T} \Sigma Y_2^2}.$$

Using reduced forms:

$$Y_1 = \frac{U_1 + \beta U_2}{1 - \alpha\beta}$$

$$Y_2 = \frac{U_2 + \alpha U_1}{1 - \alpha\beta}$$

we have

$$\text{plim } \hat{\beta} = \frac{\alpha \sigma_{11} + \beta \sigma_{22} + (1 + \alpha\beta)\sigma_{12}}{\alpha^2 \sigma_{11} + 2\alpha \sigma_{12} + \sigma_{22}}$$

$$= \beta + \frac{(1 - \alpha\beta)(\alpha \sigma_{11} + \sigma_{12})}{\alpha^2 \sigma_{11} + 2\alpha \sigma_{12} + \sigma_{22}}.$$

(b) The denominator is strictly positive (being the plim of a square) so that the sign of the bias depends on the sign of $(1 - \alpha\beta)(\alpha \sigma_{11} + \sigma_{12})$ (i.e. the size of σ_{22} does not affect the direction of bias). Also σ_{11} must be strictly positive so that prior information on parameter signs relates to α, β, and σ_{12}. Knowledge of the sign of the bias depends on knowing the signs of the bracketed terms. Taking the first term, this is definitely positive if $\alpha > 0$ and $\beta < 0$ or $\alpha < 0$ and $\beta > 0$ or $\alpha = 0$ or $\beta = 0$. We can never say that the term is definitely negative without quantitative prior information. Taking the second term—since σ_{11} is positive (of unknown magnitude) we can decide the sign when α and σ_{12} are of the same sign, i.e. the second term is definitely positive if $\alpha > 0$ and $\sigma_{12} < 0$ and definitely negative if $\alpha < 0$ and $\sigma_{12} < 0$. If $\sigma_{12} = 0$ the second term takes the sign of α, and if $\alpha = 0$ takes sign of σ_{12}. Therefore bias is positive if both terms positive or both negative, i.e. $\alpha > 0$, $\beta < 0$, $(\alpha > 0)\sigma_{12} \geq 0$; $\alpha = 0$, $\sigma_{12} > 0$. And the bias is negative if $\alpha < 0$, $\beta > 0$, $(\alpha < 0)\sigma_{12} \leq 0$; $\alpha = 0$, $\sigma_{12} < 0$.

(c) Using the reduced forms we have:

$$\text{plim}\frac{1}{T}\Sigma Y_1^2 = \frac{\sigma_{11} + \beta^2\sigma_{22} + 2\beta\sigma_{12}}{(1-\alpha\beta)^2}$$

$$\text{plim}\frac{1}{T}\Sigma Y_2^2 = \frac{\sigma_{22} + \alpha^2\sigma_{11} + 2\alpha\sigma_{12}}{(1-\alpha\beta)^2}$$

$$\text{plim}\frac{1}{T}\Sigma Y_1 Y_2 = \frac{\alpha\sigma_{11} + \beta\sigma_{22} + (1+\alpha\beta)\sigma_{12}}{(1-\alpha\beta)^2}$$

i.e. three equations in five unknowns. If we know $\sigma_{12}(\bar{\sigma}_{12})$ and can express $\sigma_{12} = K\sigma_{22}(K \text{ known})$ then we can in principle solve the three equations in the three remaining unknowns, although the actual manipulation is not necessarily easy.

9.3 Using the estimated reduced forms based on the incorrect specification:

$$\hat{Y}_1 = \hat{\pi}_{11}X_1 + \hat{\pi}_{12}X_2$$

$$\hat{Y}_2 = \hat{\pi}_{21}X_1 + \hat{\pi}_{22}X_2$$

the estimator (a) rewrites the first equation

$$Y_1 = \beta_{12}\hat{Y}_2 + \gamma_{11}X_1 + W_1$$

$$\therefore \quad \hat{\beta}_{12} = \frac{\Sigma Y_1 \hat{Y}_2 \Sigma X_1^2 - \Sigma Y_1 X_1 \Sigma \hat{Y}_2 X_1}{\Sigma \hat{Y}_2^2 \Sigma X_1^2 - (\Sigma X_1 \hat{Y}_2)^2}$$

$$\therefore \quad \hat{\beta}_{12} =$$

$$\frac{\Sigma(\beta_{12}\hat{Y}_2 + \gamma_{11}X_1 + W_1)\hat{Y}_2 \Sigma X_1^2 - \Sigma(\beta_{12}\hat{Y}_2 + \gamma_{11}X_1 + W_1)X_1 \Sigma \hat{Y}_2 X_1}{\Sigma \hat{Y}_2^2 \Sigma X_1^2 - (\Sigma X_1 \hat{Y}_2)^2}$$

$$= \beta_{12} + \frac{\Sigma W_1 \hat{Y}_2 \Sigma X_1^2 - \Sigma W_1 X_1 \Sigma \hat{Y}_2 X_1}{\Sigma \hat{Y}_2^2 \Sigma X_1^2 - (\Sigma X_1 \hat{Y}_2)^2}.$$

So if the plim of the numerator of the second term is zero (while that of the denominator is not) the estimator will be consistent. Clearly this depends on the properties of W. But,

$$W_1 = U_1 + \beta_{12}(Y_2 - \hat{Y}_2).$$

Replacing Y_2 by the correct reduced form and $\hat{Y}_2$ by the estimated reduced form we have

$$W_1 = U_1 + \beta_{12}(\Pi_{21}X_1 + \Pi_{22}X_2 + \Pi_{23}X_3 + E_2 - \hat{\Pi}_{21}X_1 - \hat{\Pi}_{22}X_2)$$

where
$$E_2 = (U_2 + \beta_{21} U_1)/(1 - \beta_{12}\beta_{21}).$$

The problem is that, however accurately $\hat{\Pi}_{21}$ and $\hat{\Pi}_{22}$ are estimated, there will always be a term in X_3 remaining in W_1. Unless X_3 is uncorrelated with X_1 and X_2, we shall have the situation where in a regression a right-hand side variable ($\hat{Y}_2$) is correlated with the error term (W_1), with the resulting bias. For a correctly specified system this is not a problem since in the limit the difference between the fitted value ($\hat{Y}_2$) and the actual value (Y_2) it replaces is purely a function of the equation errors (U_i), which are uncorrelated with the exogenous variables. The estimator (b) does not create this problem. The two instruments are $\hat{Y}_2$ and X_1, both of which are uncorrelated with U_1. Hence the IV estimator (8.51)

$$\tilde{\beta}_{12} = \frac{\Sigma Y_1 \hat{Y}_2 \Sigma X_1^2 - \Sigma Y_1 X_1 \Sigma X_1 \hat{Y}_2}{\Sigma \hat{Y}_2 X_2 \Sigma X_1^2 - \Sigma \hat{Y}_2 X_1 \Sigma Y_2 X_1}$$

will still be consistent in the face of this specification error. However we know that the variance of an IV estimator will be reduced if we can find a closer fit between the IV and variable it is replacing. The mis-specification of omitting X_3 in the second reduced form has lowered the correlation between $\hat{Y}_2$ and Y_2 and hence the variance of this estimator is unnecessarily large.

9.4 (a) Applying the order condition to the first equation there are no excluded exogenous variables and one right-hand-side endogenous variable so that the equation is underidentified (by. one). The second equation is overidentified by one.

(b) (i) Substituting in the restriction we have

$$Y_1 = \beta_{12} Y_2 + \gamma_{11}(X_1 + X_2) + U_1.$$

We now have one excluded exogenous (X_1 or X_2) to match the one right-hand-side endogenous so the equation is just identified (provided X_1 and X_2 are not perfectly multicollinear). (ii) If we first estimate β_{21} by TSLS to yield $\tilde{\beta}_{21}$ and then rewrite the first equation (using $\tilde{\beta}_{12} = \tilde{\beta}_{21}/2$):

$$(Y_1 - \tilde{\beta}_{12} Y_2) = \gamma_{11} X_1 + \gamma_{12} X_2 + V_1$$

(where $V_1 = U_1 + \beta_{12} Y_2 - \tilde{\beta}_{12} Y_2$). We see that in the limit V_1 will be uncorrelated with X_1 or X_2, since it will contain just U_1, given that $\tilde{\beta}_{12}$ is estimated consistently, and hence OLS of the composite variable $(Y_1 - \tilde{\beta}_{12} Y_2)$ on X_1 and X_2 will be consistent and the parameters thus identified. (iii) Again estimating the second equation by TSLS we derive the estimated residuals: $\hat{U}_2 = Y_2 - \tilde{\beta}_{21} Y_1$. This will form a suitable instrument for IV estimation of the first equation, being correlated with Y_2, but not with U_1 in the limit (since in the limit $\hat{U}_2$ will tend to U_2 which is uncorrelated with U_1). Hence there are enough instrumental variables to allow the first equation to be identified and estimated.

9.5 We rewrite the equations
$$Y_1 = \beta_{12}Y_2 + \gamma_{11}X_1 + \gamma_{12}X_2 + V_1$$
$$Y_2 = \beta_{21}Y_1 + \gamma_{23}X_3 + U_2$$
where
$$V_1 = U_1 + \gamma_{11}X_1^{**}.$$

(a) Now by the assumption $E(X_1 X_1^{**}) \neq 0$ so $E(X_1 V) \neq 0$. Hence the first equation requires three instruments (all uncorrelated with V_1) and has only X_3 and X_2 available (provided they are uncorrelated with X^{**})—the equation is thus underidentified by one. The second equation is just identified (we cannot use X_1 as an instrument because we do not know whether $E(X_1 U_2 = 0)$).

(b) Changing the second equation to
$$Y_2 = \beta_{21}Y_1 + \gamma_{23}X_3^* + U_2$$
and rewriting it as
$$Y_2 = \beta_{21}Y_1 + \gamma_{23}X_3 + V_2$$
where
$$V_2 = U_2 + \gamma_{23}X_3^{**}.$$

The first equation continues to be underidentified—with X_3 possibly correlated with V_1 (if X_3^{**} is correlated with U_1 or X_1^{**}) the degree of underidentification could now be two.

The second equation now needs two instruments and only has X_2 available (provided it is uncorrelated with X_3^{**}). Hence the second equation is also underidentified. However, if we assume that the various measurement errors are uncorrelated with the equation errors and with each other, the measured values can be used as instrument in equations in which they do not appear. In this case the first equation is still underidentified by one, but the second equation is just identified.

10
Dynamic Models

Much of applied econometrics on macro-economic time series is concerned with modelling and estimating dynamic relationships—that is relationships in which a change in an explanatory variable occurring at a particular point in time affects the dependent variable not just in the same period but for several periods afterwards. The particular models used raise special econometric difficulties and are best treated as a special topic even though they are largely applications of ideas we have already met.

The chapter is divided into three distinct sections. The more elementary class of models do not use lagged (previous) values of the *dependent* variable as explanatory variables, while a more difficult set of problems is raised when the model used for estimation includes some lagged values of the dependent variable. The actual estimation as opposed to formulation is dealt with in the third section.

10.1 Finite lag models

We can imagine that the effect of a unit change in an explanatory variable (for example, the change in aggregate demand) does not just affect the current value of the dependent variable (for example, investment) but instead affects it for several periods—the impact is *distributed* over several subsequent periods. The corollary of this is that at any point in time the dependent variable is affected by several past values of the explanatory variable. Suppose that for a change at time t in variable $X(t)$ (of size ΔXt) the effects on the subsequent dependent variables are:

$$\Delta Y(t) = \beta_0 \Delta X(t), \Delta Y(t+1) = \beta_1 \Delta X(t) \ldots, \Delta Y(t+K) = \beta_K \Delta X(t)$$

and that after K periods the effect of the original change in X is not felt any more. Hence at time t the equation explaining Y is

$$Y(t) = \beta_0 X(t) + \beta_1 X(t-1) + \ldots \beta_K X(t-K) + U(t) \qquad (10.1)$$

The coefficients β_0 to β_K give a *finite distributed lagged* effect for variable X on variable Y. Such models are very common in situations where there is some 'cost of adjustment' so that the economic agents prefer to smooth out their response to a change in the variable X. For the econometrician this formulation may present problems. The most extreme problem would be one in which the lag length K was greater than the number of observations available

(T)—in such a case the data would be effectively multicollinear and no estimator would exist (there would be more variables than observations). Such a situation is rather rare but it is common for a related difficulty to exist. The time series nature of most studies using such models means that often the values of these explanatory variables are highly corrrelated (X is strongly correlated with its own past values). If there are not a large number of observations we know this collinearity will mean that, even if OLS is BLUE, the standard errors of the parameters will be very large and the pattern of the dynamic response may therefore be rather uncertain.

We have already seen that when the errors are 'well-behaved' we cannot hope to improve upon OLS if we stick to the BLUE criterion, unless we have extra information available. A substantial improvement in estimation could however result from the imposition of some correct restrictions on the parameter values. The commonest form of such restrictions are the so-called *polynomial distributed lags* (Almon lag). The key feature of this approach is the additional hypothesis that the values of the parameters $\beta(i)$ for $i=0$ to K lie on a function $f(i)$ of simple shape when evaluated at the integers $i = 0 \ldots K$. An illustration can make this clear. Suppose we know that the instantaneous impact of variable X is greatest and its effect declines by equal amounts as each period passes: i.e.

$$\beta_0 = a_0$$
$$\beta_1 = a_0 + a_1$$
$$\beta_2 = a_0 + 2a_1 \qquad (10.2)$$

(where a_1 is expected to be negative). The values can be plotted on a graph with a function against the time index (figure 10.1). Of course the value of the

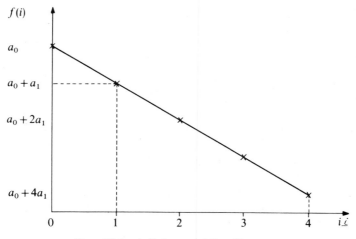

FIG. 10.1   A Polynomial Lag Pattern

function is only of interest at the $K+1$ points corresponding to the structural parameters—it is purely a device for relating the values to each other. Here a linear polynomial has been applied and we can see that if the parameters do obey such a relationship then, instead of there being $K+1$ unknowns, there are in fact only two unknowns—in fact $K-1$ restrictions have been placed on the model. A typical restriction here would be

$$\beta_1 - \beta_0 = \beta_2 - \beta_1 \tag{10.3}$$

which is obviously linear in the parameters. If we substitute (10.2) into (10.1) and collect terms with the same parameters we have:

$$Y(t) = a_0 \{X(t) + X(t-1) \ldots X(t-K)\}$$
$$+ a_1 \{X(t-1) + 2X(t-2) \ldots + KX(t-K)\} + U(t)$$

which is an equation in two composite variables. As usual we can carry out OLS since the properties of the error term will be undisturbed *if the restrictions are correct*. The properties of the estimators of the structural parameters will have the usual properties, i.e. they are best linear unbiased for all estimators which obey the restrictions. It is easy to see how to recover the structural parameters from the restricted least squares estimation. Given $\hat{a}_0$ and $\hat{a}_1$ we form

$$\hat{\beta}_0 = \hat{a}_0$$
$$\hat{\beta}_1 = \hat{a}_0 + \hat{a}_1 \quad \text{etc.} \tag{10.5}$$

The variances of the structural coefficients can also be obtained from those of the restricted model. Since

$$\hat{\beta}_K = \hat{a}_0 + K\hat{a}_1 \tag{10.6}$$

$$\therefore \quad \text{Var } \hat{\beta}_K = \text{Var } \hat{a}_0 + K^2 \text{Var } \hat{a}_1 + 2K \text{ Cov } (\hat{a}_0, \hat{a}_1). \tag{10.7}$$

The values of Var $\hat{a}_0$, Var $\hat{a}_1$ and Cov $(\hat{a}_0, \hat{a}_1)$ can be estimated by the application of standard formulae to the *restricted* model. As usual the variances of the restricted estimators (10.7) will be smaller than the variances of the unrestricted estimators, and this gain will tend to be larger the greater is the degree of collinearity among the lagged values of X.

The restriction to a first-order polynomial is rarely appropriate and often econometricians work with quadratic or higher degree polynomials. The idea might be that the impact of X could be small at first, then build up to a peak after a few periods and then decline again until no more effect is felt. Such a response needs at least a quadratic form for the restriction. We give the general treatment for a quadratic restriction: the lagged effect $\beta(i)$ is known to lie on a polynomial of degree 2 so that we can describe the general equation passing through the points as

$$\beta(i) = a_0 + a_1 i + a_2 i^2 \quad i = 0, 1 \ldots K \tag{10.8}$$

thus the restrictions are expressible by the equations:

$$\beta_0 = a_0$$
$$\beta_1 = a_0 + a_1 + a_2$$
$$\beta_2 = a_0 + 2a_1 + 4a_2 \quad \text{etc.} \tag{10.9}$$

We can now substitute in these three unknowns to replace the $K+1$ original unknowns—the restrictions are again linear and we obtain, collecting terms:

$$Y(t) = a_0\{X(t) + X(t-1) \ldots X(t-K)\}$$
$$+ a_1\{X(t-1) + 2X(t-2) + \ldots KX(t-K)\}$$
$$+ a_2\{X(t-1) + 4X(t-2) \ldots K^2 X(t-K)\} \tag{10.10}$$

This becomes a three-variable regression yielding BLU estimates of the a_i and hence BLU (subject to restrictions) estimates of the β_i obtained from substituting the estimated $\hat{a}_i$ into (10.9). The variances of the $\hat{\beta}_i$ can also be obtained by using a generalization of the method shown for the first-degree polynomial. It is very important to recover the $\hat{\beta}_i$ because, although the $\hat{a}_i$ may be significant, the implicit pattern of the $\hat{\beta}_i$ may not fit the a priori reasoning—there might be no turning point (the points just lying on a falling part of a quadratic curve or even worse on the rising part), or the values might first fall and then increase rather than vice-versa. Either case would indicate that the restriction had not found an acceptable response pattern in the data.

Testing in this model will be particularly simple because of the linearity of restrictions. We might wish either to test the length (order) of the lag or the degree of the polynomial. Using the standard F test we would wish to compare the less restrictive with the more restrictive in order to see whether the two are compatible (compatability indicates that the restriction is acceptable). Testing the degree of the polynomial requires us to go from a higher degree (less restrictive) to a lower degree (more restrictive) and this can be done by testing the incremental effect of the extra composite variables. For example in testing whether the model follows a linear or quadratic set of restrictions (for a given order of lag) we can see from (10.10) and (10.4) that the difference is in the term multiplying a_2 in (10.10). If a_2 were zero then the restriction is linear, so the test is simply:

$$H_0: a_2 = 0$$
$$H_1: a_2 \neq 0$$

(given that a_0 and a_1 are included and there are K lagged values). A simple t or F test can be used to test rival polynomial restrictions. To test the order of the lag we need to test the impact of the coefficient β_K subject to the given degree of polynomial restriction—the model can be estimated with or without the variable $X(t-K)$ and the normal F test estimated, or equivalently a 't' test on the coefficient estimated under the restriction can be used.

So far the discussion has concentrated on cases where the lag length is known (or can be specified as between two alternatives). However, it may be that we only have broadly defined prior information on the shape of the lag distribution: we may be certain that the lag weights start small, increase, and then decrease, but we may really know nothing else about the shape. In such a case imposing a low degree polynomial (a large number of restrictions) in order to save degrees of freedom is usually regarded as imposing a restriction which is approximately true (although the degree of approximation is unknown since we know nothing about the true function). From earlier discussion of specification error we know that in general any incorrect restriction will lead to bias in estimation, but that the bias will be smaller the more exact is the approximation. Hence our aim is implicitly to allow as little bias as possible while substantially reducing the variance. To this end it is clearly very advantageous to find approximating forms which are very flexible while being economical in terms of the number of parameters involved. A particularly useful idea is that of a 'rational distributed lag function'. This assumes that the weights, at the integer points 0, 1..., can be modelled at these values by a general class of functions (defined for all points) known as rational. Such a class is extremely wide, so that assuming the weights could lie on such a function is only a very weak restriction. We write such a model:

$$Y(t) = R(L)X(t) + U(t) \tag{10.11}$$

where a compact expression for the distributed lag has been introduced using the 'lag operator'

$$R(L)X(t) = r_0 X + r_1 X_{t-1} + r_2 X_{t-2} \dots \tag{10.12}$$

$$= \sum_{i=0}^{\infty} r_i X_{t-i}$$

$$= \Sigma r_i L^i X(t). \tag{10.13}$$

Here L^i is the 'lag operator' which lags the term X_t by i periods. Such a formulation can include 'infinite lags' where the terms go back (with very small weights) an infinite number of periods. Although it is not likely to be literally true that events today depend on *all* past values of a particular variable, economists and econometricians have found that some particularly interesting and appealing models do have such a lag structure.

The attraction of arguing that the weights can be described by a rational function (at the integer values) is that rational functions can be approximated by ratios of polynomial functions: the closer the approximation required, the higher will be the degrees of the polynomials needed. A particular feature is that in general while R will have a very large or infinite number of terms the polynomials will usually require relatively few terms and hence effect a large reduction in the number of parameters. This formal result that

$$R(L) \approx \lambda(L)/\mu(L) \tag{10.14}$$

over some range of the function, where $\lambda(L)$ and $\mu(L)$ are polynomials in L, has proved very attractive to econometricians because the polynomial lag functions can be treated *as if they were simple numbers*. Thus substituting (10.14) into (10.11) we have

$$Y(t) = \frac{\lambda(L)}{\mu(L)} X(t) + U(t) \tag{10.15}$$

$$\therefore \quad \mu(L) Y(t) = \lambda(L) X(t) + \mu(L) U(t). \tag{10.16}$$

If the degrees of the polynomials can be assumed we can then expand them in (10.16). Suppose both are quadratic, and that as usual we scale the equation so that the coefficient on $Y(t)$ is unity, i.e. $\mu_0 = 1$.

$$\therefore \quad Y(t) + \mu_1 Y(t-1) + \mu_2 Y(t-2)$$
$$= \lambda_0 X(t) + \lambda_1 X(t-1) + \lambda_2 X(t-2) + e(t) \tag{10.17}$$

where $e(t)$ is the composite error term from (10.16). Thus

$$Y(t) = -\mu_1 Y(t-1) - \mu_2 Y(t-2) + \lambda_0 X(t)$$
$$+ \lambda_1 X(t-1) + \lambda_2 X(t-2) + e(t). \tag{10.18}$$

This yields an equation in five unknowns which can be estimated provided we have a consistent technique available. Assuming for the moment that we do have such a technique then we can use (10.14) to reconstruct the function $R(L)$. Treating the lag operators as simple functions of L we have

$$r_0 + r_1 L + r_2 L^2 \ldots = \frac{\lambda_0 + \lambda_1 L + \lambda_2 L^2}{1 + \mu_1 L + \mu_2 L^2} \tag{10.19}$$

or

$$(1 + \mu_1 L + \mu_2 L^2)(r_0 + r_1 L + r_2 L^2 \ldots) = \lambda_0 + \lambda_1 L + \lambda_2 L^2 \tag{10.20}$$

$$r_0 + (r_1 + r_0 \mu_1) L + (r_2 + \mu_1 r_1 + \mu_2 r_0) L^2$$
$$+ (r_3 + \mu_1 r_2 + \mu_2 r_1 + \mu_3 r_0) L^3 \ldots = \lambda_0 + \lambda_1 L + \lambda_2 L^2. \tag{10.21}$$

Now if two functions are equal then the coefficients on similar terms must be equal—hence equating coefficients in powers of L we obtain:

$$r_0 = \lambda_0.$$
$$(r_1 + r_0 \mu_1) = \lambda_1 \quad \therefore \quad r_1 = \lambda_1 - \lambda_0 \mu_1$$
$$(r_2 + \mu_1 r_1 + \mu_2 r_0) = \lambda_2 \quad \therefore \quad r_2 = \lambda_2 - \mu_1(\lambda_1 - \lambda_0 \mu_1) - \mu_2 \lambda_0$$
$$(r_3 + \mu_1 r_2 + \mu_2 r_1 + \mu_3 r_0) = 0 \tag{10.22}$$

etc. Once the μ_i and λ_i are known then we can determine all the r_i in this recursive fashion. This ability to manipulate lag operators, separately from the data they are applied to, is clearly very useful in that it potentially can save large numbers of degrees of freedom when the basic lag is of high order. However, there are a number of important points to be borne in mind when considering the use of this very elegant approach.

1. We must have a consistent method of estimation for the approximation parameters.

2. The basic function $R(L)$, if it has a finite number of terms, is of course always a polynomial in L of degree the same as the lag length (a set of M points can always be exactly fitted by a polynomial of degree M) so that we are approximating a high degree polynomial by the ratio of two lower degree polynomials—this will usually reduce the number of parameters.

3. The theorems which guarantee that a rational function can be approximated by the ratio of two polynomials merely state that for an approximation of desired accuracy there exists a pair of polynomials with given weights that approximate the function in a given range (i.e. between certain values of L). The wider the range over which the approximation is to be good then the higher will the degree of the polynomials generally have to be. In general it is even possible to put limits on the size of the approximation error. The difficulty for econometricians is that we do not know enough about $R(L)$ to say how accurate our approximation will be, nor anything about the range of value of L for which it is a good approximation. Hence the choice of degrees for $\mu(L)$ and $\lambda(L)$ has to be made with care. We could try alternative degrees of approximating polynomial in order to see whether the overall goodness of fit and the estimated r_i coefficients are sensitive to our choices. The r_i are however non-linear functions of the λ_i and μ_i (unlike the Almon lags introduced earlier) so that although we can calculate standard errors for these approximating coefficients from our method of estimation, it will not be possible to calculate in a simple fashion standard errors for the r_i (more advanced methods are available to calculate the standard errors). Checks on goodness of fit would also come from the incremental effect on the standard error of estimate.

4. The model does not distinguish directly between different lag lengths for the r_i—the same λ_i and μ_i can be solved for the r_i in different ways according to the assumed length of lag in the structural equation.

Our first concern is to find a consistent method of estimation. As usual the choice of method depends on the error properties. The structural equation $U(t)$ has been transformed by the approximation polynomial to yield a new error term

$$e(t) = \mu_0 U(t) + \mu_1 U(t-1) \ldots \quad (10.23)$$

There is an obvious problem here. If the error term in the structural equation is well behaved, i.e. is not serially correlated, then this composite error term will be serially correlated:

$$e(t)e(t-s) = (\mu_0 \mu_s + \mu_1 \mu_{s+1} + \ldots)\sigma^2. \quad (10.24)$$

We already know that this means OLS is not BLUE and that we need to find the Aitken transform to remove the serial correlation. Obviously the required transform is one which has $U(t)$ as the error and hence is the original equation:

$$Y(t) = R(L)X(t) + U(t).$$

An attempt to estimate the approximation equation (10.23) by OLS, just ignoring the effects of serial correlation on efficiency, would also be unsatisfactory. As we shall show below the combination of lagged dependent variables plus autocorrelated errors produces bias when using OLS, because of the correlation between the errors and right-hand-side variables (a shock in period $t-1$ affects the current error and the lagged dependent variable). Hence if we wish to use the approximation form of the equation we cannot in general use OLS. Of course were the original errors (U) serially correlated then the approximation could achieve two ends simultaneously: (i) it would act as a low degree approximation polynomial, saving degrees of freedom; (ii) it could purge serial correlation thus opening the way for OLS to be used. The strategy is then to experiment with the denominator polynomial $\mu(L)$ until the residuals show no evidence of serial correlation, while $\lambda(L)$ is varied until a sufficiently stable fit is achieved. The problem with this approach is that we no longer have a free choice on one polynomial—it is decided by the original error structure. This means that in order to approximate a high-order polynomial we have to choose a single polynomial divided by a fixed polynomial, which clearly means that we are in a much less flexible situation than originally appeared. The chances for saving degrees of freedom are possibly reduced.

An alternative strategy, which does not tie our hands with respect to the degree of approximation, is to ignore the serial correlation effects on efficiency and find instrumental variables uncorrelated with the error terms. This would allow consistent estimators to be derived from the normal equations. The problem is to find variables correlated with the lagged dependent variables. From the logic of the model these are themselves determined by earlier lagged values of the dependent variable and lagged independent variables. Clearly only the latter are admissible as instruments, so that we could use earlier lags of the independent variables than those appearing in the approximation model. However the standard formulae for standard errors would be incorrect because of the serial correlation.

10.2 Infinite lags

As we have seen we can hope to approximate an infinite lag function by the ratio of two polynomial lags, so that were such a model to arise we have a potential technique for handling it. In fact at least two important economic models lead directly to such a situation so that it is quite common.

Consider first the so called *'partial adjustment'* model. In equilibrium the variable Y does not immediately change to a new equilibrium level but moves part of the way in each period.

$$Y_t^* = \alpha + \beta X_t + U_t \qquad (10.25)$$

and
$$Y_t = Y_{t-1} + \lambda(Y_t^* - Y_{t-1}) + V_t \qquad (10.26)$$

where Y_t^* is the 'target' or 'equilibrium' value associated with X_t. The second equation adjusts the level of Y (from its previous level) by a *percentage* of the gap between the existing value and the target value. Both equations are subject possibly to stochastic disturbances. Substituting (10.25) into (10.26) we obtain:

$$Y_t = \lambda\alpha + \beta\lambda X_t + (1-\lambda)Y_{t-1} + \lambda U_t + V_t. \qquad (10.27)$$

This of course can be converted into an *infinite* lag by repeated substitution for the lagged dependent variable: therefore,

$$Y_t = \lambda\alpha + \beta\lambda X_t + \lambda\alpha(1-\lambda) + \beta\lambda(1-\lambda)X_{t-1}$$
$$\ldots + \lambda U_t + V_t + \lambda(1-\lambda)U_{t-1} + (1-\lambda)V_{t-1} \ldots \qquad (10.28)$$

The key to estimation lies as usual with the error terms. If the original errors are well behaved then (10.27) needs no transform to get rid of serial correlation and the problems of an infinite lag are not faced: OLS will be consistent. If there is serial correlation in (10.27) the standard Aitken transform should be used (together with the two-step procedure for estimating the serial correlation coefficient). The equation (10.27) is in effect an infinite lag subject to a large number of restrictions.

A different model comes from the 'adaptive expectations' framework. The actual value of the dependent variable is related to an (unobservable) variable, but the value of this is adjusted partially to new evidence:

$$Y_t = \beta X_t^* + U_t \qquad (10.29)$$

$$X_t^* = X_{t-1} + \lambda(X_t - X_{t-1}) \qquad (10.30)$$

where X_t^* is the 'expected' value of variable X (the actual values being observable). Substituting (10.30) into (10.29) to obtain an equation solely in terms of observables we have:

$$Y_t^* = \beta\lambda X_t + \beta\lambda(1-\lambda)X_{t-1} \ldots + U_t. \qquad (10.31)$$

This model is an infinite distributed lag but with non-linear restrictions linking the coefficients. The equation can be written

$$Y_t = \beta\lambda(1 + \lambda L + \lambda^2 L^2 \ldots)(X_t) + U_t \qquad (10.32)$$

where L is the lag operator. Treating the term in brackets as a geometric series and summing we have

$$Y_t = \frac{\beta\lambda}{1-\lambda L} X_t + U_t \qquad (10.33)$$

or
$$Y_t(1-\lambda L) = \beta\lambda X_t + U_t(1-\lambda L) \qquad (10.34)$$

or
$$Y_t = \lambda Y_{t-1} + \beta\lambda X_t + e_t \qquad (10.35)$$

where
$$e_t = U_t - \lambda U_{t-1}.$$

This final form could have also been obtained by lagging (10.31), multiplying it by λ, and subtracting the result from (10.31) (the Koyck transform), thus illustrating that it is valid to treat the lag operator as a number. This transformation has produced an equation with a finite number of terms, and indeed exactly as many coefficients as there are parameters to be estimated. However the error term is now a *moving average* of the structural errors—if the U are well behaved then the e must be serially correlated. This in turn would lead to OLS, as applied to (10.35), being inconsistent because of the correlation of lagged dependent variable and error term (both being affected by U_{t-1}). The obvious Aitken transform to remove the serial correlation merely integrates the equation back to (10.31) so that this would not solve the problem. Only in the case where the structural errors followed a first-order Markov process with parameter λ

$$U_t = \lambda U_{t-1} + e_t \qquad (10.36)$$

would the Koyck transform, which imposes all the non-linear restrictions, also exactly neutralize the underlying serial correlation. In such a situation OLS would be consistent.

Since it is extremely unlikely that the serial correlation parameter in (10.36) is equal to the adaptive expectations parameter in (10.30) we need to find another method of estimation. Clearly instrumental variables could be applied to (10.35) using an earlier lagged value of X as an instrument. The high collinearity between instruments would tend to produce high standard errors, while the conventional formula for the standard errors would be incorrect because they ignore the serial correlation.

A standard approach to this, and similar problems which cannot be approximated by ratios of polynomials, is to split the equation into observable and unobservable parts. From (10.31) we have

$$Y_t = \beta \sum_0^{t-2} \lambda^i X_{t-i} + \lambda^{t-1}\left(\beta \sum_0^{\infty} \lambda^i X_{1-i}\right) + U_t \qquad (10.37)$$

i.e. we split the data into one portion which ranges from the earliest observable (X_1) back to the beginning of the process, and one which summarizes the effect on period t of values of X from periods 2 to t. Writing this as

$$Y_t = \beta X_t^* + \lambda^{t-1} X_1^* + U_t \qquad (10.38)$$

we see that X_1^* is unobservable because of the λ as well as the earlier values of X. However the expected value of Y at period 1 is equal to X_1^* so that we can write:

$$Y_t = \beta X_t^* + \lambda^{t-1} Y_1 + V_t \qquad (10.39)$$

where V_t now includes the equation errors from before the estimation period. This equation can be estimated by a grid search technique. We choose a value for λ—λ (0), say— construct X^* (which is a function of λ), and regress ($Y_t - \lambda(0)^{t-1} Y_1\}$ on $X^*(0)_t$ to yield an estimate of β. This is repeated for increasing values of $\tilde{\lambda}$ until the range including the minimum standard error of estimate has been identified. Subdivision of the critical interval for $\tilde{\lambda}$ can then be used to obtain a better approximation. Clearly the estimator attempting to use all the information available is likely to be superior to one which truncated the data at period 1 and omitted the variable $\lambda^t Y_1$ altogether.

When the parameters of an infinite distributed lag are not known a priori to follow exact restrictions then we can use the approximation of two polynomials to estimate the lag structure. Although in principle we can calculate all the coefficient of the infinite lag from equating coefficients as before from an equation of the form:

$$R(L) = \frac{\lambda(L)}{\mu(L)}.$$

This is usually done only for the first few terms, in order to obtain the general shape of the lag function. It is perhaps more useful to calculate certain summary statistics such as the mean lag and the total effect of the distributed lag, which can be done without direct use of the coefficients of the infinite lag. To obtain the total or long-run multiplier effect we need to sum all the distributed lag coefficients so that for a steady value of X we have:

$$Y^* = R(L) X^*. \qquad (10.40)$$

Hence the *total* effect of a unit change in X^* is Σr_i, which is in fact the value of the distributed lag function evaluated at $L = 1$:

$$\Sigma r_i = R(1) \qquad (10.41)$$

But
$$\frac{\lambda(1)}{\mu(1)} = R(1) \qquad (10.42)$$

so that
$$\sum_0^\infty r_i = \sum_0^\infty \lambda_i \Big/ \sum_0^\infty \mu_i. \qquad (10.43)$$

The two sums are both over a *finite* number of terms so that we can immediately obtain the total effect (the same device could be used for a finite rational lag function if desired).

The mean lag measures (for non-negative coefficients) the weighted average effect—the lags weighted by their relative importance:

$$\bar{r} = \sum_0^\infty i r_i \bigg/ \sum_0^\infty r_i. \qquad (10.44)$$

We see that from the lag function $R(L)$ we can obtain the derivative with respect to L:

$$\frac{dR(L)}{dL} = r_1 + 2r_2 L + 3r_3 L^2 \ldots$$

$$\therefore \quad \bar{r} = R'(1)/R(1) \qquad (10.45)$$

But also we have

$$\frac{dR}{dL} = \frac{d[\lambda(L)/\mu(L)]}{dL} = \frac{\lambda(L)\mu'(L) - \mu(L)\lambda'(L)}{\lambda^2(L)} \qquad (10.46)$$

$$\therefore \quad \bar{r} = \frac{\mu'(1)}{\mu(1)} - \frac{\lambda'(1)}{\lambda(1)} \qquad (10.47)$$

which can easily be evaluated given the estimated polynomial functions.

10.3 Estimation with lagged dependent variables

We have already seen that certain models will give rise to equations which contain lagged dependent variables and serially correlated errors. We have remarked that this combination will have the effect of making estimation by OLS biased because of the correlation of the lagged dependent variable—LDV—(which contains lagged errors) with the current error term (which also contains lagged errors). However it is useful to give a more systematic treatment of the problems of estimation with models including LDVs.

Suppose that we have the simple model:

$$Y_t = \alpha + \gamma Y_{t-1} + U_t \qquad (10.48)$$

where the errors have the properties:

$$E(U_t) = 0$$

$$E(U_t U_{t-s}) = 0 \qquad s \neq 0 \text{ all } t$$

$$= \sigma^2 \qquad s = 0. \qquad (10.49)$$

If we estimate by ordinary least squares we have:

$$\hat{\gamma} = \frac{\sum_2^T y_t y_{t-1}}{\sum_2^T y_{t-1}^2} \qquad (10.50)$$

$$\hat{\alpha} = \bar{Y} - \hat{\gamma} \bar{Y}_{-1}. \qquad (10.51)$$

Now clearly both numerator and denominator of (10.50) are random variables, and are not independent since they contain common elements, so that we cannot evaluate $\hat{\gamma}$ by taking the expectation of (10.50). However, we can consider the large sample estimator

$$\operatorname{plim} \hat{\gamma} = \frac{\operatorname{plim} \frac{1}{T}\Sigma y_t y_{t-1}}{\operatorname{plim} \frac{1}{T}\Sigma y_{t-1}^2}.$$

Substituting in the structural equation we obtain

$$\operatorname{plim} \hat{\gamma} = \gamma + \frac{\operatorname{plim} \frac{1}{T}\Sigma U_t y_{t-1}}{\operatorname{plim} \frac{1}{T}\Sigma y_{t-1}^2}. \qquad (10.52)$$

The numerator tends to zero as the sample size increases—although Y_{t-1} includes all the error terms of the sample in the mean value $\bar{Y}_{-1}$, the weight on each is $1/T$, so that the average of the sum of those terms goes to zero as T goes to infinity. Hence the OLS estimator is consistent. The variance of the estimator is given by

$$\operatorname{Var} \hat{\gamma} = \frac{\sigma^2}{\Sigma y_{t-1}^2} \qquad (10.53)$$

which goes to zero as the number of terms increases.

A more severe estimation problem with a lagged dependent variable arises when there is also serial correlation. Suppose that we alter the assumptions of our previous model to allow for first-order serial correlation

$$U_t = \rho U_{t-1} + e_t \qquad (10.54)$$

where e_t is well behaved. Clearly the property of the OLS estimator of γ will depend on the correlation between Y_{t-1} and U_t. There is now an additional non-zero term in (10.52) since both contain U_{t-1}. Hence at each value of t there is a non-zero term independent of sample size, with the result that the average of the sum of all these terms is finite and OLS is inconsistent. There is a very useful way of evaluating this effect. The probability limit of the least squares estimator (10.50) can be denoted by

$$\operatorname{plim} \hat{\gamma} = K_{-1}/K_0 \qquad (10.55)$$

where

$$K_{-s} = \operatorname{plim} \frac{1}{T}\Sigma y_t y_{t-s} \qquad (10.56)$$

(assuming that the limit exists). In order to evaluate this ratio we need to find an equation linking the terms to the structural parameters. We first substitute in the autocorrelated error structure in weighted difference form in order to obtain an equation with well-behaved error terms:

$$Y_t = \alpha(1-\rho) + (\gamma+\rho)Y_{t-1} - \gamma\rho Y_{t-2} + e_t. \tag{10.57}$$

Taking deviations from the mean and premultiplying by Y_{t-1} and summing we have

$$\frac{1}{T}\Sigma y_t y_{t-1} = (\gamma+\rho)\frac{1}{T}\Sigma y_{t-1}^2 - \gamma\rho\frac{1}{T}\Sigma y_{t-1} y_{t-2}$$

$$+ \frac{1}{T}\Sigma y_{t-1} e_t. \tag{10.58}$$

Taking plims and noting that

$$\operatorname{plim}\frac{1}{T}\Sigma y_{t-1}^2 = \operatorname{plim}\frac{1}{T}\Sigma y_t^2$$

$$\operatorname{plim}\frac{1}{T}\Sigma y_t y_{t-1} = \operatorname{plim}\frac{1}{T}\Sigma y_{t-1} y_{t-2} \tag{10.59}$$

(since the terms by which they differ are of decreasing importance as the sample size increases) we obtain

$$K_{-1} = (\gamma+\rho)K_0 - \gamma\rho K_{-1} \tag{10.60}$$

$$\therefore \quad \frac{K_{-1}}{K_0} = \frac{\gamma+\rho}{1+\gamma\rho}. \tag{10.61}$$

Hence we have the result that

$$\operatorname{plim} \hat{\gamma} = \frac{\gamma+\rho}{1+\gamma\rho} = \gamma + \frac{\rho(1-\gamma)^2}{1+\gamma\rho} \tag{10.62}$$

which indicates that OLS is generally inconsistent when there is serial correlation of the error terms together with a lagged dependent variable. The method of evaluation is particularly useful since it does not require us directly to evaluate the K_{-s}. Other cases, with longer lags and higher-order serial correlation, can be evaluated by an extension of the same technique. The result makes clear the problem that we encountered with the rational lag approximation by polynomial lags. If the transformation by the denominator PDL induces serial correlation, as well as creating the lagged dependent variable, then OLS will be biased and inconsistent.

These results show that it is particularly important to be able to test for the presence of serial correlation when there are lagged dependent variables in the model. The standard test against serial correlation that we introduced was the Durbin–Watson test. The statistic is defined by the formula

$$\hat{d} = \frac{\Sigma(\hat{U}_t - \hat{U}_{t-1})^2}{\Sigma \hat{U}_t^2}. \tag{10.63}$$

Suppose that we had the true values of the error terms (U_t), then the plim of the estimator $\hat{d}$ would be given by:

$$\hat{d} = 2 - 2\rho \tag{10.64}$$

when there is first-order serial correlation. However the plim has to be applied to the residuals from an OLS regression:

$$\text{plim } \hat{d} = 2\left(1 - \frac{\text{plim} \frac{1}{T}\Sigma \hat{U}_t \hat{U}_{t-1}}{\text{plim} \frac{1}{T}\Sigma \hat{U}_t^2}\right). \tag{10.65}$$

The latter terms can be solved by using

$$\hat{U}_t = Y_t - \hat{\gamma} Y_{t-1} \tag{10.66}$$

and using our result for plim $\hat{\gamma}$ we have (see problem (10.3)):

$$\text{plim } \hat{d} = 2\left(1 - \frac{\gamma\rho(\gamma + \rho)}{1 + \gamma\rho}\right). \tag{10.67}$$

Hence the estimated DWS does not have the same limit as that defined for the true residuals. This indicates that when there is serial correlation the *power* of the test may be very poor since the statistic can be biased towards the neutral value of 2 (suggesting that ρ is zero and there is lack of serial correlation). An equally serious problem is that even when there is *no serial correlation* the distribution of $\hat{d}$ in repeated samples does not follow the distribution used in the standard tables so that the usual upper and lower bounds do not apply. Attempts to use standard tables in effect are using tests of unknown size.

Hence whether or not there is serial correlation of the error terms the Durbin–Watson test cannot be used if there are lagged dependent variables. A series of more advanced tests have been designed for use in this situation. Durbin's h test is suitable for testing against first-order serial correlation for any number of lagged dependent variables. The procedure is to estimate the basic model by OLS on the assumption that there is no serial correlation. The residuals are then used to calculate an estimated coefficient for first-order serial correlation.

$$\hat{\rho}_1 = \frac{\Sigma \hat{U}_t \hat{U}_{t-1}}{\Sigma \hat{U}_{t-1}^2} \tag{10.68}$$

This is then substituted in the h statistic formula:

$$h = \hat{\rho}_1 \sqrt{\frac{T}{1 - TV(\hat{a}_1)}} \tag{10.69}$$

where $V(\hat{a}_1)$ is the estimated variance of the coefficient of the dependent variable lagged one period (in the OLS regression). In large samples under the NH of no serial correlation this statistic follows a standard normal distribution $N(0, 1)$ so that an asymptotic test for the presence of first-order serial correlation can be carried out just using the results of an OLS regression based on an equation assuming no serial correlation. This test (which is in effect an LM test) can be generalized to allow for serial correlation of higher order.

The general pattern of these tests is as follows: (i) carry out OLS on the structural equation obtaining the residuals $\hat{U}_t$; (ii) carry out OLS of $\hat{U}_t$ on all the explanatory variables of the structural model and on the M values of lagged errors ($\hat{U}(-1)$:, ... $\hat{U}(-M)$—where M is the order of the autocorrelation process); (iii) test the marginal significance of these M-lagged residuals by a conventional F statistic for a sub-set of variables.

$$F = \frac{(R^2 - R_M^2)/M}{R_M^2/(T - M - K)} \tag{10.70}$$

where K is the number of variables in the second regression and $T - M$ is the number of observations in the regression. This statistic approximately follows an F distribution for large T. If the F statistic is less than the critical value then we conclude that there is no evidence of serial correlation of up to order M. Alternatively we can take the squared multiple correlation coefficient R^2 to form the statistic:

$$TR^2 \sim \chi^2(M) \tag{10.71}$$

where T is the total number of observations. Under the null hypothesis of no serial correlation this statistic follows a $\chi^2(M)$ distribution.

These results show the importance of correcting for serial correlation in a model with lagged dependent variables. The weighted difference technique (as used in Aitken's estimator) is clearly required in order to transform the equation to a form suitable for estimation by OLS or else we must find a respecification of the basic model in which the error terms are not serially correlated.

If there is serial correlation present of a known order then we can estimate the structural equation by a two-step procedure. Consider the model:

$$Y_t = \beta X_t + \gamma Y_{t-1} + U_t \tag{10.71}$$

where

$$U_t = \rho U_{t-1} + e_t$$

where we know of the existence of the first-order serial correlation, but not the value of ρ. We write the equation in the form it would take for estimation if we knew the serial correlation parameter:

$$Y_t = \beta X_t - \beta \rho X_{t-1} + (\gamma + \rho) Y_{t-1} - \gamma \rho Y_{t-2} + e_t. \tag{10.72}$$

This can be estimated by OLS and the values obtained will be consistent. Clearly they are not efficient since there are only three unknowns (β, γ, ρ) but four estimated parameters. There is in fact a non-linear restriction between the coefficients. We can take this into account by using the unrestricted OLS coefficients on the exogenous variable (X) and its lagged value to obtain an estimate of ρ:

$$\tilde{\rho} = \frac{\widehat{\beta\rho}}{\hat{\beta}}. \tag{10.73}$$

This is then used to transform the basic variables in the standard Aitken fashion

$$Y_t^* = Y_t - \tilde{\rho} Y_{t-1}$$
$$X_t^* = X_t - \tilde{\rho} X_{t-1} \tag{10.74}$$

and a regression of Y_t^* on X_t^* and Y_{t-1}^* yields efficient estimates of β and γ.

For more complex error processes it is necessary to use other methods of estimation which utilize ML techniques. These are not described here. We conclude with an example of estimates of lag structures.

Example 10.1: Estimation of lag structures

We consider again our simple consumption function but ignore the problem of simultaneity in order to focus on the estimation of a possible lag response of consumption to changes in income.

(a) Polynomial distributed lag

We begin with the hypothesis that the level of current consumption responds to lags of up to five years. The unrestricted OLS equation is shown in Table 10.1.

TABLE 10.1. Coefficients from Unrestricted OLS

Variable	Coefficient	SE
Constant	20223	3355
Income (t)	0.753	0.124
Income ($t-1$)	-0.055	0.168
Income ($t-2$)	-0.091	0.180
Income ($t-3$)	0.168	0.181
Income ($t-4$)	0.027	0.175
Income ($t-5$)	-0.054	0.135

$R^2 = 0.990$, SEE $= 1548$, DWS $= 1.01$

We see that taken one by one only the current income term would pass a 't' test against the null hypothesis that income affects consumption. No lag, taken individually, is significant. We might wonder whether the large standard errors are due to the severe collinearity between the different lags in income and accordingly decide to restrict the pattern of the lag parameters. We choose to restrict the six coefficients to lie on a quadratic (effectively imposing three linear restrictions). The restricted regression is shown in Table 10.2 (estimated by a programme which automatically takes the restrictions into account).

TABLE 10.2 Coefficients from a Quadratic Restriction

Variable	Coefficient	SE
Constant	23017	3985
Income (t)	0.471	0.061
Income ($t-1$)	0.177	0.029
Income ($t-2$)	−0.001	0.036
Income ($t-3$)	−0.061	0.035
Income ($t-4$)	−0.008	0.029
Income ($t-5$)	0.163	0.068

$R^2 = 0.980$, SEE $= 1920$, DWS $= 1.24$

The results show, as expected, that the coefficients have smaller standard errors when the restrictions are imposed. However, lags 2, 3, and 4 would show very weak effects. Moreover, the shape of the lag response is quite unacceptable—the feature of an important initial impact with rapidly dying lags up to four years before with a dramatic increase in the effect of five years before would not fit any reasonable explanation and the model should be altered.

(b) Rational distributed lag

It is decided to use a much longer lag and so a rational distributed lag model is introduced: $C = R(L)Y$. The rational lag is approximated by the ratio of a second-order to first-order polynomial, i.e.

$$C_t + \mu_1 C_{t-1} = \alpha^* + \lambda_0 Y_t + \lambda_1 Y_{t-1} + \lambda_2 Y_{t-2} + U_t.$$

The results of the estimation are shown in Table 10.3.

TABLE 10.3 Coefficients from a Rational Lag Restriction

Variable	Coefficient	SE
Constant	6577	4948
Income (t)	0.650	0.094
Income ($t-1$)	−0.533	0.191
Income ($t-2$)	0.112	0.075
Consumption ($t-1$)	0.695	0.276

$R^2 = 0.994$, SEE $= 1284$

The individual coefficients (apart from income lagged two periods) are now much better determined but the pattern of the r_i is critical. From this equation we can reconstruct the characteristics of the rational lag response. Since

$$C = \alpha + R(L)Y$$
$$= \alpha + \frac{\lambda(L)Y}{\mu(L)}$$

$$\therefore \quad \mu(L)C = \alpha\mu(L) + \lambda(L)Y.$$

We see that the structural constant can be obtained by dividing the estimated constant by $(1 - \mu(1))$

$$\alpha = \frac{6577}{1 - 0.695} = 21564.$$

Using the relation

$$(r_0 + r_1 L + r_2 L^2 \ldots) = (\lambda_0 + \lambda_1 L + \lambda_2 L^2)/(1 + \mu_1 L)$$

we can obtain individual r_i as far back as required. First we calculate the long-run multiplier by using (10.42)

$$\Sigma r_i = R(1) = \lambda(1)/\mu(1)$$

so that here $R(1) = 0.751$ (we cannot calculate a mean lag since one of the weights is negative). Equating term by term we have from (10.22):

$$\lambda_0 = 0.65, \quad \lambda_1 = -0.533 \quad \lambda_2 = 0.112 \quad \mu_1 = -0.695$$

$$\therefore \quad r_0 = 0.650$$

$$r_1 = -0.081$$

$$r_2 = 0.056$$

$$r_3 = 0.039$$

$$r_4 = 0.027$$

etc. Again the pattern of the lag seems unrealistic with a large decline in the one period ago weight (negative) switching back to positive two periods ago, and then declining geometrically before that. It appears that further experiments would be required to establish whether there is genuinely a dynamic response to changes in income.

Problems 10

10.1 Consider the model

$$Y_t = \alpha + \sum_0^K \beta(i) X_{t-i} + U_t$$

where $E(U_t) = 0$, $E(U_t U_s) = 0 \quad s \neq t$
$\qquad \qquad \qquad \qquad \quad = \sigma^2 \quad s = t.$

The model is to be estimated subject to the restrictions

$$\beta(i) = C_0 + C_1 i + C_2 i^2 \qquad i = 0, 1 \ldots K.$$

(a) Show that the restrictions can be written in the form

$$\Delta^3 \beta(i) = 0 \qquad i = 3, 4 \ldots K,$$

where
$$\Delta \beta(i) = \beta(i) - \beta(i-1)$$
$$\Delta^2 \beta(i) = \Delta\{\Delta \beta(i)\} \quad \text{etc.}$$

(b) It is proposed to define a new variable, $\beta(-1)$. Show that imposing the restriction $\beta(-1) = 0$ is equivalent to placing a restriction on the C_i, and hence to placing another restriction on the $\beta(i)$ (for $i = 0 \ldots k$).

10.2 There is an equation

$$Y_t = \alpha Y_{t-2} + U_t \qquad |\alpha| < 1$$
$$U_t = \lambda U_{t-1} + \varepsilon_t \qquad |\lambda| < 1$$

where $E(\varepsilon_t) = 0$
$\qquad \quad E(\varepsilon_t \varepsilon_s) = 0 \qquad t \neq s$
$\qquad \qquad \qquad \; = \sigma^2 \qquad t = s$

What is the asymptotic bias of the estimator $\hat{\alpha}$ obtained from regressing Y_t on Y_{t-2}?

10.3 Consider the model

$$Y_t = \gamma Y_{t-1} + U_t$$
$$U_t = \rho U_{t-1} + e_t$$
$$E(e_t) = 0$$
$$E(e_t e_s) = 0 \qquad s \neq t$$
$$ = \sigma^2 \qquad s = t$$

Show that the probability limit of the Durbin–Watson statistic

$$\hat{d} = \frac{\Sigma(\hat{U}_t - \hat{U}_{t-1})^2}{\Sigma \hat{U}_{t-1}^2}$$

is given by

$$\text{plim } \hat{d} = 2\left\{\frac{1 - \gamma\rho(\gamma + \rho)}{(1 + \gamma\rho)}\right\}.$$

10.4 Consider the system of equations

$$Y_{1t} = \alpha_1 + \beta_{12} Y_{2t} + \gamma_{11} Y_{1t-1} + U_{1t}$$
$$Y_{2t} = \alpha_2 + \beta_{21} Y_{1t} + \gamma_{22} X_{2t} + U_{2t}$$

where

$$E(X_{2t} U_{is}) = 0 \quad \text{all } i \text{ and } s$$
$$U_{1t} = \rho U_{1t-1} + e_{1t}$$
$$E(e_{1t}) = 0$$
$$E(e_{1t} e_{1s}) = 0 \qquad s \neq t$$
$$E(U_{2t}) = 0, \; E(U_{2t} U_{2s}) = 0 \qquad s \neq t.$$

Are the equations identified? How does the presence of the serial correlation affect your answer? How would you estimate those equations which are identified?

10.5 Consider the model

$$Y_t = A X_t e^{vt}.$$

The logarithmic form of the model is subject to a dynamic adjustment process of a rational lag type

$$\ln Y_t = \ln A + R(L) \ln X_t + V_t.$$

(a) If the function $R(L)$ is expressed as the ratio of two first-order polynomials show that the equation can be written with 'differential' and 'proportional' control factors:

$$\Delta y_t = A^* + \alpha_0 \Delta x_t + (1 + \beta_1)(x_{t-1} - y_{t-1}) + V_t^*.$$

How are A^* and V_t^* related to A and V_t?
(where $y = \ln Y$, $x = \ln X$).

(b) Generalize this to the case where the numerator polynomial is second degree and show that the terms can be grouped so as to include an 'acceleration' factor $\Delta^2 X$.

Answers 10

10.1 (a) Since
$$\beta(i) = C_0 + C_1 i + C_2 i^2$$
$$\Delta \beta(i) = C_1(1) + C_2(2i-1)$$
$$\Delta^2 \beta(i) = 2C_2$$
and
$$\Delta^3 \beta(i) = 0$$

(the first difference is defined for $i \geq 1$, the second difference is defined for $i \geq 2$, and the third difference is defined for $i \geq 3$).

(b) If we invent the variable $\beta(-1)$ and insist that it is zero (i.e. the quadratic passes through the point -1 with a zero value) then $\beta(-1) = C_0 - C_1 + C_2 = 0$ and $C_0 = C_1 - C_2$. The 'end-point' restricted polynomial is

$$\beta(i) = (C_1 - C_2) + C_1 i + C_2 i^2.$$

The $\beta(i)$ are even less free to vary than before since the value on the zero lag $(C_1 - C_2)$ must equal the difference of the linear and quadratic effects. Another way to see this effect is to solve the C_i in terms of the $\beta(i)$. From the original restriction we have

$$\beta(0) = C_0$$
$$\beta(1) = C_0 + C_1 + C_2$$
$$\beta(2) = C_0 + 2C_1 + 4C_2$$

therefore solving

$$C_0 = \beta(0)$$
$$C_1 = 2\beta(1) - \frac{3}{2}\beta(0) - \frac{1}{2}\beta(2)$$
$$C_2 = \frac{1}{2}\{\beta(2) - 2\beta(1) + \beta(0)\}.$$

If we add the 'end-point' restriction we are insisting not only that the $\beta(i)$ lie on a quadratic but that $C_0 - C_1 + C_2 = 0$, i.e. $2\beta(0) + \beta(2) - 3\beta(1) = 0$. This relationship between the lag weights is most unlikely to correspond to any economic formulation.

10.2 Substitute the weighted first difference of the error-process into the

structural equation in order to produce a well-behaved error term.
$$Y_t = \lambda Y_{t-1} + \alpha Y_{t-2} - \alpha\lambda Y_{t-3} + \varepsilon_t$$
We have to evaluate
$$\hat{\alpha} = \Sigma Y_t Y_{t-2} / \Sigma Y_{t-2}^2$$
$$\therefore \quad \text{plim } \hat{\alpha} = K_{-2}/K_0.$$

Multiplying both sides by Y_{t-2} summing, averaging, and taking plims we have
$$K_{-2} = \lambda K_{-1} + \alpha K_0 - \alpha\lambda K_{-1}.$$

We need to eliminate K_{-1}, so premultiplying the equation by Y_{-1}, summing, averaging, and taking plims we also have:
$$K_{-1} = \lambda K_0 + \alpha K_{-1} - \alpha\lambda K_{-2}$$
$$\therefore \quad K_{-1} = \frac{\lambda K_0 - \alpha\lambda K_{-2}}{1-\alpha}.$$

Substituting back
$$K_{-2} = \lambda[\lambda K_0 - \alpha\lambda K_{-2}] + \alpha K_0$$
and therefore
$$\frac{K_{-2}}{K_0} = \frac{\alpha + \lambda^2}{1 + \alpha\lambda^2} = \alpha + \frac{\lambda^2(1-\alpha^2)}{1+\alpha\lambda^2}.$$

The bias is unambiguously positive.

10.3
$$\text{plim } \hat{d} = \frac{\text{plim } \Sigma(\hat{U}_t - \hat{U}_{t-1})^2}{\Sigma \hat{U}_t^2}$$

$$= 2 - 2 \left[\frac{\text{plim } \frac{1}{T}\Sigma \hat{U}_t \hat{U}_{t-1}}{\text{plim } \frac{1}{T}\Sigma \hat{U}_t^2} \right].$$

Replacing $\hat{U}_t$ by $Y_t - \hat{\gamma} Y_{t-1}$

$$\therefore \quad \text{plim } \hat{d} = 2 - 2 \left[\frac{\text{plim}\left\{\frac{1}{T}\Sigma(Y_t - \hat{\gamma} Y_{t-1})(Y_{t-1} - \hat{\gamma} Y_{t-2})\right\}}{\text{plim}\left\{\frac{1}{T}\Sigma(Y_t - \hat{\gamma} Y_{t-1})^2\right\}} \right].$$

Define
$$K_{-i} = \text{plim } \frac{1}{T}\Sigma Y_t Y_{t-i}$$

so that
$$\text{plim } \hat{\gamma} = K_{-1}/K_0$$

$$\therefore \quad \text{plim } d = 2 - 2\frac{\left\{K_{-1} - \dfrac{K_{-1}}{K_0} \cdot K_{-2} - \dfrac{K_{-1}}{K_0} K_0 + \left(\dfrac{K_{-1}}{K_0}\right)^2 K_{-1}\right\}}{K_0 - \dfrac{2K_{-1}}{K_0} \cdot K_{-1} + \left(\dfrac{K_{-1}}{K_0}\right)^2 K_0}.$$

Divide by K_0 and let $K_{-1}/K_0 = \theta$

$$\therefore \quad \text{plim } \hat{d} = 2 - 2\frac{\left(\theta - \theta\dfrac{K_{-2}}{K_0} - \theta + \theta^3\right)}{1 - 2\theta^2 + \theta^2}.$$

We need to express K_{-2} in terms of K_{-1} and K_0 and so we take the structural equation:

$$Y_t = (\gamma + \rho)Y_{t-1} - \gamma\rho Y_{t-2} + e_t.$$

Multiplying by Y_{t-2}, summing, and taking plims:

$$K_{-2} = (\gamma + \rho)K_{-1} - \gamma\rho K_0.$$

Hence

$$\text{plim } \hat{d} = \frac{2 - 2[\theta^3 - \theta\{(\gamma + \rho)\theta - \gamma\rho\}]}{1 - \theta^2}.$$

Using

$$\gamma + \rho = \theta(1 + \gamma\rho)$$

$$\therefore \quad \text{plim } \hat{d} = 2 - 2(\gamma\rho\theta)$$

$$= \frac{2\{1 - \gamma\rho(\gamma + \rho)\}}{1 + \gamma\rho}.$$

If d^* had been calculated with the 'true' errors (U_t) then

$$\text{plim } d^* = 2(1 - \rho).$$

10.4 Using the standard-order condition for identification reveals the difficulty. In the first equation there are three right-hand-side variables but only two instruments in the system (1 and X_{2t}) which are uncorrelated with U_{1t}. The presence of the serial correlation in the errors produces a covariance between the lagged dependent variable and the error term so that the former cannot be used as an instrument.

The second equation also presents a problem—it too requires three instruments and has only 1 and X_2 available. If the errors from different equations are not correlated or cross-serially correlated, i.e.

$$E(U_{1t}U_{2s}) = 0 \text{ all } s \text{ and } t$$

then Y_{1t-1} would be independent of U_{2t} and could also be used as an instrument.

The problem is to find a variable that is correlated with the endogenous or lagged endogenous variables by the logic of the model, but which is not correlated with the error terms. Since X_{2t} is correlated with Y_{2t-1} but not with U_2 we see that X_{2t-1} will be correlated with Y_{2t-1} and this is a suitable instrument. Thus providing the X_2 variable is not perfectly serially correlated, we have as possible instruments 1, X_{2t}, and X_{2t-1} and both equations are in fact identified and can be estimated by IV.

10.5 (a) Writing $R(L) = \alpha(L)/\beta(L)$ we see that there is a restriction from the structural equation that $\Sigma r_i = 1$, i.e.

$$R(1) = \frac{\alpha(1)}{\beta(1)} = 1,$$

which in turn implies

$$\alpha(1) = \beta(1)$$

or

$$\Sigma \alpha_i = \Sigma \beta_i.$$

Using first-order polynomials and the normalization that $\beta_0 = 1$, we have $1 + \beta_1 = \alpha_0 + \alpha_1$. Introducing the polynomials into the equation we have

$$y_t + \beta_1 y_{t-1} = (1 + \beta_1) \ln A + \alpha_0 x_t + (1 + \beta_1 - \alpha_0) x_{t-1} + V_t + \beta V_{t-1}$$

$$\therefore \quad \Delta y_t = A^* + \alpha_0 \Delta x_t + (1 + \beta_1)(x_{t-1} - y_{t-1}) + V_t^*$$

where

$$A^* = (1 + \beta) \ln A$$

$$V_t^* = V_t + \beta V_{t-1}.$$

(b) The new constraint is

$$1 + \beta_1 = \alpha_0 + \alpha_1 + \alpha_2$$

$$\therefore \quad y_t + \beta_1 y_{t-1} = (1 + \beta_1) \ln A + \alpha_0 x_t + \alpha_1 x_{t-1} + (1 + \beta_1 - \alpha_0 - \alpha_1) x_{t-2} + V_t^*$$

$$\therefore \quad \Delta y_t = A^* + \alpha_0 \Delta x_t + (\alpha_1 + \alpha_0) \Delta x_{t-1} + (1 + \beta_1)(x_{t-2} - y_{t-1}) + V_t^*$$

$$= A^* + \alpha_0 \Delta^2 x_t + (2\alpha_0 + \alpha_1) \Delta x_{t-1} + (1 + \beta_1)(x_{t-2} - y_{t-1}) + V_t^*.$$

This alternative form produces an 'acceleration' term as well as a lagged 'differential' adjustment factor.

11
Applying Econometric Theory

11.1 Introduction

In the earlier part of this book we have taken a very special viewpoint on the models to be estimated. In the main we have assumed that we know the true model, so that the task is merely to estimate the parameters and possibly to compare their values with certain hypothetical values of interest. In this sense the models have been completely *specified*. As an important pedagogical device we have also analysed the effects of making an incorrect assumption about the model—we have discussed this problem as if we were omniscient and thus able to evaluate the effect of mistakes made by a less than omniscient econometrician. This analysis of *incorrectly specified* models is in effect the polar case from the initial treatment.

In practice we are in neither position: we can neither be certain that we are definitely correct nor that we are definitely incorrect. Not only are we uncertain as to the status of any particular model but in any real application there is not even a single clearly defined model. For example, although economic theory tells us that consumption is related to income (via a formal analysis of consumer utility maximizing behaviour), theory does not tell us the functional form of the relationship or the lag length and structure of any dynamic adjustment. There may also be competing hypotheses concerning, say, the formulation of expectations concerning future uncertainty of monetary income receipts and the course of inflation. It soon becomes evident that when we try to estimate a model we have a good deal of latitude in initially specifying the model. This flexibility indicates a lack of precision in our model building so that the choice between rival versions of a model has to be made not on a priori grounds, but on empirical grounds. The problem is therefore to find the 'best' representation of the model from the choices available to us. In order to choose we must therefore have first a criterion or criteria by which to discriminate between models, and second a programme or methodology for deciding which models to investigate. One of the most important features of applied econometric work is that as at present there is no agreed methodology of choosing which variations to test, or for discriminating between rival versions. This largely reflects the lack of a single optimizing characteristic which could be applied to whole methodologies as well as to individual versions of a model.

This chapter is unable therefore to put forward a unified approach to actually doing econometrics and can only indicate what seem at present to be important and useful lines of practice. Section 11.2 begins the discussion with a review of methods of assessing the performance of a given model with an eye to basing the choice between models on such criteria. This difficulty is often associated with high degrees of 'collinearity' between different explanatory factors but is perhaps better thought of as lack of general variability. Since the 'multicollinearity' problem is so prevalent in economic practice, despite not involving a failure of assumptions which guarantee the optimality properties of standard estimators, section 11.3 reviews the problem and possible solutions both good and bad. The main trend of the chapter is taken up again in section 11.4 where we turn to the problems of deciding which variants of models to test and how to choose between them. The choice between models will of course be founded on some 'performance' criterion but the adequacy of the criterion will itself depend on the way in which variants are chosen.

11.2 Measures of econometric performance

There are two distinct types of measures or tests of performance. The first type, of which the multiple correlation coefficient (R^2) is the best known, accepts the specification as true or relevant and attempts to measure how well it fits the data. The second type of measure in effect is aimed at questioning the basic assumptions under which the model is estimated—the Durbin–Watson test for the absence of serial correlation is a good example of the latter. Often the model must 'perform' well on these latter types of criteria before we can use the former—for example we are often interested in the estimated standard error of a regression coefficient because it tells us something about the precision of our estimate of the effects of a particular variable. However, the use of a standard formula for that variance depends on the errors not being serially correlated etc. Hence it is likely to be good practice to demand that a model be acceptable on all such 'diagnostic' tests or checks before we assess its general standard of performance. These tests are designed to highlight possible departures from the standard conditions under which a model can be optimally estimated and classical inferential procedures (e.g. F tests) carried out. In the logical development of the book these tests are presented as if there is a very clear process in the errors, which may or may not be present, and hence which may disturb the potential optimality of our estimation procedure. In practical terms it is much more likely that the model is incorrect in some more general fashion and that this error produces the *appearance* of (say) first-order serial correlation in the residuals. Only in special circumstances will the specification error in fact exactly coincide with the simple error processes investigated. An important corollary to this realization must be that the remedy to failure in such a test is not necessarily to posit the type of error structure implied in the test (e.g. a first-order Markov process) but rather to look for a respecification of the model in which the apparent problem disappears.

(a) Diagnostic tests

The following list of diagnostics are some of the tests presented earlier in the book: they certainly do not cover the complete range of such tests available to practising econometricians, but are a representative sample of the most important of such tests.

1. *Tests for serial correlation.* In a regression model, where all the explanatory variables are exogenous and estimation can take place by OLS, the first test to be applied, to check the null hypothesis that there is no serial correlation, is the Durbin–Watson test (such restrictions rule out its use if there are lagged dependent variables). This test, which uses the residuals from OLS estimation, is a 'bounds' test in the sense that there is a range of values where the critical points might lie under the null hypothesis. The lack of precision that this implies decreases with samples size but is a problem for situations with few degrees of freedom (the table for the distribution of the test statistic nearly always assumes that the model has an intercept). From the algebraic form of the test it is clear that its power will tend to be high when there is first-order serial correlation—sample values are pulled away from the mean of 2 in proportion to the value of the serial correlation parameter ρ. However, in cases of higher-order serial correlation where, for example, an omitted variable is directly correlated with its own lagged values of more than one period before, the power of the test may be rather lower and a different test is needed.

A second approach to testing for serial correlation, which is valid in large samples even when there are lagged endogenous variables, is the LM test in which the residuals from the OLS fit are regressed on all explanatory variables and on as many lagged residuals as the serial correlation process might involve. The resulting value of TR^2 is distributed as a $\chi^2(M)$ variable where Mth-order serial correlation is allowed.

2. *Tests for heteroskedasticity.* The standard test, which assumes the minimum about the nature of any heteroskedasticity if it exists, is the two-sample F test. This separates the data at a preassigned point, calculates two separate regressions and residual variances, and tests the equality of the error variances. When it is felt that there may be a trend in the error variance the data is sometimes split into three groups and error variances from the lowest and highest groups compared (in order to maximize the chance of spotting any shifts in variance size).

A variant of such a procedure, particularly when there is no clear reason for choosing a particular point of sample separation, is to explore several different switch points to see whether any yields evidence of a shift in the error variance. Such a 'search' procedure raises methodological difficulties to which we return later.

In the case where the pattern of heteroskedasticity can be related to the size of the exogenous variable then White's LM test can be applied. This constructs the value of TR^2 on the basis of the correlation between the squared OLS residuals and the squares and cross-products of the various explanatory variables.

3. *Tests for measurement error.* An important failure for the optimality of OLS is caused by correlation between the errors and the explanatory variables. Such a correlation can be produced by:

(i) an omitted variable;
(ii) a measurement error;
(iii) the presence of a simultaneous feedback from a second equation onto the variable in question;
(iv) the presence of a lagged dependent variable together with serial correlation of the errors.

All of these produce OLS bias but if we can find an IV that is uncorrelated with the error then the parameters can be estimated consistently. The choice of instrument is likely to depend on an assessment of the probable cause of trouble. If we can obtain a separate IV estimate, then we can test for the presence of correlation between the variable and the error with Hausman's test. A significant value of this statistic indicates not necessarily that we should stick to IV estimation but possibly that respecification of the model would remove the problem.

4. *Tests for structural stability.* Most causes of specification error have the effect that if the model is estimated over very different values of the data for those explanatory variables which are included, the estimated parameters will change quite dramatically. Tests of structural stability attempt to exploit this by dividing the data into two subsets. The model is estimated, and refined if need be, using only the first set of data. The second set is then added and an F (Chow) test for structural stability is carried out. Again to achieve the maximum power for such a test the data should be divided into two dissimilar sets. In a time-series context, where most variables are growing, the final few observations (corresponding to the largest values of the variables) are often used for the second sample. A rejection of homogeneity between samples usually leads to a search for a richer formulation and not to a simple augmentation of the model by 'dummy' shift variables for the slope or intercept.

We suppose that the variant of the model under consideration has passed all these diagnostic checks and there is no evidence of any specification error. We can then legitimately use the estimated parameters and their variances to describe the model, and performance statistics can safely be based on the estimated model.

(b) Performance indicators

Performance indicators can focus either on the estimated equation as a whole or on a single coefficient.

1. *Standard error of estimate*. In many ways the most useful measures of 'goodness of fit' of a whole equation is the standard error of estimate (SEE) the average size of residuals. The fact that such a measure has the dimension of the dependent variable gives us a quantitative assessment of its importance which percentage figures, such as correlations, cannot do.

2. *Multiple correlation coefficients*. The coefficient of multiple correlation —the correlation between the actual and fitted values of the dependent variable—is a very popular measure of goodness of fit. Because it is dimensionless (being a percentage of the overall variation 'explained' by the model), values from entirely different studies are easily compared. The desire to put such measures on an equal footing led to the use of the corrected multiple correlation coefficient, where numbers of observations and numbers of variables are allowed for; such a measure can be negative despite being called a squared correlation. The problem with all correlation measures is that we cannot say what is a high or a low fit simply from the value. This need to identify a 'good' performance leads to *testing* the overall performance of an equation.

3. *Tests of goodness of fit*. The standard test of goodness of fit is to take the null hypothesis that the dependent variable is completely unrelated to any of the variables of the model. If we allow the null hypothesis to explain the dependent variable solely by an intercept, then the F statistic that all the regression coefficients are zero is a simple (increasing) function of the multiple correlation coefficients—the higher the correlation the smaller the size of test at which the model can be accepted as making some contribution to the explanation of the behaviour of the dependent variable. Such a test does not tell us which component of the model is important and which not, merely that the model as a whole performs significantly better than no model at all.

4. *Total and partial correlations*. When we turn to describing the performance of an individual component of the model it is natural to look at the correlation of that component with the dependent variable. It is not usually appropriate to use a total (simple) correlation between a particular X variable and the dependent variable because this ignores the presence of all the other X variables in the model. The more highly correlated are the X variables the less a total correlation measures the marginal impact of a given explantory variable. Partial correlations correct for the presence of other variables and can be used to assess the effect of each factor marginal to all the rest. Again there is no absolute standard for a 'good' or satisfactory value of a partial correlation so that a significance test is desirable.

5. *Student 't' (F) tests*. We must the marginal contribution of a single variable (or subset of variables) by a $t(F)$ test. It should be noted at this point

that there can be conflicts between these performance criteria. We might, for example, demand that every variable make a significant contribution to the model (conditioned on all the others being present) and indeed find that every 't' statistic was significant but that the overall regression F statistic was insignificant. More commonly we find that the overall F statistic is significant, indicating that the model as a whole is better than no model, but that no t test is significant (no variable adds significantly to the other variables). This failure to be able to separate out the individual factors' performance is very common and presents a real difficulty to the econometrician. The basic cause of the failure comes from the lack of variabiility in the data either within a series or between series. As we have pointed out, lack of variability does not disturb the optimality of the Gauss–Markov theorem and hence of OLS (except in the extreme case of no variability) so that we cannot hope to deal with the problem within the standard least squares context. Section 11.3 reviews the problem and some possible approaches to circumventing it.

11.3 Multicollinearity: lack of data variability

We return to the very simplest model of all:

$$Y_t = \beta X_t + U_t \tag{11.1}$$

where

$$E(U_t) = 0$$

$$E(U_t U_s) = 0 \quad s \neq t$$

$$E(U_t^2) = \sigma^2$$

X_t fixed in repeated samples. We saw that, provided $\Sigma X_t^2 > 0$, the OLS estimator

$$\hat{\beta} = \Sigma X_t Y_t / \Sigma X_t^2 \tag{11.2}$$

does correspond to the unique minimum of the goodness of fit criterion (RSSQ). The variance of this estimator is

$$\text{Var } \hat{\beta} = \sigma^2 / \Sigma X_t^2. \tag{11.3}$$

Now the test for the 'significance of the model' is

$$H_0: \beta = 0$$

$$H_1: \beta \neq 0. \tag{11.4}$$

The value of the 't' statistic for this hypothesis is

$$t = \frac{\hat{\beta}}{(\hat{\sigma}^2 / \Sigma X_t^2)^{\frac{1}{2}}} \tag{11.5}$$

where
$$\hat{\sigma}^2 = \frac{1}{T-1}\Sigma \hat{U}_t^2. \qquad (11.6)$$

The critical value of 't' depends solely on the degrees of freedom ($T-1$) and in fact has the general shape shown in figure 11.1, where we have taken a one-sided test of size 5 per cent. The statistic has a minimum value (1.65) and approaches this rapidly; thus, after obtaining say 10 degrees of freedom, extra observations do not do much to lower the critical value. However, looking at (11.6) we see that the possibility of finding a 'significant' model depends on three factors:

   (i) the actual goodness of fit as measured by the error terms;
   (ii) the number of observations;
   (iii) The average squared value of each observation (the 'spread' of the data).

Combining those factors we can see that the accuracy of the model (σ^2) relative to the data spread (X^2) is something that may depend on the specification of the model. A better model may have a smaller unexplained component relative to the 'explaining' component. A first solution to poor model performance defined in this way is simply to attempt to improve the model.

The second factor of importance is the number of observations: more data from the same economic structure would reduce the variance further. Such an effect is clearly most beneficial when the original data set is small—an extra 5

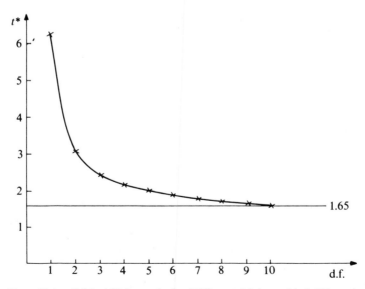

FIG. 11.1 Critical Values of t for Different d.f. (one-sided 5% test)

data points is much more valuable if initially there were only 10 points rather than 50.

The third factor is the variability of the typical observation. The more variable (the further from zero) is the typical X observation the less is the uncertainty (the variance) of the estimated parameter and the more likely are we to be able to accept a model as being 'significant'. This problem is not usually acute in the case of a single variable: there is usually enough spread in a single series to give the model a chance to be significant. The real problem arises with the multi-variable case. We recall the two-variable model:

$$Y_t = \beta_1 X_{1t} + \beta_2 X_{2t} + U_t \qquad (11.7)$$

where U_t is well behaved. By OLS:

$$\hat{\beta}_1 = \frac{\Sigma Y X_1 \Sigma X_2^2 - \Sigma Y X_2 \Sigma X_1 X_2}{\Sigma X_1^2 \Sigma X_2^2 - (\Sigma X_1 X_2)^2} \qquad (11.8)$$

and

$$\text{Var } \hat{\beta}_1 = \frac{\sigma^2 \Sigma X_2^2}{\Sigma X_1^2 \Sigma X_2^2 - (\Sigma X_1 X_2)^2} \qquad (11.9)$$

$$= \frac{\sigma^2}{\Sigma X_1^2 (1 - \tilde{r}_{12}^2)} \qquad (11.10)$$

where $\tilde{r}$ is the correlation between X_1 and X_2 in 'raw' form. Lack of variation in X_1 will continue to lead to imprecision in the estimation of β_1 (similarly with X_2 and β_2), but also if X_1 and X_2 are highly correlated then, even when X_1 is well spread, the variance of $\hat{\beta}_1$ (and of $\hat{\beta}_2$) will be so large that the variable is 'insignificant'. The same can happen for β_2. The data has too little variation, in the sense that the two series do not vary in separate ways, so that we cannot distinguish between their impacts on the dependent variable. In macro-economic time-series modelling many variables are growth or cycle dominated and hence very highly correlated. As extra variables are added to a model there can be correlations between combinations of variables which produce the same effect.

The problem of lack of data variability arising either from correlation between variables or too little variation from an individual variable can thus produce such imprecise estimates of the marginal impact of any given factor that the model is unacceptable if the normal procedures are followed. The problem for the applied econometrician is to find, if possible, a way of improving the precision of estimation in order to reach a sharper view of the correctness of the model. The first point to notice, and one which we have referred to many times is that OLS is still BLUE (providing that there is not perfect collinearity/complete lack of variation in the data). No technique of estimation which does not add more information can improve on OLS, unless we change our criteria of estimator optimality. We review some of the major solutions that have been tried at various times.

1. *Dropping a variable.* If we omit one of the highly correlated variables it is

likely that the *estimated* standard errors of the coefficients of the remaining variable will decline dramatically (compare 11.10 and 11.3). This, if the original model is correct, will in fact create the specification error of incorrect exclusion and lead to a bias of unknown amount in the remaining coefficients.

2. *Adding variables*. Although the denominator of the estimated variance will become smaller as we add extra variables to a regression we must remember that better specification can reduce the estimated residual variance and also change the value of the $\hat{\beta}$ involved. It is quite possible to improve the significance of an individual variable by adding other variables, as well as reducing bias if the variable has been wrongly excluded.

3. *Transforming the equation*. There are two distinct ways in which transformations are made to equations prior to estimation. The first essentially redefines the exogenous variables in such a way as to produce separate variables with good marginal significance. Consider the model

$$Y_t = \beta_1 X_{1t} + \beta_2 X_{1t} + U_t.$$

This as estimated may suffer from high collinearity between X_1 and X_2. The resulting large values of the estimated variances of $\hat{\beta}_1$ and $\hat{\beta}_2$ make it impossible to accept that X_1 has an important incremental effect once X_2 is included. This might lead us to conclude that solely X_2 should be included. We could however rewrite the equation:

$$Y_1 = \beta_1(X_1 - X_2) + (\beta_2 + \beta_1)X_2 + U \qquad (11.11)$$

and regress Y_1 on $(X_1 - X_2)$ and X_2. These variables might be much less correlated so that the estimated variances of $\hat{\beta}_1$ and $\widehat{(\beta_2 + \beta_1)}$ were both small and the variables both significant. We conclude that if X_2 is included the extra effect of the difference between X_1 and X_2 is also significant (which is not the same as arguing that X_1 is significant). If we attempted to recover the original parameters and variances we should of course finish with exactly the same results as if we had carried out the original regression. This approach looks for a grouping of variables that has an economically interesting interpretation and which at the same time removes the collinearity problem.

The second approach to equation transformation is much less satisfactory. This applies (usually) a linear transform to the whole equation, e.g. first differences:

$$\Delta Y_t = \beta_1 \Delta X_{1t} + \beta_2 \Delta X_{2t} + \Delta U_t. \qquad (11.12)$$

Although the correlation between ΔX_1 and ΔX_2 is much lower than that between X_1 and X_2, a regression of ΔY and ΔX_1 and ΔX_2 does not in fact improve our estimators of β_1 and β_2. This is simply because the errors have also been transformed and are no longer 'well behaved' (here they are serially correlated). Conventional formulae for variances will be incorrect, and applying the Aitken transform to obtain the optimal estimating form merely returns us to the original equation.

4. *Adding more data.* The obvious way to improve the performance of the equation is to add more data if that is possible. Even if the new observations on the X variables are as highly correlated as the old data, the effect of enlarging sample size on ΣX_1^2 in (11.10) will reduce the variance.

The usual reason for not adding more data is either that none is available, or that the data comes from a period in which it is feared the values of the parameters were entirely different. Using dummy shift variables would not help if all the parameters were thought to have shifted—the estimated parameters and variances for the original sub-period would then be unchanged.

5. *Use of restrictions.* The most attractive solution to the problems caused by lack of information in the data is to supply some information via a prior restriction on some of the parameters. Such a technique, even though the overall goodness of fit is worsened, results in smaller estimated parameter variances. The problem with this technique is simply that of finding the correct restriction. If the restriction (exact) is correct or the stochastic restriction is unbiased then we can preserve our unbiasedness criteria for estimation while reducing variance. If however the restriction is incorrect or biased then the use of RLS will be biased. The general use of restrictions, particularly in the field of dynamic adjustment (Almon polynomial lags and rational lags) is to try to choose forms for the restrictions which are 'flexible'—i.e. will be approximately true in the circumstances of the model, so that any bias arising from their use is small.

6. *Alteration of optimality criteria.* Many of the techniques we have reviewed may result in biased estimation. While we insist on unbiasedness, at whatever cost in variance, these techniques are therefore ruled out. It would therefore appear very attractive to consider relaxing our optimality criteria and allow some bias—the trade-off between bias and variance could be formalized by using the mean square error criterion. The difficulty with this criterion is that although we can often show, particularly when collinearity is high, that other estimators apart from OLS have better MSE performance, inevitably we need to know something about the true parameters of the model in order to calculate the value of such an estimator. Since, by definition, we do not know the parameter values this route is scarcely practical: knowing that such an estimator exists is not enough.

7. *Alteration of significant levels for testing.* If there is too little information in the data to reject the null hypothesis using a test of a fixed size (usually 5 per cent), then we can reconsider the choice of size of test. If we have demanded that the size be kept constant, then it follows that the power of the test has declined as the collinearity has increased—it becomes harder to accept the alternative hypothesis when true.

From the point of view of choice between the two types of errors it may well be that in the face of poor data we wish to maintain reasonable power (to accept a model if true) and so we can 'trade off' size. Setting tests at 10 instead of 5 per cent in situations where the data is poor in variation may be superior to other methods of proceeding because the implications are clear: we are more likely to reject the NH when it is true.

This review of the problems caused by lack of variability in the data makes it clear that there is no easy solution, and that we may finally have to accept that although the model is the 'best' we can find and contains no evidence of specification error, there is just not enough evidence to allow us to assess the validity of all its components.

11.4 Modelling strategy

We return now to the larger question of how to choose which models to test. When so many variants of a basic economic model can be represented econometrically, and we have no a priori view as to which is the correct model, it is necessary to have a strategy for examining these variants. One school of thought is to examine all possible combinations of specification and choose the 'best'. This suffers from a number of difficulties and is scarcely used in practice—however, it does make a useful starting point for the discussion. A second line of thought is to start with very simple models, check them for adequacy, and then generalize them only if they fail one of the diagnostic or model-fitting tests. A third and polar approach is to specify the most general model that encompasses all our possible hypotheses and then try to simplify it as much as possible until further simplification produces a variant which fails a diagnostic or model-fitting test. We briefly review the merits and problems of these procedures.

1. *All possible combinations.* Suppose that we have six possible explanatory variables, which may or may not be mutually exclusive from the point of economic specification (this might include different definitions of a variable, different lag lengths, and different variables). A strategy which evaluates all (feasible) combinations clearly gives them all equal weight a priori—it does not take any one as a standard from which we depart only if there is good reason. The other procedures will only prefer one version to another if it is 'significantly' better while this procedure takes any improvement in performance as an indication of superiority. This technique is best suited to situations where there is little to choose a priori between variants of the model. A difficulty with this procedure used to be its computational burden where large numbers of variables were involved but this objection is much less important with modern techniques.

A more important difficulty is to decide on the choice criteria. The simplest are the $\bar{R}^2$ or SEE criteria, both of which measure the goodness of fit of the total model allowing for the degrees of freedom. This is merely one dimension of performance: correct parameter signs and satisfactory diagnostic checks are both required in practice. A common approach would be to take the best fitting ($\bar{R}^2$) version of the model out of all those versions where all the diagnostic and fitting tests are passed.

There is a danger inherent in this process which aims to maximize the chances of finding a good fit. The twin factors of (i) treating all variables equally so that new variants can be brought in at any stage, (ii) choosing by maximizing a performance criterion, whether or not it is subject to certain side conditions, means that we can almost guarantee to achieve any desired level of goodness of fit if we are persistent enough in thinking up new versions of the model to test. Even if every version is in fact incorrect there will always be some which accidentally fit well for the particular period. This is the danger of 'data mining'. In fact the problem is in effect that the size of tests used to check goodness of fit becomes very large—we are more and more likely to reject the null hypothesis when true if we test in various ways and treat each one *as if it were a unique test*. Checking an F test (or R^2) for overall model adequacy against a *constant* critical value F^* (based on a 5 per cent test) gives us a very much greater than 5 per cent chance of finding at least one version which is more extreme than the critical value. Suppose all their variants were statistically independent (which is in fact not likely to be true) and in every case the null hypothesis of no economic relationship were true, then the chance of finding no 'significant' model if we use a test of size ρ and have N tests would be

$$(1-\rho)^N \qquad (11.13)$$

and so the chance of finding at least one 'significant' model would be

$$1-(1-\rho)^N. \qquad (11.14)$$

This rises quite strongly with values of N. Of course we could guard against this increasing type I error (which is in effect allowing the power to increase) by adjusting ρ so that (11.14) is always equal to 5 per cent, whatever the value of N. The difficulty with this is that the correct formula depends on the degree of correlation between the various alternatives. For example, if alternatives were virtually perfectly correlated then all would be rejected or none so that the chance of finding a significant model would be independent of the number of hypotheses tried. The dangers of claiming too much for a model arrived at by data-mining have led econometricians to prefer much more structured approaches to the question of search.

2. *Simple to general modelling.* Many econometric studies start with simple versions of a model—this is estimated and then tested. If it fails the tests, the tendency has then been to think of other variables or variations of existing factors to put into the model. This procedure is continued until a version is found which satisfies all the diagnostic and performance criteria.

The first point to note is that the methodology has a preference for certain variations built into it. The choice of the initial model, then what to try next, and so on combined with a rule that stops when an adequate level of performance is attained, does imply, to an omniscient viewer of the process, that there will be a certain pattern to the results. One important feature of this process is that of 'pre-testing' bias. If we only add variables if a significance test is failed at our earlier stage (but not otherwise) and that variable is genuinely important, then our estimate of its coefficient taken over repeated samples is biased (we set many sample values equal to zero rather than equal to the values they would have attained had the variable been included). The offsetting virtue to this potential bias is the lower variance created by the lack of collinearity made possible by the variable's exclusion from the equation. In essence this technique is unlikely to allow us to reach an unbiased estimate of complex models and some simpler version will appear adequate.

A second disadvantage with this approach is that it does not encourage the user to attempt a very thorough specification of the model initially. Given that we recognize that we will search further if the initial model proves inadequate it is tempting not to start by thinking of the most complex situations. This means that the 'best' model may be completely missed because our step-by-step process does not get round to considering it before our sequence of trials (which is not based on full information) finds a model which satisfies all the diagnostic and testing criteria.

3. *General to simple modelling.* This more recent approach recognizes the weaknesses of the simple to general procedure and hence demands that sequential steps in model building should normally be in the direction of simplification. This means that before we start estimation we should attempt to specify the complete list of alternatives that we would be willing to contemplate. A general model involving all elements is then estimated and checked. Variables are likely to be insignificant and can then be removed, lag lengths shortened, and so on. This procedure continues until any further simplification fails to make economic sense, or else introduces a diagnostic or testing failure. There can also be some ambiguity about the sequence of moves but the choice is made explicit by the test failures rather than implicit as in simple to general modelling where the next variable to be entered is arbitrary. The two important features of this procedure are the requirement to specify the model in its broadest terms at the outset, plus the fact that it is less likely to overlook an important version of the model since all should be included in the initial general formulation.

11.5 Final thoughts

This book has introduced the main techniques of estimation and inference in econometrics. There are of course many other techniques and variations suitable for special situations which are not appropriate for a first course. Also the full generality of the many-variable model has been avoided by the decision not to use matrix algebra. Nevertheless the biggest gap, between the theory discussed in a book of this nature and what econometricians actually do, lies in the imagination needed to handle any real problem. Despite the substantial history of the subject and the very large number of its applications there is certainly at present no agreement as to the best way to 'do' econometrics. Experience counts for a great deal in learning to handle new applied problems and this cannot be taught. The reader is strongly urged, if at all possible, actually to try the complete process of specifying a model, collecting data, estimating and evaluating the model. It is not enough to know that a technique exists or even how and why to use it in theory: it is equally crucial to be able to choose the right techniques for the situation.

Index

adaptive expectations 286
Aitken's theorem 180–2, 184, 191, 194, 202–3, 209, 210, 266, 285, 286, 293, 311
asymptotic bias 158, 161–2, 297
autocorrelation; *see* serial correlation

best fitting line 11
best linear unbiased estimator (Blue) 42, 48–9, 54–5, 58, 59, 60, 79, 89, 100, 101, 102, 157, 170, 178, 181, 201, 203, 222, 279, 285, 310
beta coefficient 83
bias 42, 71–2, 220–2, 224–6, 242, 245, 247, 248, 274, 282, 285, 289, 315

catching-up 145, 150
Cauchy–Schwarz inequality 27, 28, 29, 31, 34, 35, 49, 70, 75, 82–83, 105, 110, 208, 209, 225
central limit theorem 116, 160, 162
Chi-squared distribution 123–4, 165, 166, 167, 189, 197, 293
Chow test 306
Cobb–Douglas production function 89
Cochrane–Orcutt transform 194, 233, 246
coefficient of variation 74
completing the square 31
confidence interval 135–8, 141–2, 144, 146, 147, 157
consistency 158, 160, 186, 251
consumption data 7; function 2, 6, 8, 10, 15, 25, 86, 96, 98, 126, 146, 153, 167–9, 190, 197, 199–200, 236, 247, 269, 294, 303
contemporaneous covariance 201, 249, 259–60, 264
central factors; differential 298; proportional 298
correlation 21–3, 33, 35, 84, 96, 225; in raw form 28; between estimators 59, 69, 100, 106; multiple correlation 81–3, 84, 99, 129, 134, 146, 167, 182, 189, 197, 293, 307; partial 84, 86, 88, 99, 105, 134, 307; correction for degrees of freedom 82–3, 146
covariance; between parameters 45, 51–3, 86, 88
Cramer's theorem 159
critical region 119, 174, 309; best critical region 120
cross-section 145, 179

data; in deviation form 14, 27; in raw form 27

data mining 130, 314
degrees of freedom (df) 20, 50, 81, 165, 282
diagnostic tests 304–6
differences 206, 299
dummy variable 138–9, 140–1, 155, 179, 201, 306
Durbin's h test 292–3
Durbin–Watson test 195–6, 197–8, 233, 271, 292, 298, 305

efficiency 59, 61, 64, 71, 75, 204, 207, 210, 218
end point restrictions 299
endogenous variable 10, 154, 247, 253
Engel's Laws 179
equations; additive 6
error; equation error 9
errors and hypothesis tests; type I 122; type II 122
estimates 14, 15
estimation 1
estimators and intercept 27; linear 42–3, 45–7, 48, 49–50, 66, 71, 74, 77, 92, 93, 103, 107, 150, 154–7; least squares 15; optimal, 14; restricted 89–96, 123, 125, 150; unbiased 42, 44, 45–7; unrestricted 118, 123, 139
exogenous variable 10, 247, 253
expectation 41
experimental design 3
explained sum of squares (ESSQ) 22–3; in raw form 28
extraneous information 210–12

Fisher's F test 123–30, 132–4, 137, 138–41, 142, 144, 146–7, 149, 150, 151, 152, 157, 162, 169, 172, 174, 184, 187–9, 194, 198, 203, 232–3, 235, 264, 281, 293, 305, 306, 307, 314
fitted value 22, 24, 25
forecasting 2, 53, 57–58, 61; and multiple regressions 97–8; and hypothesis tests 141–3, 152–3; and serial correlation 198–200; and simultaneous equations 267–8, 270–1
full information maximum likelihood 265

Gauss–Markov theorem 40–50, 54, 58, 77–81, 92, 98–9, 107, 143, 155–7, 176–8, 180–1, 183, 191, 211, 308
goodness of fit 20–3, 28, 30, 32, 37, 81–6, 88, 141, 307
graphs 7–9, 26–7

grouped data 204, 208–10
Hausman's test 235–6, 306
heteroskedasticity 171, 178–90, 193, 202, 210, 305
homoskedasticity 41, 180, 187, 202, 204, 208

identification 251, 270, 273, 298; and under-identification 109, 251, 273; and overidentification 261
income 7, 15, 86, 96, 98, 128, 146; incomes policy 145, 150
independence of; of errors 42; of variables 41
inference 2
instrumental variables 66, 224–9, 235–6, 238–9, 242, 244, 249–50, 252–3, 258, 261, 262, 264, 268, 276, 285, 287, 301, 306
intercept term 10, 26–8, 29, 30, 37–8, 56–7, 60, 71, 91, 180–1, 183, 229
iteration 186, 194

Keynesian model 247
kink 145
Koyck transform 287

lag; Almon 279–81, 283–4, 294, 312; finite 278–85; infinite 285–9; mean 288–9, 296; polynomial (see Almon lag); rational 282–5, 295, 312; order of 281; operator 282–3, 287
lagged dependent variable 285, 289–97
Lagrange multiplier test 166–7, 169, 170, 173, 175, 189, 190, 197, 198, 234, 293, 305
Lagrangian 37, 58, 64, 90, 100, 164
least squares 10, 13, 16, 29, 99; and intercept-tion 27; and scaling 18; and slope coefficient 28; and diagrammatic representation 23; and forecasting 53–8, 74
least squares; ordinary 14, 19, 22, 25, 39–40, 42–9, 50–2, 76–7, 78–81, 87–9, 93, 98–9, 100, 101, 125, 131, 140, 144, 151, 152, 154–7, 161–2, 163, 176–8, 191, 230, 241, 246, 285, 290; restricted 90–1, 93, 95, 96, 97, 100, 102, 109, 110, 112, 114, 139, 163, 164–5, 167, 168, 173, 281, 312; generalized 181, 192, 194, 203, 209, 233; indirect 259–60, 261, 265, 267, 268–9
L'Hôpital's Rule 104
likelihood; function 159–61, 163, 164
likelihood ratio test 164–5, 168, 170, 171–2, 175, 197, 234
linear; estimator 42–3, 45–50, 58, 59, 60, 61, 63, 66, 67, 71, 72, 74, 77, 78, 92, 93, 103, 107, 154–7, 176, 180–1, 192–3, 201–3; model 10, 43; predictor 54–5, 67, 73, 97, 113, 143; relation 6, 30; restriction 89–96, 131, 137, 174; transformation 19–20, 34, 180–1

logit model 171

Markov process 191, 198, 287
maximum likelihood estimator 185, 194, 196, 204, 234, 294
mean square error 43, 60, 64, 71–3, 102, 112–13, 236, 240–1, 312
measurement error 17, 194, 273, 277, 306
mixed regression 211
moving average 206, 287
multicollinearity 29, 32, 80, 87–9, 101, 111, 112, 140, 148, 151–2, 157, 245, 254–5, 274, 279, 308–13
multiplier 268; impact, 268

non-linear equations 20, 29, 34, 185, 194
normal distribution 116–18, 129, 141–2, 160, 167, 293; and estimators 122, 162
'normal' equations 13, 31, 37, 68, 77, 79, 87, 90, 102, 104, 161–2, 163, 165, 168, 170, 228, 244, 249, 251, 253
null hypothesis 118, 119, 126–9, 136, 139, 142, 147, 148

order condition 253–4, 256–8, 261, 265, 276

partial adjustment 286
power of test 121, 129, 144, 148, 188, 223, 236, 292, 306, 312, 314
pre-testing bias 315
prior information 252
probability limit 155, 158–9, 230–2
probit model 171
proxy variable 237–8, 241–2, 243

quarterly data 140, 145, 150–1

random term 10
random variables 41, 58, 61, 155–7, 224, 230–2
rank; variable 227; condition 255–6, 258
recursive system 265
reduced form; coefficients 248, 260, 262–3, 267–8; equations 248, 258, 261, 275
regression 14
regulation and control 2
repeated sampling 40, 44, 78, 115, 119, 135, 154–7
residual sum of squares (RSSQ) 20, 81, 95–6, 182; and intercept 27, 106; and F test 123, 146
residuals 20, 22, 24, 25–7, 29, 81, 104, 227
restrictions 89–96, 100, 101, 102, 108–12, 164, 166, 167, 189, 234; and forecasting 97; and identification 252–60, 262, 264; and distributed lags 279–81, 286, 294, 297

scaling 17, 29, 32, 83, 100, 105, 241

scattergram 8, 11, 25, 85
search procedure 185, 194, 288
second order conditions; and least squares 14, 29, 31, 36, 69, 87, 103, 108, 111, 112
seemingly unrelated regressions 200–4, 207, 217, 265–6
serial correlation 191–200, 203, 205–6, 233–4, 240, 246, 284, 293–4, 301, 305
significance; level 119, 147, 312–13
size of hypothesis test 119, 121, 129, 137, 143, 144, 172, 177, 223
slope term 28–9, 37
Slutsky's theorem 159
stacked equations 138–9, 201, 204, 266
standard error; of estimate (SEE) 20, 33, 50, 81, 95, 185, 307; and intercept 27
stochastic term 10, 23–5
structural change 138–41, 144, 146, 152–3, 179, 306, 234–5
structural equation 248, 293
student's t test 113, 129, 130–3, 134–5, 135–8, 142, 143, 147, 148, 149, 150, 152, 157, 173, 198, 232–3, 235, 241, 264, 270, 281, 307, 309
summations 13, 14; and double sum 43

test; one-sided 120, 123, 127, 131, 150; two-sided 120, 124, 126, 127–8, 131, 147; unbiased 120; uniformly most powerful 120, 123
Theil–Goldberger estimator 211
three stage least squares 266–7
tolerance interval 142, 143
total sum of squares (TSSQ) 22–3, 146, 182; in raw form 28
transformations 17, 150, 170, 180, 311; of variables 6, 19, 30, 32, 33, 35, 100, 106, 110, 195
trend variable 155, 227
two-stage least squares 220, 239, 262–7, 268–9, 273–4

unemployment 85–6, 96, 98, 126, 146
united variable 28, 32, 155, 236, 252

variance; of error term 41, 50, 51–3, 226, 239, 251, 266; of estimators 43–5, 46–7, 50–3; of predictor 55–6, 62, 97–8; of product 61, 74; of sum 131

Wald test 166, 168, 170, 172–3, 175
White test 187, 189, 190, 234, 305